HIGHER
GCSE Mathematics for Edexcel

TEACHER'S RESOURCE

ALAN SMITH

SERIES CONSULTANT: JEAN LINSKY

CONTRIBUTOR: ANNE PULLAN

Hodder Murray
www.hoddereducation.co.uk

Acknowledgements
The Publisher would like to thank the following for permission to reproduce copyright material:
p.6 © Science Museum / Science & Society Picture Library

This high quality material is endorsed by Edexcel and has been through a rigorous quality assurance
programme to ensure that it is a suitable companion to the specification for both learners and
teachers. This does not mean that its contents will be used verbatim when setting examinations, nor
is it to be read as being the official specification – a copy of which is available at
www.edexcel.org.uk

Orders: please contact Bookpoint Ltd, 130 Milton Park, Abingdon, Oxon OX14 4SB.
Telephone: (44) 01235 827720. Fax: (44) 01235 400454. Lines are open 9 am to 5 pm, Monday to
Saturday, with a 24-hour message answering service.
Visit our website at www.hoddereducation.co.uk

First published in 2006 by
Hodder Murray, an imprint of Hodder Education,
a member of the Hodder Headline Group
338 Euston Road
London NW1 3BH

Impression number   10   9   8   7   6   5   4   3   2   1
Year   2011   2010   2009   2008   2007   2006

Cover illustration © David Angel @ Début Art
Illustrations © Barking Dog Art

Typeset in Great Britain by Tech-Set Ltd
Printed and bound in Great Britain by Hobbs The Printers, Totton, Hants

A catalogue record for this title is available from the British Library.

ISBN-10: 0340 913606
ISBN-13: 978 0340 913604

CONTENTS

Introduction

This *Teacher's Resource* contains a variety of materials to support teachers using Hodder's *Higher Tier GCSE Mathematics for Edexcel Student's Book*.

Teaching Notes

The *Teacher's Resource* begins with a set of Teaching Notes; these give a few sentences of commentary on the purpose of each Starter, Exercise and Internet Challenge. Some of the chapters are supported by supplementary Worksheets, and brief comments are given about the goal of these, too.

The Teaching Notes also warn of many common pitfalls throughout the exercises. While such advice should be helpful to all users of the book, it is likely to be of particular benefit to teachers new to the profession, or those who are not mathematics specialists.

Worksheets

The next section comprises a number of photocopiable Worksheets. These are intended to provide extra support and consolidation for students needing further practice at the core topics underpinning much of the grade C and B material, and they therefore tend to concentrate on the first two-thirds of the *Student's Book*.

The Worksheets are numbered to match the numbering of sections in the *Student's Book*: thus Worksheet 8.4, for example, provides further work on the material in section 8.4 and Exercise 8.4 of the *Student's Book*.

Speed-up Sheets

These photocopiable Speed-up Sheets save the time that students would otherwise spend constructing basic grids and diagrams before starting to answer questions from the *Student's Book*. The sheets are of particular help for graphical or geometrical questions.

Answers

The *Teacher's Resource* concludes with answers to all the Starters, Exercises and Worksheets. Where appropriate, answers to Internet Challenges are also given, but some of these are open-ended in nature, while for others the answers may change over time (e.g. the largest known prime, the current rate of inflation etc.). It is recommended that teachers try the Internet Challenges for themselves before going through the answers with students.

Coursework

The two written Edexcel examinations account for 80% of the GCSE award, with the remaining 20% being provided by two coursework tasks. These are an Algebraic Task (10%) and a Data Handling Project (10%). Detailed advice and information about coursework is provided in *Foundation GCSE Mathematics for Edexcel Assessment Pack*.

CHAPTER 1
Working with whole numbers

Starter 1

This exercise is an opportunity to practise simple arithmetic and the hierarchy of operations in a fun way. Students will find it starts easily, but becomes tricky later. Try using $4 \div 0.4 = 10$ to help make the higher numbers.

Other useful calculations include $4 \div 0.\dot{4} = 9$ and $\sqrt{4} \div 0.\dot{4} = 3$, although some students might consider this to be cheating!

Exercise 1.1

This first exercise is not hard: the idea is to try and find quick ways of solving the problems, where possible. After completing this exercise, you might consider a brief discussion to see what short-cut strategies the students have used.

Exercise 1.2

There is another decomposition method, where, for example, 492 is written as $400 + 90 + 2$. It is strongly recommended that students do not use this method. It does, admittedly, provide a nice visual image of the distributive law at work, but it invariably leads to numbers followed by lots of zeros within the working, so errors occur more often than with the two methods shown in the Student's Book.

Exercise 1.3

When doing a long division, say by 13, many students find it helpful to write out the first 9 multiples of 13 (13, 26, 39, 52, 65, 78, 91, 104, 117) at the beginning, to save time later on.

W Worksheet 1.3

This worksheet provides additional work on adding, subtracting, multiplying and dividing without a calculator.

Exercise 1.4

This topic is easier to understand if the students realise that a minus sign can mean two different things:
- a place on the number line -4 means a place on the number line 4 below 0
- an instruction to move -4 means 'move 4 to the left'.

W Worksheet 1.4

This worksheet provides additional work on positive and negative numbers.

Exercise 1.5

Students enjoy making factor trees. The actual order of factorising does not matter, but it is a good idea to write the final answer with the smaller prime factors first.

Exercise 1.6

Students are likely to find higher common factors (HCFs) using intuitive methods quite easily, but some practice is required to use the prime factor method accurately. In an exam students should take care to read the questions carefully: candidates often give an HCF in response to a question asking for a lowest common multiple (LCM).

Exercise 1.7

For any pair of numbers, the product of their HCF and LCM is always the same as the product of the two original numbers. This provides a handy check.

Able students might want to see under what circumstances a similar result might hold for the HCF and LCM of *three* numbers.

Internet Challenge 1

Some of these questions are about the mathematical definition of prime (cannot be factorised) while others are about the everyday use of the word prime (first, as in Prime Minister, Prime Meridian).

This illustrates an important feature of mathematics: we take words that already have a (perhaps fairly broad) meaning in everyday usage, and attach very precise mathematical meanings to them.

CHAPTER 2
Fractions and decimals

Starter 2

▶▶ A speed-up sheet is available for question 3 of this starter.

These questions are not as simple as they look. Question 2 could be done using the knowledge that 'of' means 'multiply', anticipating the work on the multiplication of fractions later in the chapter.

Many students will instinctively try to solve question 3 by cutting into two *reflected* halves, which cannot be done. If they need a clue, suggest that they cut it into two parts that map to each other by *rotation through 90°*.

Exercise 2.1

Students should find cancelling fractions straightforward, but encourage them to make sure that they have done this fully. Some of the addition/subtraction problems towards the end of the exercise are a little harder than would typically be encountered in a GCSE exam.

Exercise 2.2

Some students will feel more comfortable doing all the cancelling at the end. While this is perfectly valid, it can lead to arithmetic with very large numbers, so students

should be gently encouraged to cancel early rather than late. They need to realise, however, that division problems must be converted to multiplications before any cancelling is attempted.

Once again, some of the problems towards the end of the exercise are a little harder than would typically be encountered in a GCSE exam.

W Worksheets 2.1 and 2.2

In these exercises students must decide which is the correct operation to use in order to solve the given problem. Worksheet 2.1 requires either addition or subtraction and could be used after section 2.1; Worksheet 2.2 requires either multiplication or division, and should be left until section 2.2 has been completed.

These problems should be solved without the use of a calculator, as students will meet this type of question on the non-calculator paper.

Exercise 2.3

This is a routine exercise based squarely on the standard methods and examples shown in the Student's Book. Be vigilant for students who, having multiplied both numbers by 10 or 100 prior to dividing, mistakenly believe it necessary to divide by 10 or 100 at the end.

Exercise 2.4

Recurring decimals are considerably more awkward than terminating decimals, and intuition can be misleading: the formal methods shown in the Student's Book must be adhered to.

Watch for common misunderstandings, including the common (but incorrect) belief that 0.3 and 0.33 are the same as $\frac{1}{3}$.

Exercise 2.5

Rounding and approximation should be quite straightforward, though poor practice is frequently encountered in everyday life. For example, an internet site lists the top speed of a cheetah as 70.00 mph, implying an unlikely degree of precision. You could discuss with your students why 70.00 is not the same thing as 70; this will help them understand the idea behind question 20.

This topic is developed again more fully in relation to upper and lower bounds in Chapter 18.

W Worksheet 2.5A

The sort of questions presented in this worksheet are becoming increasingly common on the GCSE papers, but some students find them rather difficult to get to grips with and plenty of practice is advisable.

This worksheet can be used in conjunction with the other work done on estimation whilst at the same time developing the theme.

W Worksheet 2.5B

This worksheet is intended to enable students to assess whether the result they have obtained from a calculation, usually involving a decimal, is reasonably correct or not.

In the exam students will also be presented with calculations and asked to estimate the final result.

Questions 1 to 10 simply require students to round numbers so that they are easier to work with: some discussion as to what might be the most suitable estimate for a particular value might be appropriate here. Questions 11 to 20 are calculations involving only one operation: you might like to extend the earlier discussion here to consider how accurate the estimate is likely to be. The final five questions involve multiple operations.

W Worksheet 2.5C

Students can find upper and lower bounds, which are introduced in Exercise 2.5, quite challenging: introducing it at this early stage (and then developing it further in Chapter 23) should help them tackle the topic more successfully. Grade C students should be able to cope with the ideas involved, especially if they are approached through the use of a number line. As students gain an understanding of what is required, however, it should be possible to produce results without a number line.

At this stage the aim is for students to become familiar with the idea of a range within which is located the given value; it is not necessary at this stage to apply the idea of bounds to practical problems such as perimeter and area.

Internet Challenge 2

This challenge is quite difficult. The point behind question 7 is that it is much easier to judge $\frac{1}{2}$ of one pile plus $\frac{1}{8}$ of another, compared with trying to judge $\frac{5}{8}$ of a single pile.

CHAPTER 3
Ratios and percentages

Starter 3

This starter is a little off-topic, since it is not, strictly speaking, about percentages.

A good strategy is to solve it backwards. Ending with the T in the centre, there are six different approach boxes containing N's. Each N may be approached from one of two E boxes, so now we have $2 \times 6 = 12$. Continue in this way, building up more multipliers, until the final answer is achieved.

Exercise 3.1

Many students find ratios intuitively obvious, but weaker students can become fretful. A calculator fraction key can simplify ratios for them, but only for ratios of two numbers, not three.

 **3 Ratios and percentages**

Ratios are sometimes written in the form $1 : n$ or $n : 1$, where n need not be a whole number. The worksheet provides practice at this way of writing ratios.

Exercise 3.2

Resist impulsive thinking: these problems are all quite straightforward if tackled systematically. Some older calculators have percentage keys: these should *not* be used, as they merely add more confusion.

Question 19 is best done by expressing both test scores as percentages: this is one very valuable use of percentages, and is used on food packaging.

Exercise 3.3

The multiplying factor method is by far the best way of dealing with percentage increase and decrease. Not only is it elegant and efficient, but it also makes compound interest and reverse percentage problems almost trivially easy.

Multiplying factor methods assume calculator access, however, so other methods need to be mastered as well, specifically for use in the non-calculator GCSE paper.

Worksheet 3.3

This worksheet provides further practice of percentage increase and decrease. It should be attempted without a calculator, as students will be required on the non-calculator exam paper to achieve results using either fraction work or decimal work.

Exercise 3.4

Again, the method of choice is multiplying factors. Multiply by the factor for a forwards increase/decrease, and divide for reverse problems (question 1b), for example.

Exercise 3.5

Simple interest may be calculated using common sense, though some students prefer to learn the formula given in the Student's Book.

Compound interest is best done using multiplying factors. To perform repetitive steps on a calculator, type in the initial amount and press $\boxed{=}$.
Then type $\boxed{\text{ANS}}\ \boxed{\times}\ \boxed{1}\ \boxed{.}\ \boxed{0}\ \boxed{5}\ \boxed{=}$ and just keep pressing $\boxed{=}$ to see this increase by 5% compound.

Worksheet 3.5A

This worksheet provides further examples of simple interest calculations.

In the examination, candidates would expect to have to calculate the interest as in questions 1 to 4 and question 12. The worksheet also contains some more demanding calculations in questions 5 to 11; these are provided to help consolidate learning of the simple interest formula $I = PRT/100$ in its various forms.

Worksheet 3.5B

This worksheet provides further examples of compound interest calculations. They can sometimes be rather demanding and plenty of practice is advisable. Question 1 involves simply finding the multiplying factors for a number of rates of interest, since it is important that students are able to do this easily.

Internet Challenge 3

This challenge provides a real-life example of percentages in use. Some students think that you cannot have more than 100% of anything but, while this is true in some situations (how many percent did you score in an exam?), it is clearly not true of financial applications.

CHAPTER 4
Powers, roots and reciprocals

Starter 4

We take the modern, base ten (Arabic) number system for granted, but this starter can be used to explain to students that other cultures have done arithmetic differently in the past.

As an extension, you could ask your students to try some arithmetic problems with Roman numerals: for example, work out XXIV multiplied by XI. Students should try to do this using Roman numerals, rather than by converting back into 'normal' numbers.

Exercise 4.1

The squares and cubes on page 58 must be learned thoroughly; students may be expected to recognise these in the non-calculator exam. When using a calculator, care should be taken to round the answers in exactly the way specified in the question, as carelessness here loses exam marks.

Worksheet 4.1

This worksheet requires students both to remember squares and square roots and to use them in simple calculations. This style of question may be encountered on the non-calculator exam paper; frequent practice will help students remember the required results.

Such questions are also useful since they reinforce the relationship between squares and square roots by demonstrating that they are directly related.

Exercise 4.2

Calculator power and root keys will do all the hard work later in the exercise, but different calculators implement these functions in slightly different ways. Each student must master the keys fully on their own model, and practise regularly before the exam.

Exercise 4.3

These problems are best done by taking a root first (the bottom of the fraction) then raising to a power (the top of the fraction). A calculator could be used as a check afterwards.

Exercise 4.4

Negative powers will confuse many students, especially when combined with a fractional power. This is a topic that will benefit from further revision and practice at regular intervals during the GCSE course.

Exercise 4.5

Calculators are deliberately excluded here for two reasons: these questions are often examined on the non-calculator paper at GCSE and, in addition, mastery of the rules for combining indices will be needed for the algebra work to follow in Chapter 5.

W Worksheet 4.5

This worksheet makes use of the base number two as a means of calculating multiplication and division problems, and requires the use of the rules of indices for both of these operations.

Students should recognise that this method provides a short cut for what can be troublesome questions, as it avoids the need to perform a multiplication and/or division.

Exercise 4.6

Standard index form (or scientific notation) is regularly used in other subjects, and students will find that their calculator automatically adopts standard form when the display would otherwise overflow. This exercise provides simple practice at converting large and small numbers into and out of standard form.

Exercise 4.7

Standard form calculations with a calculator are easy, although some calculators require a non-intuitive way of inputting and reading out the numbers. Non-calculator methods include adjusting all the numbers so that they have the same power of ten, or even converting them out of standard form entirely.

Internet Challenge 4

▶▶ A speed-up sheet is available for this challenge.

This challenge uses some examples of very large (or very small) numbers used by scientists. Note that some values found on the internet may be slightly different from the values offered in the options here, but it should be quite clear how the answers match up.

CHAPTER 5
Working with algebra

Starter 5

Some of these are obvious, but questions 4, 5 and 6 illustrate some common pitfalls. Some TV game shows are guilty of overlooking BIDMAS in chains of number calculations. It is important to get this right before working with algebra.

Exercise 5.1

Students should try to do questions 1 to 12 without a calculator, perhaps just checking on a calculator afterwards. A calculator is required for questions 13 to 15.

Watch for common errors with expressions such as $4f^2$, where students may be tempted to multiply by 4 before squaring instead of after.

W Worksheet 5.1

This worksheet provides further practice of substituting numbers into formulae.

Exercise 5.2

The first six questions are easier than typical Higher Tier questions and are provided to help students warm up before tackling the more substantial questions later in the exercise. In the latter part of the exercise, watch for confusion between cases such $x^2 \times x^3 = x^5$ and $(x^2)^3 = x^6$.

W Worksheet 5.2

This worksheet provides further practice of the laws of indices.

Exercise 5.3

Many students will have no problems with this exercise until question 12, where care must be taken over the minus sign between the brackets. Some find it helpful to write a factor of (-2) in brackets in front of the second bracket, as a reminder that both terms in the second bracket need to be multiplied by -2.

W Worksheet 5.3

This worksheet provides further practice of expanding brackets.

Exercise 5.4

The Student's Book shows three different ways of writing out expansions of brackets, all of which lead to the same final result. Students should use whichever method they are most comfortable with; weaker students often find the table method a reliable one to use.

 **5 Working with algebra**   GCSE Maths for Edexcel: Higher Teacher's Resource © Hodder Murray 2006

W Worksheet 5.4

This worksheet provides further practice of multiplying two brackets together.

Exercise 5.5

The factorising methods practised here and in the next two exercises are all quite standard, but care must be taken not to confuse them. Exam papers typically have one long question on factorising, in which all of the various types are tested.

W Worksheet 5.5

This worksheet provides further practice of factorising expressions using the common factor method.

Exercise 5.6

The standard method of factorising quadratics, shown in the Student's Book, involves some trial and improvement. Although it is a little tedious, checking by expanding the brackets is a valuable exercise.

There are other methods involving splitting the middle term: some students like these methods but others find them confusing and, moreover, they do not always generalise well to problems where the coefficient of x^2 is not 1.

W Worksheet 5.6

This worksheet provides further practice of factorising quadratic expressions.

Exercise 5.7

The principle behind the factorising here is exactly the same as in the previous exercise, but when the coefficient of x^2 is not 1 there are more possibilities to consider. Extra care should be taken to verify that the proposed answer does indeed multiply out correctly.

These examples set the scene for the solution of quadratics, which will be covered in Chapter 25.

Exercise 5.8A

The difference of two squares is easily learned, but easily forgotten, too. In some cases it is necessary to take out a common factor first (see question 5, for example).

Exercise 5.8B

In this exercise students must recognise which type of factorisation is required, before actually carrying it out.

Exercise 5.9

Some exam questions require candidates to generate algebraic formulae from verbal information. Students tend to find this quite difficult, perhaps because it is their first real experience of mathematical modelling.

W Worksheet 5.9

This worksheet provides further practice of generating formula from given information.

Exercise 5.10

Rearranging formulae is a hard topic to teach, since every problem is different.

Some students (and teachers) like the reverse flow chart method. This can be used to make a the subject of $v = u + at$ (question 3), for example, as follows.

Start with the intended subject, a, and list the steps you would use to work out v.
Write them in a list:

Start with a
Multiply by t
Add u
End up with v

Now write the opposite processes alongside these:

Start with a End up with a
Multiply by t Divide by t
Add u Subtract u
End up with v Start with v

To complete the solution, read out the second list but in reverse order, to obtain:

$$\frac{v - u}{t} = a.$$

This method is not perfect, and cannot be used when the variable appears twice. It does, however, provide an alternative approach for simple problems that nervous students might like to use for a while, even if they outgrow it later.

Internet Challenge 5

▶▶ A speed-up sheet is available for this challenge.

This wordsearch is a light-hearted way of highlighting some important mathematical vocabulary. You could conclude the activity by having a class discussion about the everyday and mathematical meanings of some or all of the words.

CHAPTER 6
Algebraic equations

Starter 6

These arithmagons are not difficult, but students will probably use some kind of trial and improvement method to solve the last example. Later in the chapter you could come back to this one, and solve it by writing x in the top circle; this forces $13 - x$ and $17 - x$ into the two lower circles. The bottom row may then be expressed as the equation $13 - x + 17 - x = 16$, which may then be simplified and solved to obtain the value of x.

Exercise 6.1

The words *expression*, *equation* and *formula* are often used incorrectly; this exercise aims to reinforce using them accurately. Many students fail to make the important distinction between an equation and an *identity*.

You could think of an equation as a filter: when possible x values are fed into it, the equation is either true or false, and the values for which it is true are the roots (solutions) of the equation. Then an identity is simply an equation where every possible x value turns out to be a solution.

Exercise 6.2

These questions are all essentially one-stage equations, so only one process is needed to extract the solution. You should insist that your students do solve them formally, despite the fact that they can be done by inspection; the idea is to build up a formal procedure that can be used when the equations contain two or more stages.

Exercise 6.3

As a general guide, students will need to add or subtract first, then do a division or multiplication to complete these two-stage equations. You could think of this as BIDMAS in reverse, since the arithmetic that has been done to the x terms is essentially being undone.

Remind students that it is not always necessary to collect the x terms on the left-hand side: if they wish to keep the final x coefficient positive then collecting on the right-hand side is a good idea in some cases (see question 13, for example). This establishes good practice before meeting inequalities in Chapter 9.

W *Worksheet 6.3*

This worksheet requires students to use given information to connect together algebraic expressions to form an equation. The equations can then be solved using the basic algebra rules/methods without the need to make use of a calculator.

The three sets of questions relate to the sides of a square, the perimeter of a triangle and the angles of a triangle respectively. Some students will find it useful to draw a diagram for each question.

Exercise 6.4

By multiplying out the brackets, these problems come to resemble the ones encountered in the previous exercise. Watch that negative signs are handled correctly, especially when in front of a bracket (see question 7, for example). Students may need to be reminded of this point.

W *Worksheet 6.4*

This worksheet may be used as a starter to Exercise 6.4.

This worksheet increases in difficulty: the first ten questions all have positive whole number values for x; questions 11 to 20 all have whole number negative solutions; questions 21 to 30 give rise to solutions which are mixed numbers or fractions, either positive or negative; and the final ten questions are a mixture of more demanding examples of all three types.

Exercise 6.5

The best way of dealing with algebraic fractions in equations is simply to multiply the whole equation by a large enough integer to clear them away. Avoid

methods that involve adding two algebraic fractions to make another fraction: this topic is on the GCSE specification, but is dealt with in Chapter 26.

W *Worksheet 6.5*

This worksheet is intended to be completed before Exercise 6.5.

Exercise 6.6

Some students are reluctant to show the trials, feeling that they are somehow 'incorrect' until the last one. Emphasise that the examiner will want to see evidence of trials that missed to begin with being gradually refined: full details of the trials used must be shown. The exam question will always specify the level of accuracy required (typically 1 decimal place).

W *Worksheet 6.6*

The questions in this worksheet are easier than those in Exercise 6.6 and it could be used as an introduction for weaker students.

Internet Challenge 6

Gauss was one of the greatest mathematical minds of all time, and his work included much on the theory of equations. The banknote image at the end reminds us that he was such a prominent figure that he appeared on the currency of his home country; those of us who can remember UK £1 notes will know the same is true of our own Sir Isaac Newton.

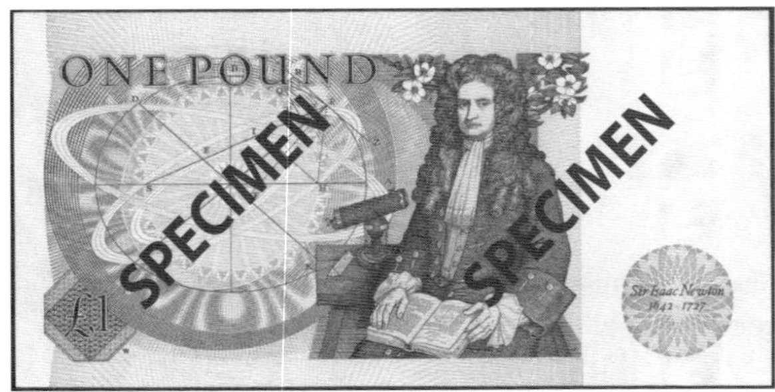

Notice, incidentally, that this £1 banknote contains a mathematical error: the Sun should be placed at one focus of the ellipse representing a planetary orbit, not in the exact centre.

CHAPTER 7
Graphs of straight lines

Starter 7

Matchstick puzzles are good fun, and although they may not seem to have anything to do with graphs of straight lines, there is a connection with question 3 of Exercise 7.1. Much of the GCSE specification requires students to learn and practise standard methods; puzzle solving is a refreshing change from this routine.

Exercise 7.1

▶▶ A speed-up sheet is available for questions 3 and 4 of this exercise.

These questions provide straightforward practice at using coordinates in all four quadrants. Questions 3 and 4 require special attention to detail; they are also self-checking in that the final diagram is unlikely to work if a slip has been made. Question 4, the maze puzzle, is actually quite easy, and students might like to devise their own, harder versions.

Exercise 7.2

▶▶ A speed-up sheet is available for this exercise.

This exercise builds confidence in plotting graphs of straight lines. The use of three points rather than two helps to guard against errors in calculating the y coordinates. There is a speed-up sheet for the whole exercise; in the GCSE exam, grids will always be provided.

Exercise 7.3

This is a very simple exercise on calculating gradients and intercepts: it provides practice in anticipation of developing the link between the algebra of $y = mx + c$ and the corresponding graph in the next section.

Exercise 7.4

For mathematics teachers the connection between numerical values of gradient and intercept and the equation of a line as $y = mx + c$ is very obvious, but students meeting it for the first time need a while for the ideas to sink in. This exercise helps reinforce the links between the algebraic and graphical representations of straight lines.

Exercise 7.5

If two lines are parallel then they must have the same gradient: this is easy to check once both lines have been written in the form $y = mx + c$.

Higher Tier GCSE students also need to know the link between the gradients of perpendicular lines: this is more sophisticated, and is covered in Chapter 31.

Review Exercise 7

▶▶ A speed-up sheet is available for this exercise.

Internet Challenge 7

This challenge features some amusing material on the theme of parallels. The optical illusion is particularly striking and students may wish to extend the challenge by searching for other optical illusions involving parallel lines.

CHAPTER 8
Simultaneous equations

Starter 8

The first equation is easy and, done in sequence, so are the next three. The last two equations are much harder, since they need to be solved as a pair.

By comparing the two equations, students should see that the grapes are worth three more than a banana. By halving the last equation, they should see that grapes and a banana add up to 11. These ideas underpin the concept of solving two equations in two unknowns, which forms the main focus of this whole chapter.

Exercise 8.1

GCSE examiners are not likely to set questions quite as easy as these, but the ability to spot a solution by identifying small changes between two equations is a useful supporting skill for this whole topic.

Exercise 8.2

Algebraic elimination is the standard method for solving linear simultaneous equations, and is used throughout this exercise.

Substitution methods could also be used, but really come into their own when the equations are more complicated – when, for example, one is linear and one quadratic. Such instances are covered in Chapter 26.

W *Worksheet 8.2*

This worksheet is intended for use in conjunction with the work contained in exercise 8.2; it could also be used in preparation for the non-calculator paper. In each question students are required to decide which is the most suitable method to use in order to obtain a solution.

Exercise 8.3

▶▶ Speed-up sheets are available for this exercise.

Questions on the graphical solution of simultaneous equations will usually have whole number or simple fractional answers. Exam questions often provide one line, and ask candidates to add a second line on the same grid.

W *Worksheet 8.3*

▶▶ A speed-up sheet is available for this worksheet.

The worksheet offers additional questions on solving simultaneous equations by a graphical method, and can be used in conjunction with Exercise 8.3.

All the questions can be solved without using a calculator. The early questions have positive whole number answers, while the solutions to the later questions include some fractional and negative values.

Exercise 8.4

This exercise requires some skill at turning verbal information into precise mathematical equations. Stronger students find this quite easy, but others may benefit from further practice.

Worksheet 8.4

This worksheet can be used to supplement the questions in Exercise 8.4. The questions are graded, with easier ones at the beginning and some quite demanding ones towards the end.

Review Exercise 8

▶▶ A speed-up sheet is available for questions 13, 14 and 15 of this exercise.

Internet Challenge 8

Magic squares would be too hard to formulate using simultaneous equations – there are just too many variables. Instead, there are methods (algorithms) based on underlying symmetry. It is an interesting fact that odd squares are very easy to make, and even squares very difficult; thus it is far easier to make an 11 by 11 magic square than a 6 by 6, for example. You might get your students to make enormous magic squares as classroom displays.

CHAPTER 9
Inequalities

Starter 9

One of the ideas behind the starter is for students to locate all points satisfying the condition $x + y = 16$, albeit in an informal and intuitive way at this stage.

Exercise 9.1

The four symbols $<$ $>$ $\leqslant$ and $\geqslant$ underpin this whole chapter, so the first exercise confirms that students are comfortable working with them. Note that the integer solution sets listed in the examples are all in increasing order, but students may feel it is more natural to write $2, 1, 0, -1, \ldots$ rather than $\ldots, -1, 0, 1, 2$.

Exercise 9.2

Inequalities may be manipulated in similar ways to equations, but beware of multiplying or dividing by a negative number at the end, since this would reverse the direction of the inequality. This difficulty can be avoided altogether by always collecting the x terms on the side of the inequality that will ensure a positive coefficient.

Do check that your students do not use any equal ($=$) signs when working with inequalities as this would mean they would not gain any method marks in the exams.

Worksheet 9.2

The worksheet is intended to review and revise the questions asked in Exercises 9.1 and 9.2. All the questions are intended to be solved without the use of a calculator, and the results are positive or negative values with some simple fractional values.

Exercise 9.3

When a diagram is asked for in the examination, students will normally be given a section of number line printed on the paper; this can give a strong hint as to whether an answer is right or wrong. In this exercise, students are asked to draw their own number lines from scratch, making the questions a little more challenging.

Worksheet 9.3

This worksheet supplements Exercises 9.1, 9.2 and 9.3 and presents inequalities in practical contexts rather than as a purely theoretical algebraic exercise. It could be used simply as an introduction to the work or as reinforcement.

Exercise 9.4

▶▶ Speed-up sheets are available for this exercise; adapt these for use with Worksheet 9.4A.

Some of the inequalities in this section are strictly $<$ or $>$, while others use $\leqslant$ or $\geqslant$. This has no significant effect on the shading of a region R, but when asked to mark points that satisfy given inequalities, students must check carefully whether the boundaries form part of the allowed solution set.

In examinations, candidates often do not draw the required line long enough, thereby often losing marks. Encourage students to draw all required lines the full length of the given graph area.

Worksheet 9.4A

▶▶ Speed-up sheets are available for this worksheet.

The worksheet gives students practice in finding a region on a graph when the inequality is given in terms of two variables.

Worksheet 9.4B

▶▶ Speed-up sheets are available for this worksheet.

This worksheet is more demanding than Worksheet 9.4A, involving more than one inequality in each case; however, only question 10 involves inequalities in terms of two variables. Students are also required to select coordinate values which satisfy the given conditions: similar questions have appeared on a number of past GCSE question papers.

Review Exercise 9

▶▶ A speed-up sheet is available for questions 21, 22b and 225c of this exercise.

Internet Challenge 9

▶▶ A speed-up sheet is available for this challenge.

The goal of this challenge is to stimulate some reading about the history behind the development of the symbols we now take for granted, and to realise that they were added over a period of time rather than all at once. In particular, division was a late developer, and in the 16th century was only taught at university level in some countries!

It could be very time consuming for students to try to search for each symbol individually. Instead, if they enter 'mathematical symbol history' or similar into a search engine textbox, then they are likely to find pages that will discuss all the symbols together.

CHAPTER 10
Number sequences

Starter 10

This starter uses three different number sequences, all generated by the same progression of diagrams. Note that the third sequence (1, 2, 4, 8, ...) does not continue to double in the way that most students would expect.

Exercise 10.1

This is a mixed bag of sequences, which are all related in some way to the standard sequences described in the Student's Book. When you go through the answers, you may wish to flag up the ones that go up or down in equal steps (questions 1, 2, 3, 4, 9). These are arithmetic sequences, and are studied more closely later in the chapter.

Ⓦ *Worksheet 10.1*

This worksheet provides further practice at mixed problems on number sequences.

Exercise 10.2

Terms in a number sequence may be defined by an algebraic rule. They may also be defined inductively, by saying how each term is obtained from the one before. This second type of rule, known as a term-to-term rule, will not be examined using formal algebraic notation in the GCSE, but candidates might be asked to interpret such a rule given in verbal form, as in questions 1 and 2 here.

Exercise 10.3

Exam questions on arithmetic sequences often ask for, say, the 30th term in an early part of the question, and a general rule for the nth term as the last part of the question. Strong candidates may prefer to establish the general rule for the nth term first, then use this rule to find the 30th term.

Internet Challenge 10

There are many fascinating number sequences in mathematics, but the GCSE specification focuses largely on a small number of standard ones, especially those that are arithmetic sequences. Thus some of the more interesting sequences tend to be overlooked.

Many situations in the real world are well modelled by the Fibonacci sequence and its close relative, the Golden ratio. This challenge provides a short introduction to a very rich vein of mathematical material, which students may wish to probe in greater depth.

CHAPTER 11
Travel and other graphs

Starter 11

Some interesting general knowledge about the speeds of different animals is presented here, along with simple calculations involving division or ratio. This begins to prepare the students for more detailed calculations involving speed and distance, which they will meet in the main section on travel graphs.

Exercise 11.1

▶▶ A speed-up sheet is available for this exercise.

All the questions in this exercise are about how a quantity varies over time. Care should be taken with the water container questions, as the container may be filling or emptying. The key point is that if water is flowing in (or out) at a constant rate, then the water level will change most rapidly when the container is at its narrowest.

Ⓦ *Worksheet 11.1*

This worksheet provides further practice of interpreting real-life graphs.

Exercise 11.2

▶▶ A speed-up sheet is available for this exercise.

Exam questions on travel graphs are usually quite easy, and can carry a lot of marks. Speeds and distances are often more conveniently calculated using proportional ideas, rather than conventional formulae, especially in a non-calculator paper. Do encourage your students to think in this way: for example, 40 miles per hour means 40 miles in 60 minutes, 20 miles in 30 minutes, 10 miles in 15 minutes, 2 miles in 3 minutes and so on.

Note that many exam questions will use multiples of 5 minutes along the time axis.

Ⓦ *Worksheet 11.2*

This worksheet provides further practice of interpreting distance–time graphs.

Exercise 11.3

▶▶ A speed-up sheet is available for question 4 of this exercise.

Students will be expected to know that the area under a velocity–time graph corresponds to the distance travelled, and that the gradient of the graph gives the accelereation. Questions will be set on line segments only, not curves.

Review Exercise 11

▶▶ Speed-up sheets are available for questions 2, 3, 4, 6, 7 and 9 of this exercise.

Internet Challenge 11

In contrast to the starter, which was about speeds achieved by creatures in their natural element, this investigation looks at the speeds of various man-made objects.

Different internet sources may give slightly variable answers, but they should be sufficient to distinguish the correct rank order from slowest to fastest. You could perhaps get the class to vote on what they think is the correct order, prior to doing the internet research.

CHAPTER 12
Working with shape and space

Starter 12

Alphabet soup is a fun way of revising basic angle facts, including that angles at a point add up to $360°$ and that angles on a line add up to $180°$. Students should be familiar with these ideas from Key Stage 3.

Exercise 12.1

Students usually prefer to use the informal names of F-angles and Z-angles, which help to reinforce the shape they are looking for. Once they are using these results with confidence, however, it is better to insist on the formal mathematical names of, respectively, corresponding angles and alternate angles.

Exercise 12.2

Students should already know that angles in a triangle add up to $180°$, and those in a quadrilateral add up to $360°$, so this exercise should be quite straightforward. One or two questions can be done more efficiently by using the triangle result that the exterior angle is equal to the sum of the interior opposite angles; this result is given, and proved, in the examples in this section of the book.

W *Worksheet 12.2*

This worksheet gives more practice at calculating angles using the rules introduced in the first two sections of the chapter.

Exercise 12.3

Angles inside an n-sided polygon add up to $180 \times (n - 2)°$, while the external angles of any polygon add up to $360°$. Hence the external angle of a *regular* n-sided polygon must be $360° \div n$. Conversely, the number of sides of a regular polygon can be found by dividing $360°$ by the size of an exterior angle.

Exercise 12.4

The numerical questions at the beginning of the exercise should be straightforward, but exam candidates often struggle with algebraic questions like numbers 18, 19, 20: further practice of this type of question might be advisable.

W *Worksheet 12.4*

This worksheet gives more practice at calculating the area of 2-D shapes; the final question also covers perimeter.

Exercise 12.5

Students should be reminded to state their units carefully when answering questions about area or volume; failure to do so can cost marks in the exam. The formula for the volume of a prism is given on the formula sheet for the Higher Tier exam (and the Foundation one too).

Internet Challenge 12

The four-colour theorem is an amusing geometrical result that students might not have encountered before. In practice map-makers might want to use more than the minimum four colours necessary, because they usually shade different regions belonging to the same country in the same colour, for example Alaska and the contiguous USA, or Denmark and Greenland.

CHAPTER 13
Circles and cylinders

Starter 13

Here are some famous, and not so famous, approximations to pi. Interested students might like to do some internet research to find out when these various approximations were first obtained. The best one $(355 \div 113)$ is astonishingly accurate, and has been known for a surprisingly long period of time.

Exercise 13.1

The most common mistake here is to fail to distinguish between radius and diameter when applying the formulae: students should check which is given. They should also be encouraged to round their answers carefully, to avoid unnecessarily losing marks in the exam.

  **13 Circles and cylinders**   GCSE Maths for Edexcel: Higher Teacher's Resource © Hodder Murray 2006

Exercise 13.2

At first sight this topic looks daunting, but the problems are quite straightforward provided the basic circle formulae are well understood. Emphasise that perimeter calculations must include the two radii as well as the curved arc length.

Exercise 13.3

Working back from circumference to radius (or diameter) is not hard, but working back from area is trickier, since the operations are less simple. Note that there is no direct route from area to diameter; students should work back from area to radius first, then double the radius at the end.

Exercise 13.4

Consider making a simple model, out of strong paper or thin card, to demonstrate the diagram on page 233; this makes it easier for students to learn the correct formulae. Exam questions on this topic often include density, so that the volume may be converted into mass at the end.

Exercise 13.5

These questions are easier than the calculator ones. Encourage students to think of pi as a mathematical symbol rather than 3 point something. Students with sound algebraic skills should have no difficulty here.

Internet Challenge 13

The Earth is approximately (but not exactly) spherical, so questions 3 and 4 allow the use of the formula for the circumference of a circle to be used to find the distance all the way round the world.

We perhaps imagine that people who travel 'round the world' do so by following a longest possible route (a Great Circle), but actually only astronauts can do this! The route followed by a UK round-the-world sailor, for example, actually comprises a long journey south, a short (but difficult) circuit around the southern ocean, and then a long northerly return home.

CHAPTER 14
Constructions and loci

Starter 14

This simple starter is intended to make sure that candidates can draw circles accurately and consistently using compasses. Once the central circle has been drawn, the first outer circle should be drawn using an arbitrary point on the circumference of the initial circle. Further circles are constructed by progressively moving around the initial circle, and the final circle should intersect with two other distinctive intersections, as a check on the final accuracy of the construction.

Exercise 14.1

Before attempting a detailed construction, a rough sketch is always a good idea. In the exam, candidates will usually be given a measured line segment to start them off; this also puts each candidate's drawing in the same place on the script, to help maintain consistency in the marking.

Exercise 14.2

▶▶ A speed-up sheet is available for this exercise.

Construction questions sometimes look easier if the page is rotated so that a given line segment is horizontal. Emphasise that all construction lines should remain visible. Examiners usually allow a tolerance of ± 2 mm.

Exercise 14.3

▶▶ Speed-up sheets are available for questions 1 to 8 of this exercise.

Locus constructions are simply a matter of applying the abstract constructions of Exercises 14.1 and 14.2 to real-life problem. The six standard loci shown on pages 257 and 258 should be studied and learnt thoroughly. Remember that examiners sometimes attach bearings to this topic also; these are revised at the end of the chapter, immediately before the Review Exercise.

Review Exercise 14

▶▶ Speed-up sheets are available for questions 1, 2, 3, 6, 8, 10, 11 and 13 of this exercise.

Internet Challenge 14

▶▶ A speed-up sheet is available for this challenge.

After a fairly straightforward chapter, this internet challenge is quite demanding.

Students should use the strong diagonal lines to construct two vanishing points on the horizon; these should be level, horizontally. These vanishing points can then be used to project the two buildings so that they are both at the same depth from the observer. The Answers section of this Teacher's Resource shows one way of doing this.

CHAPTER 15
Transformation and similarity

Starter 15

The goal of this activity is to start the students thinking (informally at this stage) about congruence and the properties that are preserved under translation, rotation and reflection. These ideas are developed later in the main chapter.

Exercise 15.1

▶▶ Speed-up sheets are available for this exercise.

The first question in the exercise checks that students can spot lines and planes of symmetry by eye. This is quite intuitive, and many students will not require much further practice.

The remainder of the exercise gives practice at identifying mirror lines. Emphasise the advice in the textbook: diagonal mirror line problems are best tackled by turning the page so the mirror line is vertical. (The human eye/brain finds this much easier to recognise.)

Exercise 15.2

▶▶ Speed-up sheets are available for questions 1 to 6 of this exercise.

Many students will solve rotations problems using tracing paper or by drawing radii from the centre of rotation: both methods are effective. Strong students may realise that a vector method is applicable to clockwise rotations of 90° and 270°: count the (x, y) offsets from the centre of rotation and replace them with $(y, -x)$ or $(-y, x)$ respectively.

When describing rotations, make sure that students record whether they are clockwise or anticlockwise (except in the case of 180°).

Exercise 15.3

▶▶ Speed-up sheets are available for this exercise.

It is important the students read this sort of question carefully and carry out the instructions in the correct order: some questions ask for A → B then B → C, while others ask for A → B and A → C.

Exercise 15.4

▶▶ Speed-up sheets are available for questions 1 to 5 of this exercise.

Emphasise that the lengths of construction lines must always be measured from the centre of enlargement. As a check, take a line segment from the original object (say 2 squares long) and multiply this length by the enlargement factor, then check that this matches the corresponding length in the enlarged image.

Exercise 15.5

This topic benefits from constant reinforcement. In the examination, questions may refer to objects by mass: ensure that students appreciate that mass scales in the same way as volume, provided the objects are of constant density.

Review Exercise 15

▶▶ Speed-up sheets are available for questions 1, 4, 5, 6, 10 and 12 of this exercise.

Mathematics is a language, and, as such, has its own specialised vocabulary. This challenge encourages students to develop confident use of correct mathematical definition.

Internet Challenge 15

The chapter contains a lot of specialist geometrical vocabulary; the challenge builds on this. Some of these definitions are central to the Higher GCSE, while others are provided as enrichment.

CHAPTER 16
Pythagoras' theorem

Starter 16

This starter revises squaring and square rooting, since these will be needed for work on Pythagoras. Some whole numbers, such as 9, 16 and 25, have exact whole-number square roots: these are the **perfect squares**. Some GCSE questions will be designed so that the answer is a whole number; others require a calculator approximation. The instruction 'Give your answer correct to 3 significant figures' indicates that the latter is the case.

Exercise 16.1

This exercise begins by looking at what is, strictly speaking, the converse of Pythagoras' theorem, rather than the theorem itself. That is, if the three sides of a triangle fit the condition $a^2 + b^2 = c^2$, then the triangle must be right angled; otherwise it is not. The corresponding calculations are simple, since no square rooting is necessary, making this a nice way of introducing the topic.

Exercise 16.2

This is a straightforward exercise on using Pythagoras' theorem to find the hypotenuse. A few questions involve perfect squares (sometimes concealed as decimals) and thus will give exact values.

Exercise 16.3

Here Pythagoras' theorem is used to find one of the short sides of a right-angled triangle (that is, not the hypotenuse), so subtraction must be used instead of addition.

Exercise 16.4

Pythagoras' theorem in 3-D is used for the first time in this section. The approach is to formulate the problem using two separate 2-D triangles, and solve each by Pythagoras. Stronger students may discover for themselves the 3-D generalisation of Pythagoras theorem to $a^2 + b^2 + c^2 = d^2$; this result is mentioned briefly in the Internet Challenge, and is then picked up again in Chapter 27 Further trigonometry.

Internet Challenge 16

Many Pythagorean triples (such as 3, 4, 5 and 5, 12, 13) can be generated by the formula sequence n, $\frac{1}{2}(n^2 - 1)$, $\frac{1}{2}(n^2 + 1)$ and simple multiples thereof. But there are others, for example 8, 15, 17.

Fermat's last theorem is, in effect, an astonishing generalisation about Pythagorean triples; it is intriguing to know that the truth of this conjecture has been suspected for centuries, but was only proved rigorously in 1994 (the proof was published in 1995).

CHAPTER 17
Introducing trigonometry

Starter 17

The triangular spiral exercise makes the point that as the triangle becomes longer and skinnier, so the size of the apex angle shrinks; this anticipates the behaviour of the sine function. Since the hypotenuses form a progression of $\sqrt{5}$, $\sqrt{6}$, $\sqrt{7}$ and so on, some of them will be exact square roots: the hypotenuse of the fifth triangle, for example, is $\sqrt{9} = 3$.

Exercise 17.1

In this exercise the sine function is used to find the length of an unknown side. Ensure that students know that to find an unknown short side, multiplication is required; to find an unknown hypotenuse, division is required. This distinction is summarised and developed in section 17.4.

Exercise 17.2

Here the cosine function is used, together with multiplication or division, to find an unknown side. Work on unknown angles is deferred until later in the chapter.

Exercise 17.3

Finally, the tangent function is used, together with multiplication or division, to find an unknown side.

Exercise 17.4

In the GCSE exam, students will need to decide for themselves which trigonometrical function to use, and whether to couple it with multiplication or division. By writing the familiar mnemonic SOHCAHTOA as three 'physics triangles' students can easily pick the right function, together with the correct process.

Exercise 17.5

Students often find inverse trigonometry problems surprisingly difficult. Emphasise that if they expect the answer to be an angle, then they must press, for example, [INV] [SIN] (depending on their calculator) rather than just [SIN] on their calculator. Once again, the 'physics triangles' method helps to identify which ratio is to be used.

W *Worksheet 17.5A*

This worksheet provides more practice in calculating lengths in practical contexts using the trigonometrical ratios: there is roughly the same number of questions using each of the three ratios. The earlier questions have the unknown in the numerator of the ratio, whilst in the later, harder questions, the unknown appears in the denominator.

W *Worksheet 17.5B*

This worksheet concentrates on the calculation of angles in practical contexts using the three trigonometrical ratios, with two sides or lengths given. Students should be encouraged to use the information to draw a diagram for each question.

W *Worksheet 17.5C*

This worksheet requires students to position the information given correctly and then to calculate either a side or an angle. Some questions involve the use of Pythagoras' theorem, rather than the trigonometrical ratios. Questions 1 to 10 have the unknown in the numerator or the ratio; questions 11 to 20 are slightly harder, with the unknown in the denominator.

Exercise 17.6

GCSE questions are often multistage – the answer to one part of the question provides a length or angle that is needed to solve a later part. Students should be careful not to round off answers too early in this kind of question.

Internet Challenge 17

Everyone has heard of Pythagoras, but there are many other mathematicians whose names are associated with important results in geometry. This challenge invites students to track down some of the more famous ones.

CHAPTER 18
2-D and 3-D objects

Starter 18

This exercise uses the concept of nets, in case students have not already met them or need to be reminded. You could use this as a theoretical exercise, in which they have to visualise how the nets would fold up. Alternatively, some students may prefer to make little models, perhaps as a homework exercise, in order to see which ones work.

Exercise 18.1

Although a plan view is clearly defined, it is not always obvious what the difference is between a front elevation and a side elevation; these two terms are sometimes used in an unexpected way. Question 6 of the Review Exercise, for example, chooses to refer to one specific view as a front elevation, whereas many students would instinctively call this a side elevation. The definition of 'front' is thus arbitrary, although the front and side elevations must be taken from perpendicular directions.

Exercise 18.2

▶▶ A speed-up sheet is available for question 3 of this exercise.

The x, y and z axes are at right angles to each other, but the orientation of these axes may vary from one book/teacher to another. Here we have the x and y axes in the plane of the page, with the third (z) axis coming out of the page; this corresponds to Edexcel exam usage.

Exercise 18.3

The formulae for the surface area and volume of a cone and a sphere are all given in the Edexcel exam formula sheet, though candidates will need to learn the rule for the volume of a pyramid (one-third times the area of the base times the height).

Exercise 18.4

Conversion of square and cube units is a simple process, and students will probably have no difficulties with this exercise. In the GCSE exam, however, it is regularly done poorly, indicating that candidates have forgotten (or failed to recognise) what to do. This topic benefits from regular revision.

Exercise 18.5

▶▶ A speed-up sheet is available for questions 1, 2, 4 and 5 of this exercise.

Some students find it helpful to replace all the dimensionless numbers with 1, so that, for example, $2a + 3b$ is treated as if it were $a + b$.

Curiously, students tend either to understand this topic immediately or to struggle with it for some time.

Review Exercise 18

▶▶ A speed-up sheet is available for questions 2 and 9 of this exercise.

Internet Challenge 18

Constructing models of polyhedra can be great fun! Specialist publishers such as *Tarquin* provide helpful further resources.

Teaching Notes: Chapter 20

CHAPTER 19
Circle theorems

Starter 19

▶▶ A speed-up sheet is available for this starter.

When talking about some of the more advanced theorems (for example, the intersecting tangent and chord theorem) it is helpful for students to understand and use the associated vocabulary correctly. This starter checks that the key terms are properly understood.

Exercise 19.1

This section looks at some tangent properties, before the circle theorems that apply inside a circle are introduced. The fact that a tangent and radius (or diameter) meet at right angles is surprisingly awkward to prove; students might enjoy seeing a proof based on a limiting case of a series of bisected parallel chords.

Exercise 19.2

The first circle theorem states that the angle at the centre is twice the angle at the circumference. The other two theorems here follow as special cases, so it is worth spending some time looking carefully at how to prove that first theorem.

Exercise 19.3

The remaining circle theorems are introduced in this section, including the intersecting tangent and chord theorem (also known as the alternate segment theorem). Candidates may be required to know how to prove this result.

Note that questions in this exercise, and the review exercise which follows, often require the use of more than one circle theorem. You may want to draw students' attention to question 10, and suggest drawing an extra pair of lines forming an angle of $\frac{1}{2}p$ in the segment at the top of the circle.

W *Worksheet 19R*

Mixed questions on the various circle theorems are included in this revision sheet.

Internet Challenge 19

The proof of the nine-point circle theorem is a nice illustration of the circle theorems used in reverse, which very able students will enjoy. Others might find the accurate construction of such a diagram a challenge in itself: it is not trivial! A large diagram is a good idea.

CHAPTER 20
Collecting data

Starter 20

▶▶ A speed-up sheet is available for this starter.

Tally charts are very simple, but care must be taken about the number of groups being used – not too few (Enrico) nor too many (Kwame). A certain amount of judgement is necessary to get this just right.

Exercise 20.1

In order to decide whether a particular sampling method is fair or not, students need to have an idea of what the background population is.

Exercise 20.2

▶▶ A speed-up sheet is available for this questions 5 and 6 of exercise.

Exam questions on questionnaire design can appear vague and confusing, but they usually expect students to improve on a poor question by:
- making the question more precise
- removing bias
- using response boxes (tick boxes).

Review Exercise 20

▶▶ A speed-up sheet is available for question 7 of this exercise.

Internet Challenge 20

The National Census is used to collect very basic information about the people who live in England and Wales. By law every single household is required to submit a reply; this is intended to remove bias issues that might arise if only a sample of households were used. Thus the penalty for non-response is quite high.

CHAPTER 21
Working with data

Starter 21

Suppression of the origin (as in the medical research diagram) is a commonly used device adopted to create a misleading impression of rapid growth. Such diagrams are often found in newspapers, typically with no scale on the y axis.

Exercise 21.1

The stem and leaf diagram is a brilliant idea: it allows discrete data to be grouped (like a tally chart) but without losing sight of the ungrouped, raw values. Emphasise that 23 should be written as 2 | 3 and **not** 20 | 3, since the latter may be mistaken for 203, an error that several otherwise reputable textbooks make. Students should always add a key, since this removes any possible ambiguity.

Exercise 21.2

Finding the mean from a frequency table is a favourite examination question, and the examiner will usually provide one (or, for grouped data, two) extra columns printed on the paper: this is a strong hint of the method required. Students should note the slightly different way in which discrete and continuous calculations are done; this also has an impact on the corresponding diagrams to be drawn in the next section.

Exercise 21.3

▶▶ Speed-up sheets are available for this exercise.

Vertical line graphs are used for discrete data, whereas grouped discrete and continuous data may be displayed using a frequency polygon. The frequency polygon does not need to be extended down on to the x axis, and thus is usually open: the name 'polygon' is rather misleading in this context.

For continuous data it is just as easy to draw a histogram as a frequency polygon, yet the GCSE examiner rates histograms as a higher graded topic. A favourite exam question is to provide an incomplete frequency table and an incomplete (and uncalibrated) histogram. The idea is to identify one group that is complete in both the table and the diagram, and use this to calibrate the frequency density axis. The table and diagram may then be completed quite simply.

Exercise 21.4

▶▶ Speed-up sheets are available for questions 2, 3 and 4 of this exercise.

Cumulative frequency diagrams should not cause any problems, though students occasionally plot the cumulative frequencies against the midpoints of the groups rather than the upper bounds, introducing a systematic error into the values of the quartiles and median.

For the interquartile range it is not sufficient to say, for example, 'from 32 to 48': candidates must calculate the difference, in this case $48 - 32 = 16$.

Exercise 21.5

▶▶ Speed-up sheets are available for questions 1, 2, 4 and 5 of this exercise.

Scatter diagrams are easy, and Higher Tier students should be fully familiar with them from work at Key Stage 3. Examination questions usually provide a printed grid with some of the points already plotted.

The line of best fit should pass through the (hypothetical) mean point, computed by finding the mean of all the x values and the mean of all the y values. This is not required at GCSE, but more able students might find it helpful to calculate this and use it as a pivot when they construct the line of best fit.

Review Exercise 21

▶▶ Speed-up sheets are available for questions 6 to 13 of this exercise.

Internet Challenge 21

Correlation coefficients are not on the specification, but they are useful in many real life situations. A value close to 0 indicates no correlation in the population, whereas a value towards 1 (or -1) indicates positive (or negative) correlation.

The precise borderline value where the evidence becomes significant will depend on the sample size, and

may be found from statistical tables of critical values, such as those published by A-level boards for use in Statistics module examinations.

Product–moment correlation is only applicable to straight-line analysis (more specifically, where there is evidence of an underlying joint bivariate normal distribution, which gives a characteristic elliptical cloud of points). Spearman's is applicable in a more general range of situations, including non-linear correlation, which is why subjects such as Geography and Biology tend to use it a lot.

CHAPTER 22
Probability

Starter 22

One dice should give an approximately uniform distribution: that is, each number should occur about 10 times.

Two dice should, theoretically, give a triangular distribution, with 7 the most likely score, followed by 6 and 8, then 5 and 9 and so on. Of course, for a small sample of values, the actual distribution of the data will be a little different.

Exercise 22.1

Probabilities should be written as decimals or fractions between 0 and 1, so percentages should not be used. The GCSE examiners often ask for a predicted frequency, obtained by multiplying the probability by the number of trials.

W *Worksheet 22.1*

This worksheet offers additional probability questions of an introductory nature. It could be used either to review earlier work on probability or to reinforce the techniques covered in the first section of the chapter.

Exercise 22.2

Some of the formal language here – for example 'mutually exclusive' and 'exhaustive' – is not strictly necessary for the exam paper, but it does help to make some of the underlying principles clearer.

Rules like 'Or means Add' and 'And means Multiply' should be used with some caution: they work only under certain conditions.

W *Worksheet 22.2*

Ideally, this worksheet should be tackled without a calculator, although some students may need one for question 8.

Exercise 22.3

Here students will be multiplying two probabilities to obtain the probability of a combined event. The multiplicative rule works only if the two separate events are independent; this is either assumed (for example, in the case of two dice throws) or stated in the question.

Exercise 22.4

▶▶ A speed-up sheet is available for questions 1 and 3 of this exercise.

Tree diagrams are a popular device with the GCSE examiner, and are a source of easy marks. In all these problems the probabilities on the second set of branches are independent of those on the first set, but in harder problems (see Chapter 32) the probabilities on the second set of branches may change, depending on the outcome of the first set.

W *Worksheet 22.4*

This worksheet provides further questions of the sort seen in exercises 22.3 and 22.4 and should also help to reinforce the fact that probabilities should be stated as a fraction, decimal or percentage. Students are required to decide whether or not a probability tree is required and, in the questions involving a probability tree, whether the events are independent or not.

Review Exercise 22

▶▶ A speed-up sheet is available for questions 5, 7, 11 and 12 of this exercise.

Internet Challenge 22

Some students will be surprised to learn that it is better to switch doors. Sticking to the original choice means that the probability of winning remains at its original value of one-third, whereas switching increases the probability of winning to two-thirds.

CHAPTER 23
Using a calculator efficiently

Starter 23

Popular suggestions here will be 431^2 and 321^4 but also consider stacked powers like $4^{3^{21}}$ (which has 6×10^9 digits) and $3^{4^{21}}$ (which has 2×10^{12} digits)!

Exercise 23.1

This exercise confirms that candidates can use their calculator's power and root keys effectively. Older models still use an $\boxed{x^y}$ key, but some more recent machines use a caret ($\boxed{\wedge}$) as a power. Remind students that it is unwise to borrow an untried calculator immediately before a major exam, as implementation of these keys varies so much from one model to another.

Exercise 23.2

In a question like number 12, some candidates prefer to work out the top (29) and the bottom (25) separately, and write them down, before going on to complete the division. While this is, perhaps, not especially efficient, GCSE examiners like it because they can see exactly

23 Using a calculator efficiently   GCSE Maths for Edexcel: Higher Teacher's Resource © Hodder Murray 2006

what is going on. Candidates who break their calculations up in this way usually have a higher success rate.

Exercise 23.3

A calculator fraction key is very versatile, helping with questions on ratio as well. An interesting addition to the calculator line-up is Casio's Natural Display range; such calculators can be set up to look for rational answers first, with decimal equivalents available at the touch of a toggle key. These calculators also have a natural-looking screen input mode for intricate calculations, similar to the appearance of Microsoft's Equation Editor for Word.

Exercise 23.4

The $\boxed{\text{ANS}}$ key offers an extremely useful shortcut for compound interest and similar calculations, but it is under-utilised by many students at GCSE.

Exercise 23.5

Upper and lower bounds can cause some confusion, since the interval formed by these bounds, while symmetric, is closed at the bottom and open at the top. Past GCSE questions have asked such things as 'A length is measured as 12 cm to the nearest centimetre. What is the greatest possible length?', to which, strictly speaking, there is no answer; the limit of 12.5 cm is an upper bound (or supremum), but it is not attainable and therefore is not a maximum in the truest sense of the word.

Reassure your students that such distinctions are not important at GCSE; when the examiner asks for a maximum or greatest value he simply expects the value of the upper bound. Answers such as 12.4 or 12.49 would be penalised, since they are definitely too low.

Internet Challenge 23

▶▶ A speed-up sheet is available for this challenge.

This crossword provides some interesting and amusing insight into words used in computing and calculating.

The number 10^{100} is a googol, after which the internet search engine was supposed to be named. An unfortunate spelling mistake turned out to be a good thing because the domain name google.com was available, whereas googol.com was not.

CHAPTER 24
Direct and inverse proportion

Starter 24

The five puzzles are not to be taken too seriously! Puzzles 2 and 3 may be solved by using the unitary method, which students may well know from Key Stage 3, though this method will not be explicitly examined at GCSE.

Exercise 24.1

The best way to succeed with this topic is to be very formal, and write down an algebraic equation. Using the initial values, the constant of proportionality, k, may be worked out, enabling any other value to be computed with ease. Students should be discouraged from using shortcut alternative methods, since these may not work when extended to inverse proportion later in the chapter.

Exercise 24.2

Once again, algebraic equations should be used as much as possible in the formulation of solutions to inverse proportion problems.

Exercise 24.3

▶▶ A speed-up sheet is available for this exercise.

This topic is quite easy: direct proportion is revealed by a straight-line graph through the origin, and inverse proportion by a hyperbola of the form $y = \dfrac{k}{x}$. No other graph corresponds to either direct or inverse proportion.

Internet Challenge 24

▶▶ A speed-up sheet is available for this challenge.

This challenge shows that the laws of mathematical proportion may be extended way beyond the bounds of the Earth.

CHAPTER 25
Quadratic equations

Starter 25

Simple substitution reveals that some equations have more than one solution, and, in fact, the results from this table should support the quotation (which is a simplified form of Gauss's Fundamental Theorem of Algebra): polynomials of degree n have n roots.

Exercise 25.1

This exercise is divided into three sections, each more difficult that the last. Students who find questions 11 to 20 difficult might need some further, similar practice examples before moving on to the harder questions 21 to 30. The final ten questions underline the point that all the terms of a quadratic must be collected together on one side of the equation before it can be factorised.

W *Worksheet 25.1*

This sheet provides further practice at solving quadratic equation by factorising.

Exercise 25.2

Although the formula requires division by $2a$, the value of a is simply 1 in the first two questions. Watch out for

students who consequently slip into the habit of just dividing by 2 in every problem, instead of by $2a$.

The formula is used when an exact answer by factorisation cannot be found. A useful clue that the formula is necessary is when an exam question says 'Give your answer to 3 significant figures.'

Exercise 25.3

GCSE exam questions often set a problem in a context that leads to a quadratic equation. Typically such a question will say something like 'Show that x must satisfy the equation $2x^2 + 11x - 30 = 0$' and then require the candidates to solve the equation.

Stronger candidates enjoy this sort of question. Weaker candidates should be reassured that, if they cannot see how to do the 'show that…' part of the question, they should rejoin the question part way through, and solve the given quadratic, thus still earning most of the marks.

Exercise 25.4

Completing the square is a technique that many GCSE students find unpalatable; it can appear rather dry and pointless. Certainly it is not a neat way to actually solve a quadratic; the formula does this in a much tidier and more familiar way.

The real benefit of completing the square is that it allows the axis of symmetry of the parabola to be determined, and likewise the coordinates of the vertex, so it is useful in helping to construct a sketch of the graph of a quadratic function.

Internet Challenge 25

▶▶ A speed-up sheet is available for this challenge.

There are four conic sections – circle, ellipse, parabola, hyperbola – and this challenge introduces just a few of their properties.

The curve whose envelope is drawn in question 3 seems to resemble a rectangular hyperbola at first, but it is in fact a parabola.

With care the string and two pins method can be made to work – it can form a successful whiteboard demonstration, but some practice beforehand is advisable!

CHAPTER 26
Advanced algebra

Starter 26

Here is a justification for the given formula.

If the grid contains m squares by n squares, then it contains $m + 1$ by $n + 1$ lines. Let any rectangle contained in the grid be specified by the position of its

two opposite corners. The first corner can be located on any of the $m + 1$ and $n + 1$ lines, so may be placed in $(m + 1)(n + 1)$ ways. The opposite corner must not lie on the same horizontal or vertical line, so may only be done in mn ways. Thus there are $m(m + 1)n(n + 1)$ combinations altogether.

Further thought shows that each particular rectangle is counted four times by this process. Thus the true number of rectangles is $\dfrac{m(m + 1)n(n + 1)}{4}$.

Exercise 26.1

GCSE questions will not require excessive manipulation of surds, but will require factorising and simple rationalisation. These questions will feature in the non-calculator paper, since some calculators like the Casio Natural Display models can manipulate exact surd expressions.

Exercise 26.2

Some students like to use a shortcut method based on cross-multiplying, but this can lead to difficulties with questions like number 12 where this is a repeated factor, so should be discouraged.

The answers to questions 13 to 20 can be checked by substituting back into the left-hand and/or right hand-sides of the given equations.

Exercise 26.3

Students find these questions quite enjoyable, but some try to cancel too far. Once there are no further common factors then the cancelling process ceases.

Exercise 26.4

Simultaneous equations, one linear and one quadratic, are encountered here purely in an algebraic sense. The topic is revisited in Chapter 31 Introducing coordinate geometry, in order to find the points of intersection of a straight line with a curve.

Exercise 26.5

Some students may have developed a liking for the reverse flow diagram method of rearranging an equation, but that method only works when the new subject appears once. If the symbol occurs twice, then it is necessary to expand, rearrange, and then take out a common factor so the symbol ends up appearing only once. This is a fairly predictable topic, and is also tested by many exam boards in their first AS pure mathematics module.

Exercise 26.6

Exponential growth and decay is a relatively minor topic at GCSE, though it does occasionally merit a reasonably long question. Candidates who understand the use of multiplying factors and compound interest have no difficulty at all in grasping the key ideas.

Internet Challenge 26

The formulae presented in this challenge feature regularly in algebraic questions; they are all drawn from

26 Advanced algebra

famous applications in mathematics or science. Keen students might like to write an extended list of famous formulae, perhaps trying to get 40 or 50 without having to use too many obscure ones.

CHAPTER 27
Further trigonometry

Starter 27

This starter looks at how to find the distance to an inaccessible point by surveying. The same idea has been used over the years to measure the widths of rivers, the heights of mountains, and the distances of stars.

Exercise 27.1

The sine rule is a simple but effective way of finding the value of an unknown length or angle. Note that the sine rule connects two sides and two angles; three of these four quantities must be known in order to calculate the remaining one.

Exercise 27.2

Since $\sin \theta = \sin (180 - \theta)$ the sine rule has the possibility of being ambiguous when used to find an unknown angle. Students should remain alert to this possibility.

Exercises 27.3A

The cosine rule connects three sides and one angle; again, three of these four quantities must be known in order to calculate the remaining one. Note that $\cos \theta$ is positive for an acute angle and negative for an obtuse angle, so there is no ambiguous case with the cosine rule.

Exercise 27.3B

This exercise includes questions using both the sine rule and the cosine rule, since students need to be able to decide for themselves when each is required.

Exercise 27.4

The $\frac{1}{2}ab \sin C$ formula is a quick method for finding a triangle's area provided you know two sides and an included angle. Some exam questions give you the area and two sides, and require working backwards to deduce the angle so that other properties of the triangle (such as its perimeter) may be deduced; a few questions of this type are included in the Review Exercise at the end of the chapter.

Exercise 27.5

This exercise develops some of the three-dimensional calculations first encountered in Chapter 16 Pythagoras' theorem. It helps the examiner to understand students'

work if they extract the relevant triangle and draw it out flat on the paper.

Exercise 27.6

There is a rather cumbersome formula for the volume of a frustum of a cone, but it is much simpler to calculate it as the difference of two (mathematically similar) cones.

Internet Challenge 27

Although Heron's formula is not on the GCSE specification, it is a useful technique for students to add to their mathematical toolbox, especially if they are going to take the subject on to a higher level in the future. The result has been known and used for a surprisingly long time.

CHAPTER 28
Graphs of curves

Starter 28

▶▶ A speed-up sheet is available for this starter.

The making waves activity introduces students to the cyclical nature of the sine function, a useful prelude to the chapter.

Exercise 28.1

▶▶ Speed-up sheets are available for questions 1 to 8 of this exercise.

Students should already be familiar with plotting simple quadratic graphs from Key Stage 3. This section revises the drawing up of tables of functions and also includes cubics, exponentials and reciprocals.

Exercise 28.2

▶▶ A speed-up sheet is available for questions 1 to 4 of this exercise.

In this section graphs are used to solve equations by reading off intersection with the x axis, or a simple horizontal line such as $y = 3$. Questions involving the intersection with an oblique line, such as $y = 2x + 3$, are no longer on the Edexcel GCSE specification.

Exercise 28.3

Students need to understand the symmetric properties of the sine and cosine function, and appreciate that their graphs repeat after a period of 360°. The corresponding behaviour of the tangent function, and the fact that it is disconnected at intervals of 180°, also need to be known.

Exercise 28.4

▶▶ A speed-up sheet is available for questions 1 to 4 of this exercise.

Transformation of graphs is probably one of the hardest topics on the entire Higher Tier specification; indeed, the same work appears spread across several of the Core

Pure modules in A level. Many students will find this tricky; in particular, the fact that $f(x) \to f(x + 2)$ corresponds to a translation of 2 units left, not right, is counter-intuitive to some.

Review Exercise 28

▶▶ A speed-up sheet is available for questions 1, 2 and 8 of this exercise.

Internet Challenge 28

These nine weird and wonderful curves have equally weird and wonderful names! Most of them have equations that are best expressed using a polar coordinate system; you might encourage your most able students to investigate polar coordinates on their own.

CHAPTER 29
Vectors

Starter 29

A knight's move (for example two across and one up) is an example of a displacement vector. In this starter, students are adding 64 such displacement vectors together, to end up with a zero vector – though they won't realise this at the time.

Exercise 29.1

These questions are rather simplistic compared to typical exam questions; they are intended merely to confirm basic understanding before moving on to the next level.

Exercise 29.2

Much of the work on this topic requires a clear understanding of how vector addition and subtraction works, including the application to geometric proof later in the chapter.

Exercise 29.3

This section establishes an important result: if one vector is a multiple of another, then the two vectors are parallel. This result is used repeatedly in work on geometric proof later in the chapter.

Exercise 29.4

Proving geometric results using vectors is an elegant topic, but one which students find hard. They should persevere: exam questions often consist of many parts, and some easy marks may be scored at the beginning even if the question becomes hard towards the end.

Internet Challenge 29

There are many fascinating chessboard problems: these are only the tip of the iceberg. Some students will be able to solve the eight Queens problem surprisingly quickly; there is a methodical approach that yields a quick solution, based on separating a string of Queens by a knight's move.

CHAPTER 30
Mathematical proof

Starter 30

At first sight all the steps look correct, but since $a = 2b$ it follows that $2b - a$ is zero. Since division by zero is not permitted, the algebra breaks down at Step 6; in effect the previous line is saying that $1 \times 0 = 2 \times 0$ (which is true), but it does not follow that therefore $1 = 2$.

Exercise 30.1

Students need to proceed carefully with this exercise, checking that matching lengths are included between the same angles in each case. If a question specifies two of the angles in a triangle, then it is often a good idea to work out the third one too.

Exercise 30.2

Question 6 shows one way of proving Pythagoras' theorem; there are literally hundreds of different ways of achieving this. Here is a nice alternative proof.

Begin with a right-angled triangle, with sides a, b, c (where c is the hypotenuse).

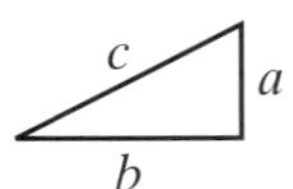

Now create three enlargements of this triangle, using enlargement factors a, b and c respectively.

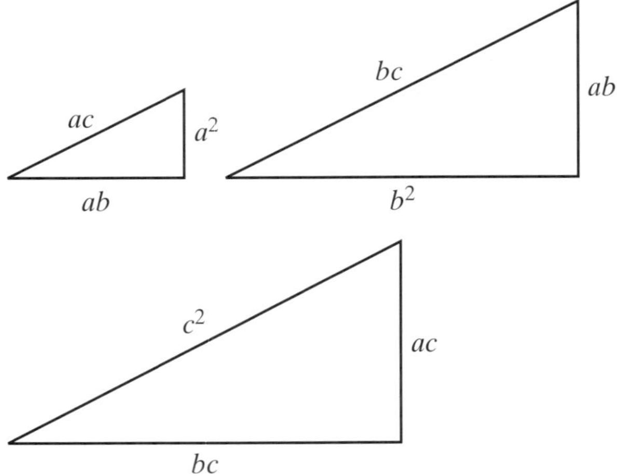

Now rotate and fit them together.

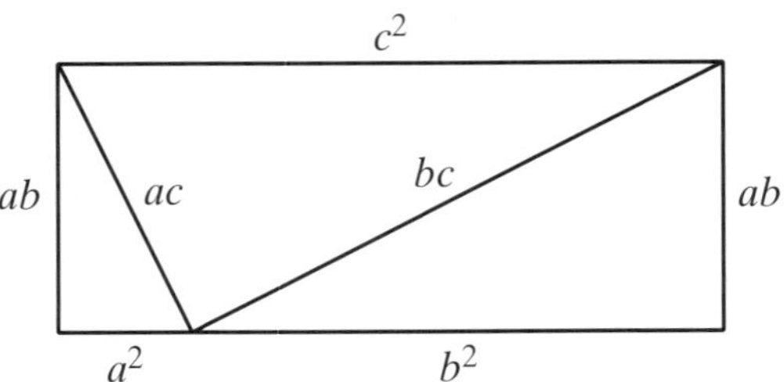

30 Mathematical proof   GCSE Maths for Edexcel: Higher Teacher's Resource © Hodder Murray 2006

This rectangle has length $a^2 + b^2$ along the bottom, but c^2 along the top.

Thus $a^2 + b^2 = c^2$ and Pythagoras' theorem is proved.

Exercise 30.3

In this section students need only find one counter-example to show that a generalised statement is false, and most will find this quite easy.

The formula in question 10 is of special interest; it fails to be prime when $n = 41$ because $41^2 + 41 + 41 = 41(41 + 1 + 1) = 41 \times 43$. Thus 41 is a factor of a result, and so the result cannot be prime.

The formula also fails for $n = 40$, as $40^2 + 40 + 41 = 40(40 + 1) + 41 = 40 \times 41 + 41$, which clearly has a factor of 41.

In all the cases from $n = 0$ to 39, however, the result is prime.

Internet Challenge 30

Students may be surprised to learn that there is a $1 million prize for proving the Riemann hypothesis – a strong incentive to study mathematics beyond GCSE level!

CHAPTER 31
Introducing coordinate geometry

Starter 31

The chapter begins with a simple coded message of encouragement. Even without being told the coding system, substitution codes are easy to crack, using the relative frequency of letters. (E should be most common, then T, and so on.)

Exercise 31.1

The idea of using Pythagoras' theorem to find the distance between two points on a coordinate grid appears both on GCSE and AS specifications, so this is a bridging topic between GCSE and A level. The concept is perhaps best explained using the formula in words to begin with, but the algebraic form should be introduced once the principle is understood.

Exercise 31.2

Equations of circles are another bridge between GCSE and A level, but A level students will meet them in the form $(x - a)^2 + (y - b)^2 = r^2$. At GCSE, circles will always be centred on the origin, giving the simpler equation $x^2 + y^2 = r^2$.

Exercise 31.3

Students should be familiar with the gradients of parallel lines from earlier work at Key Stage 3, but perpendicular lines are a new topic at this level. Once again, the idea that the gradients multiply together to make -1 is a bridging concept that is used throughout the early part of an A level course.

Internet Challenge 31

This rather unusual challenge is based on Pick's theorem, which was developed about a century ago.

CHAPTER 32
Further probability and statistics

Starter 32

This chapter begins to introduce some of the ideas of conditional probability, which, although not fully developed until A-level, can lead to some unexpected results. Students should be surprised at the outcome of this starter too: Ben spends less per present than Joe on Monday and again on Tuesday, yet spends more per present when the two days are combined.

Exercise 32.1

The key to success with combined means is always to work with totals when combining the two groups.

Exercise 32.2

▶▶ A speed-up sheet is available for this exercise.

Moving averages are a very powerful statistical tool for studying long-term behaviour, but are under-utilised at GCSE simply because of time pressure in the exam. The internet challenge at the end of this chapter shows how moving averages can be used to 'smooth out' a long run of data, and how they enable periodic variations to be seen very clearly.

Exercise 32.3

▶▶ A speed-up sheet is available for question 1 of this exercise.

Students should already be very familiar with tree diagrams from the work in Chapter 22, which looked only at independent events. Here the probabilities on the second set of branches depend on the outcome of the first trial, so the figures on the upper set of branches are different from those on the lower set.

These questions typically arise when sampling without replacement (taking two socks from a drawer, for example), so students should use fractions, not

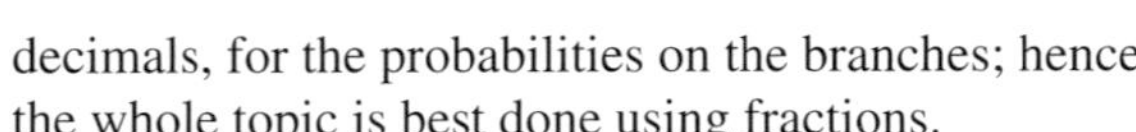

decimals, for the probabilities on the branches; hence the whole topic is best done using fractions.

Internet Challenge 32

This challenge demonstrates the advantages of plotting moving averages; the effect can be seen by comparing the raw graph with the one based on three-point moving averages. The averaging process acts, if you like, as a filter, suppressing high frequency 'flickering' in the data so that low frequency changes can be seen more clearly; this is not unlike some digital enhancement techniques that are used to clean up flawed or damaged images (scratch filters, for example).

The table stops with data for 2004; keen students might want to use the internet to bring the data set fully up to date. Care must be taken; there are several slightly different systems in use for counting sunspot numbers.

Teaching Notes: Chapter 32

  **32 Further probability and statistics** GCSE Maths for Edexcel: Higher Teacher's Resource © Hodder Murray 2006

1 Annabelle and Beatrice pick five barrows of apples.
   Each barrow holds 120 apples.
   The apples are packed into boxes; each holds 20 apples.
   The boxes are sold for £3 each.
   If all the boxes are sold, how much money do Annabelle and Beatrice make?

2 Three schools are planning a skiing holiday.
   At the three schools there are 120, 150 and 90 students, respectively, who want to go.
   There are also 25 teachers going.
   The coaches that are to be hired cost £175 per coach and will each hold 55 people.
   a) How many coaches are needed?
   b) What will be the total cost of hiring them?

3 An aeroplane flying at 6585 feet suddenly has to descend 2550 feet to miss an oncoming aircraft.
   It then climbs 7845 feet.
   What height is the plane flying at now?

4 A new office needs $612 \, \text{m}^2$ of industrial carpet.
   If the carpet costs £17 per m^2, work out the total cost of the flooring.

5 24 cans of lemonade cost £3.84.
   a) Work out the cost in pence of one can of lemonade.
   Fizzy orange costs 19p a can.
   Simran buys two cans of lemonade and three cans of orange.
   b) **(i)** How much does Simran spend?
       (ii) How much change does she get from £1?

6 **a)** Luke is paid £175 per day for 65 days. Anna is paid £1020 per week for 13 weeks.
       Work out the total amount that each is paid.
   b) How much does a doctor who earns £62 400 per year get paid:
       (i) per month?
       (ii) per week?

1 Work out the difference between each of these pairs of temperatures.
 a) $-5°$ and $10°$ b) $-3°$ and $4°$
 c) $0°$ and $7°$ d) $-1°$ and $9°$
 e) $-6°$ and $-11°$ f) $5°$ and $-4°$

2 Arrange these temperatures in order, lowest first.
 a) $-7°, 3°, -9°, -1°, 4°, -4°$ b) $-1°, 2°, -2°, 0°, -8°, 7°$

3 Arrange these numbers in order, largest first.
 a) $2, -1, 0, -3, 1, 6$ b) $-8, -4, 3, -2, 1, 0$

4 The following temperatures were taken from a garden thermometer.
 Complete the table with the missing numbers.

Day	Maximum temperature	Minimum temperature	Change in temperature
Monday	5	-2	
Tuesday	8		9
Wednesday		-3	6
Thursday	0	-4	
Friday	11		13

5 Mr Rich earns £2250 every month after tax.
 At the beginning of April, he owes the bank £400.
 During April he spends £2000.
 a) What is his bank balance at the end of April?
 b) What will his bank balance be at the end of May if he spends £2150 during this month?

      GCSE Maths for Edexcel: Higher Teacher's Resource © Hodder Murray 2006

Work out the answers to these questions about adding and subtracting with fractions. Do not use a calculator.

1 Simon walks $2\frac{2}{3}$ km, rests for a while, and then walks on a further $1\frac{3}{4}$ km in the same direction.
   How far is he now from his starting point?

2 Emily is $1\frac{5}{8}$ m tall and stands on a wall $2\frac{3}{5}$ m high.
   What is the combined height of Emily and the wall?

3 The distances between four petrol stations on a main road are $\frac{4}{9}$ km, $4\frac{5}{36}$ km and $3\frac{1}{3}$ km.
   If a motorist drives from the first petrol station to the fourth one, how far has he travelled?

4 I travel $24\frac{3}{4}$ km before my car breaks down. My total journey is measured as $27\frac{15}{16}$ km.
   How far am I from my destination?

5 A plank of wood measures $3\frac{4}{7}$ m. I cut off $1\frac{4}{5}$ m to make a shelf.
   What length of wood do I have left?

6 Chris works $4\frac{5}{7}$ hours on the day shift and $5\frac{8}{21}$ hours on the night shift.
   How many hours longer did he work on the night shift? Simplify the fraction as far as you can.

 **Worksheet 2.2**

Work out the answers to these questions about multiplying and dividing with fractions. Do not use a calculator.

1 The side of a square measures $3\frac{3}{4}$ cm.
   What is its area?

2 A journey takes $2\frac{2}{5}$ hours by train.
   How long does it take me to travel $\frac{3}{4}$ of the journey?

3 A piece of wood is $4\frac{3}{4}$ m long. I want to saw it into three equal sections.
   Calculate the length of each section.

4 Each cut with a pair of scissors is $9\frac{1}{5}$ cm long.
   How many cuts are needed for a length of $73\frac{3}{5}$ cm?

5 How many centimetres are there in three quarters of two and a half metres?

6 Gina measures her stride to be $\frac{5}{8}$ m. She walks for a distance of $1\frac{3}{7}$ km.
   How many strides does she take? Give your answer to an appropriate whole number.

Answer each of these questions using the estimation method described in the Student's Book. Do not use a calculator.

1 $39 \times 98 = 3822$

    a) $3.9 \times 98 = \ldots\ldots\ldots$

    b) $39 \times 9.8 = \ldots\ldots\ldots$

    c) $0.39 \times 9.8 = \ldots\ldots\ldots$

2 $19 \times 59 = 1121$

    a) $19 \times 590 = \ldots\ldots\ldots$

    b) $1.9 \times 5.9 = \ldots\ldots\ldots$

    c) $190 \times 0.59 = \ldots\ldots\ldots$

3 $71 \times 29 = 2059$

    a) $7.1 \times 29 = \ldots\ldots\ldots$

    b) $0.71 \times 29 = \ldots\ldots\ldots$

    c) $71 \times 2.9 = \ldots\ldots\ldots$

4 $78 \times 18 = 1404$

    a) $7.8 \times 18 = \ldots\ldots\ldots$

    b) $78 \times 180 = \ldots\ldots\ldots$

    c) $7.8 \times 1.8 = \ldots\ldots\ldots$

5 $39 \times 41 = 1599$

    a) $39 \times 4.1 = \ldots\ldots\ldots$

    b) $3.9 \times 41 = \ldots\ldots\ldots$

    c) $3.9 \times 4.1 = \ldots\ldots\ldots$

Now use your calculator to check your answers.

Round each of these numbers correct to 1 significant figure.

1 9.8		**2** 153		**3** 289		**4** 8.97	
5 47.8		**6** 0.049		**7** 79.6		**8** 104.8	
9 1794		**10** 7.09					

Round each of the numbers in these calculations to 1 significant figure, and use your rounded values to find an estimate for the answer to each calculation. You do not need to work out the exact numbers.

11 29×39	**12** $78.8 \div 19.7$	**13** $498 \div 24.9$	**14** $59.7 + 81.2$
15 3.84×5.07	**16** $88.8 - 27.8$	**17** $521.3 + 603.4$	**18** 1012×685
19 0.98×0.49	**20** $6.07 \div 2.79$		

By using suitable rounding, estimate the answers to these calculations. You should show all the steps in your working.

21 $\dfrac{2.07 \times 9.84}{1.96}$

22 $\dfrac{19.7 \times 3.87}{7.85}$

23 $\dfrac{4.96 + 34.7}{6.13 + 3.98}$

24 $\dfrac{12.01 - 3.96}{0.47 \times 7.83}$

25 $(29.2 - 10.07) \times (4.06 + 5.91)$

Worksheet 2.5C

Find the lower and upper bounds of each of these numbers, which are written correct to 1 decimal place.

1 2.7	**2** 3.8	**3** 4.9	**4** 16.2
5 12.1	**6** 100.4	**7** 7.0	**8** 50.0
9 0.7	**10** 9.9		

Find the lower and upper bounds of each of these numbers, which are written correct to 2 decimal places.

11 6.34	**12** 1.82	**13** 2.50	**14** 10.32
15 0.87	**16** 1.07	**17** 0.01	**18** 5.00
19 12.34	**20** 30.00		

21 An athlete runs a race in 75 seconds, measured to the nearest second.
Write down the lower and upper bounds for the actual time he took to run the race.

22 A rectangle measures 16 cm by 20 cm, each measurement being correct to the nearest centimetre.
   a) Work out the perimeter of the rectangle, assuming these values to be exact.
   b) Work out the smallest value that the perimeter could actually be.

23 The length of a lizard's head is measured as 2.4 cm. This measurement is correct to 1 decimal place.
   a) Write down the smallest length that the lizard's head could be.
   b) Write down the greatest length that the lizard's head could be.

24 A square has sides of length 12.5 cm, correct to 1 decimal place.
Work out the smallest possible area the square could have.

25 A town has a population of 25 000 people, to the nearest thousand.
What is the smallest number of people who might actually live in the town?

26 A baby weighs 12 kg, correct to the nearest kg.
Write down the greatest value that the baby's weight could be.

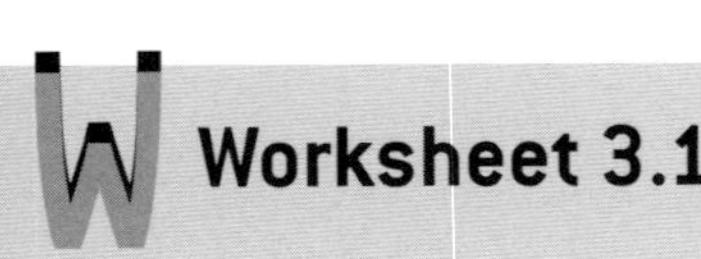

Worksheet 3.1

1 Write the following ratios in the form $1 : n$
 a) $10 : 15$ b) $40 : 50$ c) $10 : 13$ d) $5 : 4$
 e) $20 : 45$ f) $8 : 11$ g) $5 : 3$ h) $4 : 7$
 i) $8 : 1$

2 Write the following ratios in the form $n : 1$
 a) $7 : 2$ b) $11 : 4$ c) $8 : 5$ d) $15 : 6$
 e) $3 : 4$ f) $3 : 5$ g) $11 : 5$ h) $40 : 50$
 i) $12 : 5$

3 In a Year 11 class there are 20 boys and 14 girls.
 Express the ratio of boys to girls:
 a) as a ratio of whole numbers, in the simplest possible form
 b) as a ratio in the form $1 : n$

Worksheet 3.3

The non-calculator paper will require you to work out simple percentages of an amount without a calculator.
It may be helpful to work out 10% and 1% of the amount, and use these to build up the required percentage.

1 Work out 10% of £450

2 Work out 20% of £300
 Hint: Work out 10% first.

3 Work out 11% of £700
 Hint: Work out 10% and then 1%

4 Work out 12% of £900

5 Increase £300 by 8%

6 Increase £240 by 6%

7 Decrease £250 by 4%

8 Decrease £450 by 8%

9 Increase £70 by $12\frac{1}{2}$%
 Hint: Work out 50%, then 25%, then $12\frac{1}{2}$%

10 My weekly wage of £150 is increased by $7\frac{1}{2}$%
 What is my new weekly wage?

11 In a sale a shop offers 15% off an item costing £250
 What is the sale price?

12 Calculate the area of a rectangle measuring 12 cm by 15 cm
 If the area is reduced by 20%, what is the area of the new rectangle?

13 A tree increases its height by 10% each year.
 If the tree is 4 m high at the start of one year, what height would the tree be at the end of the second year?

14 A computer loses 20% of its value each year.
 If the computer costs £600 to buy new, how much is it worth after two years?

15 A picture frame is 20 cm long by 10 cm wide. The length is increased by 10% and the width decreased by 20%
 What is the area of this new picture frame?

16 I decrease 1000 m by $12\frac{1}{2}$% and then increase this new length by $12\frac{1}{2}$%
 What is the difference between the length I started with and the final length?

Solve these problems about simple interest. If you wish, you may use this formula:

$$I = \frac{P \times R \times T}{100}$$

where I = interest earned, P = principal amount invested, R = rate of interest (%) and T = time.

1 £400 is invested for 5 years at 3% per annum. Find the simple interest earned.

2 £250 is invested for 3 years at 4% per annum. Find the total amount after 3 years.

3 £70 is invested for 4 years at 5% per annum. Find the simple interest earned.

4 An amount of £350 is invested for $2\frac{1}{2}$ years at 4% per annum. Find the total amount at the end of this time.

Sometimes you need to find the Principal, Time or Rate and this can be done by rearranging the formula above into one of these three forms:

$$P = \frac{100 \times I}{R \times T} \qquad R = \frac{100 \times I}{P \times T} \qquad T = \frac{100 \times I}{P \times R}$$

Use the appropriate formula to answer these questions.

5 Harry earns simple interest of £40 by investing an amount at 5% per annum for 2 years.
How much did he originally invest?

6 I put a sum of money in a savings account. The account pays simple interest at 8% per annum.
At the end of 4 years I have earned £120 interest.
How much was my original investment?

7 Carol invests £300 in a building society account. The account pays simple interest at 5% per annum.
At the end of the investment period Carol has earned £60 in interest.
How long was the investment period?

8 Jacob invests £250 and earns simple interest at 8% per annum. At the end of the investment period Jacob has earned £40 in interest.
How long was the investment period?

9 Sam receives £400 on her birthday and invests it for 5 years. She earns £50 in simple interest.
What is the rate of interest she receives?

10 Regency Building Society pays £300 simple interest on £750 invested for 10 years.
Calculate the annual rate of interest.

11 How much interest do I receive if I put £350 in a bank account that pays simple interest at 10% per annum for $4\frac{1}{2}$ years?

12 Tony invests £500 at 6% per annum simple interest for 2 years. Gordon invests £350 at 4% per annum simple interest, also for 2 years.
a) Work out the interest earned by each investor.
b) Find the difference between the amount of interest earned by each investor.

Worksheet 3.5B

1 Compound interest problems are best done using a multiplying factor method.
Write down the multiplying factors for each of these rates of interest.

    a) 4%   **b)** 9%   **c)** 12%

    d) 15%   **e)** 3.5%   **f)** $4\frac{3}{4}\%$

    g) 2.8%   **h)** 5.65%

2 Jayne invests £400 at 6% per annum compound interest.

    a) Write down the multiplying factor for the amount after 1 year.

    b) Use your multiplying factor to find the value of Jayne's investment after:

        (i) 1 year   **(ii)** 2 years   **(iii)** 5 years

3 I invest £500 at a compound interest rate of 3% per annum for 4 years.
How much interest have I earned at the end of 4 years?

4 Armin puts £200 in a building society at a compound interest rate of 6% per annum for 3 years.
Find the value of his investment at the end of this time.

5 Jessica receives £450 on her fifteenth birthday and decides to invest it at a compound interest rate of 5% per annum for 3 years, until her eighteenth birthday.
How much interest will she receive at the end of the 3 years?

6 a) Calculate the compound interest on £500 for each of these offers.

        LFC Bank   5% for 3 years

        Magnet Building Society   3% for 5 years

    b) What is the difference between the amount or interest earned in each savings plan?

7 Matthew is saving up for a car costing £4500. He reckons that if he invests £3000 for 4 years at 9% compound interest he will have enough money to buy the car.

    a) Show that Matthew is not correct.

    b) Work out how much extra money he will need.

Worksheet 4.1

Work out these without using your calculator.

1 $4^2 + 5^2$ **2** $2^2 \times \sqrt{25}$ **3** $14^2 - 12^2$

4 $15^2 \div 5^2$ **5** $\sqrt{49} + \sqrt{81}$ **6** $\sqrt{16} \times \sqrt{36}$

7 $7^2 - 8^2$ **8** $\sqrt{(3^2 + 4^2)}$ **9** $\sqrt{169} \times \sqrt{16}$

10 $2^2 + 3^2 + 4^2$ **11** $14^2 - 13^2$ **12** $\sqrt{4} \times \sqrt{9} \times \sqrt{16}$

13 $\sqrt{(13^2 - 5^2)}$ **14** $12^2 \div \sqrt{9}$ **15** $(\sqrt{4} + \sqrt{9})(\sqrt{9} - \sqrt{4})$

16 $(9^2 - 8^2) \times 3^2$ **17** $\sqrt{4^3}$ **18** $\dfrac{10^2 + 5^2}{5^2}$

19 $\dfrac{4}{3^2} + \dfrac{3}{4^2}$ **20** $\sqrt{(\sqrt{49} + \sqrt{81})}$

Worksheet 4.5

This exercise is all about powers of 2. The answer to each question should be left as a power of 2.
Here is a reminder of some of the more frequently used cases.

$$2^0 = 1 \qquad 2^1 = 2 \qquad 2^2 = 4 \qquad 2^3 = 8 \qquad 2^4 = 16$$

1   **a)**   Write 4 as a power of 2.
    b)   Write 16 as power of 2.
    c)   Using the rule $2^a \times 2^b = 2^{a+b}$ write 4×16 as a power of 2.

Now work out the values of the following expressions in a similar way, giving each final answer as a power of 2.
Do not use a calculator.

You should use these rules wherever possible:

$$2^a \times 2^b = 2^{a+b} \qquad 2^a \div 2^b = 2^{a-b}$$

2 $32 \div 4$ **3** 8×64 **4** $512 \div 16$

5 $16 \div 64$ **6** 256×4 **7** $1024 \div 64$

8 64×8 **9** $128 \div 128$ **10** $256 \div 32$

These questions involve both multiplication and division. Once again, use the rules:

$$2^a \times 2^b = 2^{a+b} \qquad 2^a \div 2^b = 2^{a-b}$$

to express each one as a single power of 2.

11 $(256 \times 16) \div 128$ **12** $(512 \times 64) \div 1024$

13 $\dfrac{128 \times 256}{64}$ **14** $\dfrac{8 \times 16 \times 64}{64 \times 128}$

15 $1024 \times 8 \div 512$

Worksheet 5.1

1. If $a = -3$, $b = 2$ and $c = 5$, find the value of:
   - **a)** $b + c$
   - **b)** $a + b$
   - **c)** ab
   - **d)** $cb + a$
   - **e)** $c - a$
   - **f)** $(a - b)^2$
   - **g)** $ac + b$
   - **h)** $b(a - c)$

2. Temperatures can be converted from Fahrenheit to Centigrade using the formula:
   $$C = \frac{5(F - 32)}{9}$$
   - **a)** Work out C when the temperature is:
     - **(i)** 50°F
     - **(ii)** 32°F
     - **(iii)** 95°F
   - **b)** On Monday the temperature in Kenya was 28°C. On Tuesday it was 86°F.
     Which day was hotter and by how much in °C?

3. The cost of hiring a car, £C, is given by the formula $C = 100 + 25d$, where d is the number of days it is hired for.
   - **a)** Find the cost of hiring a car for four days.
   - **b)** How much would it cost to hire a car for two weeks?

4. Work out the value of each of these expressions:
   - **(i)** when $x = 3$
   - **(ii)** when $x = 10$
   - **a)** $4x - 10$
   - **b)** $x^2 + 8$
   - **c)** $2(x - 1)$
   - **d)** $5(7 - x)$
   - **e)** $4(9 - x)$
   - **f)** $7(1 - x)$
   - **g)** $(x + 1)(x - 1)$

5. If $p = 3$ and $q = 2$, find the value of:
   - **a)** $pq + p + q$
   - **b)** $p^2 - q^2$
   - **c)** $(p - q^3)^2$

Worksheet 5.2

Simplify these expressions using the index laws $x^a \times x^b = x^{a+b}$ and $x^a \div x^b = x^{a-b}$

1. $a^2 \times a^3$
2. $y^3 \times y^3$
3. $2y^2 \times y$
4. $\dfrac{b^{10}}{b^4}$

5. $\dfrac{e^7}{e^2}$
6. $\dfrac{4f^5}{f^3}$
7. $(h^2)^3$
8. $(n^3)^2$

9. $(3b)^2$
10. $\dfrac{g^2 \times g^3}{g}$
11. $6p^2 \times 5qr^3$
12. $4p^2 \times 3pr^2$

Worksheet 5.3

Multiply out the brackets and simplify the results.

1. $5(2x + 3)$
2. $3(k - 3)$
3. $z(z + 4)$

4. $9(n + 1) + 2n$
5. $6(3a + 1) - 4a$
6. $5b(b - 2)$

7. $3(p + 2) + 2(p - 1)$
8. $2q(q + 4) + 3(q - 1)$

Worksheets 5.1/5.2/5.3

Worksheet 5.4

Expand and simplify these products of brackets.

1 $(a + 2)(b + 4)$ **2** $(c + 3)(c + 2)$ **3** $(y + 4)(y - 1)$

4 $(x - 1)(x + 3)$ **5** $(q + 4)(q - 2)$ **6** $(x + 5)^2$

7 $(m - 8)^2$ **8** $(2p + 1)(p + 1)$ **9** $(3n + 1)(n + 2)$

Worksheet 5.5

Factorise these expressions.

1 $4y - 12$ **2** $16x + 4$ **3** $5x + 25y$

4 $3p^2 + 6p^3$ **5** $10s^3 - 60s$ **6** $t^2 - t$

7 $2a^2b + 4ab^3$ **8** $14p^2q - 21p^2q^2$ **9** $6x^3 - 3x^2 + 9x$

Worksheet 5.6

Factorise these quadratic expressions.

1 $x^2 - 16$ **2** $y^2 - 100$ **3** $t^2 - 9$

4 $x^2 + 3x + 2$ **5** $y^2 + 5y + 6$ **6** $a^2 + 7a + 12$

7 $m^2 + 7m + 10$ **8** $p^2 + 9p + 18$ **9** $q^2 + 2q - 8$

10 $u^2 - 25$

Worksheet 5.9

1 Sandra has £*m*. Terri has £10 more then Sandra.
  a) Obtain an expression for the amount of money that Terri has.
  b) If Terri has £38, how much does Sandra have?

2 Vanessa has two brothers, William and Yann. Vanessa is *H* cm tall.
  a) If Yann is 9 cm taller than Vanessa, obtain a formula for Yann's height.
  b) If William is 12 cm shorter than Yann, obtain an expression for William's height.
  c) If William's dad is twice as tall as William, obtain an expression for William's Dad's height.
  d) Obtain an expression for the mean height of the three children.
  e) If Vanessa is 95 cm tall, how tall is:
    (i) William?   **(ii)** Yann?   **(iii)** their dad?

3 A rectangular field is *W* metres wide and *L* metres long.
  a) What is the formula for the perimeter of a rectangle?
  b) What is the formula for the area of a rectangle?
  The field is 50 metres wide and 200 metres long.
  c) Find the perimeter of the field.
  d) Find the area of the field.

4 Ray repairs washing machines. He charges a £60 call-out fee plus £25 per hour he works on the machine.
  a) Write a formula for the total cost, £*C*, of Ray repairing a washing machine if it takes him *H* hours.

$$C = \text{................} + \text{................}$$

  b) Shannon's machine takes $2\frac{1}{2}$ hours to mend. How much should she pay Ray?

5 The charge for a single room at a hotel Riviera is £60 per night. The charge for a double room is £100.
  Write a formula for the total amount, £*T*, taken by the hotel if *x* single rooms and *y* double rooms are occupied.

$$T = \text{................} + \text{................}$$

 **Worksheet 5.9**

1. A square has a width of $2x + 1$ cm and a height of $x + 4$ cm.
   a) Draw a sketch to illustrate this information.
   b) Explain why $2x + 1 = x + 4$.
   c) Solve your equation to find the value of x.
   d) Work out the dimensions of the square.

2. A square has sides $4x - 1$ cm and $x + 8$ cm.
   a) Illustrate this information on a sketch.
   b) Work out the dimensions of the square.

3. A square has sides $5x + 4$ mm and $2x + 10$ mm.
   a) Illustrate this information on a sketch.
   b) Work out the dimensions of the square.

4. A square has sides $4 + x$ cm and $5x - 18$ cm.
   a) Illustrate this information on a sketch.
   b) Work out the dimensions of the square.

5. A square has sides $9 - x$ mm and $17 - 3x$ mm.
   a) Illustrate this information on a sketch.
   b) Work out the dimensions of the square.

6. A triangle has sides of lengths $x + 2$ mm, $3x$ mm and 7mm.
   a) Illustrate this information on a sketch.
   The triangle has a perimeter of 17 mm.
   b) Explain why $x + 2 + 3x + 7 = 17$.
   c) Solve this equation to find the value of x.
   d) Hence work out the length of each side of the triangle.

7. A triangle has sides of length $2x$ cm, $3x - 1$ cm and 5 cm. Its perimeter is 9 cm.
   a) Illustrate this information on a sketch.
   b) Work out the length of each side of the triangle.

8. A triangle has sides of length $2x + 1$ m, $x + 2$ m and $x + 3$ m. Its perimeter is 10 m.
   a) Illustrate this information on a sketch.
   b) Work out the length of each side of the triangle.

9. A triangle has sides of length $5x - 2$ cm, $2x$ cm and $x + 4$ cm. Its perimeter is 14 cm.
   a) Illustrate this information on a sketch.
   b) Work out the length of each side of the triangle.

10. A triangle has sides of length $3x - 4$ m, $4x$ m and $x + 5$ m. Its perimeter is 25 m.
    a) Illustrate this information on a sketch.
    b) Work out the length of each side of the triangle.

11. A triangle has angles of $2x°$, $4x°$ and $54°$.
    a) Explain why $2x + 4x + 54 = 180$.
    b) Solve this equation to find the value of x.
    c) Work out the value of each angle in the triangle.

12. A triangle has angles of $x + 15°$, $2x°$ and $45°$. Find the value of each angle in the triangle.

13. A triangle has angles of $x + 30°$, $x + 40°$ and $x + 50°$. Find the value of each angle in the triangle.

14. A triangle has angles of $3x - 20°$, $x + 60°$ and $50°$. Find the value of each angle in the triangle.

15. A triangle has angles of $x + 20°$, $2x + 10°$ and $2x + 10°$. Find the value of each angle in the triangle.

Multiply out the brackets and solve each of these equations.
The answers are all positive whole numbers (positive integers).

1 $2(x + 3) = 8$
3 $3(x - 4) = 12$
5 $2(x + 1) + 3x = 12$
7 $4(x + 1) - 2x = 6$
9 $5x + 3(x - 2) = 34$

2 $5(2x + 1) = 25$
4 $4(3x + 1) = 52$
6 $4x + 3(x - 1) = 18$
8 $3x - 2(x + 1) = 0$
10 $4(2x + 3) + 3x = 56$

Multiply out the brackets and solve each of these equations.
The answers are all negative whole numbers (negative integers).

11 $2(x + 3) = 4$
13 $2(x - 3) = -8$
15 $3x + 2(x + 4) = -7$
17 $3(4x + 7) - 2x = -7$
19 $2(3x + 2) - 3(2x + 3) = -5$

12 $3(2x + 1) = -9$
14 $3(2x + 4) = 0$
16 $4(x + 5) + 3x = 6$
18 $5(x - 1) + 3(2x + 9) = 0$
20 $7x - 2(4x - 3) = 10$

Multiply out the brackets and solve each of these equations.
The answers may include fractions.

21 $4(x + 2) = 14$
23 $3(4x - 1) = -21$
25 $2(4x - 3) + 2x = 19$
27 $2(3x + 1) - 9x = -2$
29 $5(3x + 2) - 5x = 5$

22 $2(3x - 1) = 1$
24 $5x + 2(x + 1) = 5\frac{1}{2}$
26 $3(x - 4) + 2(x - 1) = 8\frac{1}{2}$
28 $2(4x - 1) - 2(8x - 1) = 5$
30 $\frac{5}{2}(4x - 2) - 3(x + 1) = 4$

Finally, here are some more difficult equations to solve.

31 $2(2x + 3) - 14 = 2(x - 1)$
33 $4(x - 1) - 11 = 3(2 - x)$
35 $16 - 2(x - 4) = 4(x - 1)$
37 $4(x - 3) - 3(x - 2) + 2(x - 1) = 7$
39 $4x = 12 - 3(2x - 4)$

32 $5(2x - 1) - 6 = 3(x + 1)$
34 $8x = 14 + 2(2x - 9)$
36 $2(x + 1) + 3(x + 2) + 4(x + 3) = -7$
38 $3(4x - 1) - 2(2x - 1) = 4$
40 $\frac{3}{4}(3x - 1) - \frac{1}{2}(5x - 1) = 2$

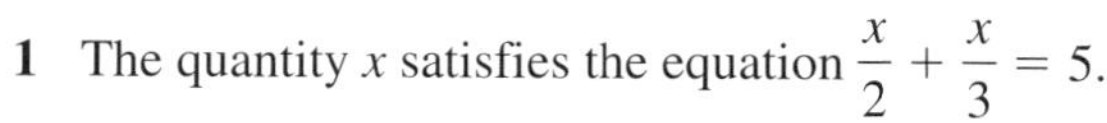

1 The quantity x satisfies the equation $\dfrac{x}{2} + \dfrac{x}{3} = 5$.

    a) Write down the lowest common multiple (LCM) of 2 and 3.
    b) Hence show that $3x + 2x = 30$.
    c) Simplify your answer to part b) and hence find the value of x.

Using a similar method, solve each of these equations.
You should begin by multiplying through by a suitable LCM in each case.

2 $\dfrac{x}{4} + \dfrac{x}{2} = 3$ **3** $\dfrac{x}{5} + \dfrac{x}{3} = 8$ **4** $\dfrac{x}{2} + \dfrac{x}{5} = 7$

5 $\dfrac{x}{5} + \dfrac{x}{2} = 14$ **6** $\dfrac{y}{3} - \dfrac{y}{5} = 4$ **7** $\dfrac{a}{3} - \dfrac{a}{4} = 1$

8 $\dfrac{b}{4} - \dfrac{b}{6} = 1$ **9** $\dfrac{y}{2} - \dfrac{y}{8} = 3$ **10** $\dfrac{y}{2} - \dfrac{y}{4} = 3$

11 The quantity y satisfies the equation $\dfrac{y}{5} + \dfrac{y}{2} = 7$.

    a) Write down the lowest common multiple (LCM) of 5 and 2.
    b) Hence show that $2y + 5y = 70$.
    c) Simplify your answer to part b) and hence find the value of y.

Using a similar method, solve each of these questions.
You should begin by multiplying through by a suitable LCM in each case.

12 $\dfrac{3x}{2} + \dfrac{x}{4} = 28$ **13** $\dfrac{3a}{4} + \dfrac{a}{2} = 5$ **14** $\dfrac{3b}{4} + \dfrac{3b}{2} = 9$

15 $\dfrac{4x}{5} + \dfrac{x}{10} = 18$ **16** $\dfrac{3a}{4} - \dfrac{a}{2} = 1$ **17** $\dfrac{2x}{5} - \dfrac{x}{10} = 3$

18 $\dfrac{2b}{3} - \dfrac{b}{4} = 20$ **19** $\dfrac{5x}{8} - \dfrac{3x}{4} = -1$ **20** $\dfrac{y}{5} - \dfrac{3y}{10} = 2$

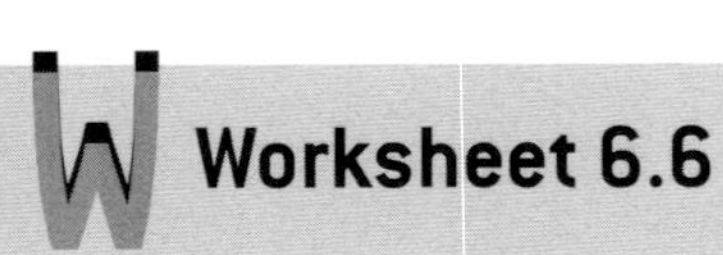

Worksheet 6.6

1 The equation $x^2 + x = 15$ is to be solved using a trial and improvement method.
   The incomplete table below shows some x values to be used in the trials.

x value	Value of $x^2 + x$	Result
1	2	too low
2	6	too low
3		
4		

   a) Copy the table and complete the missing values.
   b) Continue the table, to find a solution of the equation of $x^2 + x = 15$.
       Obtain your solution to an accuracy of 1 decimal place.

2 The equation $x^3 + 2x = 80$ is to be solved by a trial and improvement method.

x value	Value of $x^3 + 2x$	Result
0		
1		
2		
3		
4		
5		

   a) Copy the table and complete the missing values.
   b) Continue the table, to find a solution of the equation of $x^3 + 2x = 80$.
       Obtain your solution to an accuracy of 1 decimal place.

3 The equation $x^2 - 3x = 3$ has a solution between 0 and 5.
   Use a trial and improvement method to find this solution correct to 1 decimal place.
   Show the results of all your trials in a table similar to the ones in questions 1 and 2.

4 Show that the equation $x^3 + x^2 = 50$ has a solution between 3 and 4, and use a trial and improvement method to
   find this solution correct to 1 decimal place.
   Show the results of all your trials in a table.

5 Show that the equation $x^2 + 12x^2 = 35$ has a solution between 2 and 3, and use a trial and improvement method
   to find this solution correct to 1 decimal place.
   Show the results of all your trials in a table.

6 Show that the equation $5x + \dfrac{1}{x} = 12$ has a solution near 2, and use a trial and improvement method to find this
   solution correct to 1 decimal place.
   Show the results of all your trials in a table.

7 Show that the equation $3x^3 - 2x^2 = 200$ has a solution between 4 and 5, and use a trial and improvement method
   to find this solution correct to 1 decimal place.
   Show the results of all your trials in a table.

8 The equation $x^2 - 5x + 1 = 0$ has one positive solution.
   Use a trial and improvement method to find this solution correct to 2 decimal places.
   Show the results of all your trials in a table.

Worksheet 8.2

Use the most suitable method to find the value of x and y in each of the following pairs of simultaneous equations.
Do not use a calculator.

1 $x + y = 5$
    $2x - y = 1$

2 $x - y = -2$
    $2x + y = 5$

3 $4x + y = 4$
    $y = 2x + 1$

4 $x = y + 6$
    $3x + y = 10$

5 $4x + y = 11$
    $2x + 2y = 7$

6 $2x - 3y = 5$
    $y = 2x + 1$

7 $5y - 2x = 13$
    $2y - x = 5\frac{1}{2}$

8 $y = 2x + 3$
    $x = 2y + 3$

9 $2x + y = 1\frac{1}{2}$
    $3x - y = 8\frac{1}{2}$

10 $2x + 4y = 1$
    $x + y = 0$

11 $y = 2x - 2$
    $x + 2y = 16$

12 $4x + y = 8$
    $3x + 4y = -7$

13 $3x + y = -3$
    $3x - y = -9$

14 $2y = x + 1$
    $x + y = 3\frac{1}{2}$

15 $4x + 3y = -1$
    $y = \frac{1}{2} - x$

16 $2x + 3y = 22$
    $2y - 3x = -7$

17 $y = x + 1\frac{1}{2}$
    $x + y = 3\frac{1}{2}$

18 $2x - 3y = 1$
    $3x - 2y = -1$

19 $2x + y = 13$
    $x = 2y - 11$

20 $4x - 5y = 7$
    $3x = 2y$

21 $4x + 2y = 1$
    $3x = 2 - y$

22 $3x + 2y = 20$
    $x - 2y = -4$

23 $4x + y = 2$
    $12x + 4y = 7$

24 $4x - y = 7$
    $8x = 4 - 3y$

25 $5y + x = 10$
    $2x + y = 11$

Worksheet 8.3

For each of these questions draw a coordinate grid in which x and y can range from -10 to 10.
Draw the lines corresponding to each equation, and hence solve the simultaneous equations graphically.

1 $x + y = 5$
    $4x + y = 8$

2 $x + y = 4$
    $x - y = 2$

2 $y = 2x - 2$
    $x + y = 7$

4 $2x + y = 4$
    $2y = x - 2$

5 $2y + x = 9$
    $x - y = 3$

6 $2y - x = 5$
    $x + 2y = 3$

7 $2y - x = -1$
    $x + y = -5$

8 $2x + y = 5$
    $y = x - 7$

9 $2x + y = 8$
    $y = 3x - 4\frac{1}{2}$

10 $x + y = 2\frac{1}{2}$
    $2x - 2y = 11$

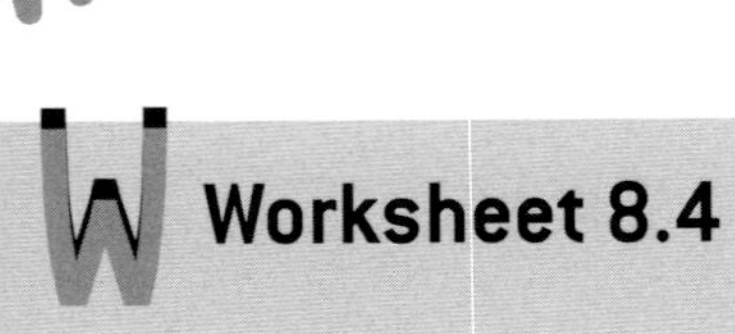

Worksheet 8.4

Use simultaneous equations to help you solve the following problems. Remember to show all your working carefully. Do not use a calculator.

In some questions you may find it helpful to draw a diagram.

1 At Landing Football Club a bacon roll and a cup of tea cost £1.70. A supporter orders two bacon rolls and three cups of tea. He pays £3.90 in total. What is the cost of:
   a) a cup of tea?   **b)** a bacon roll?

2 Mandy has 16 coins which are either 1p or 2p. The total value of the coins is 25p. Let x be the number of 1p coins and y the number of 2p coins.
How many of each does Mandy have?

3 In a sports match a team is awarded x points for a win and y points for a draw. Team A wins five matches and draws two, earning a total of 31 points. Team B wins two games and draws three to earn 19 points.
How many points are awarded for a win?

4 At a cinema, adult tickets cost £x each and child tickets cost £y each. If there are two adults and three children the total cost is £33. If there are four adults and two children the total cost is £46. Find the cost of:
   a) an adult ticket.   **b)** a child ticket.

5 The average person walks at x miles per hour and runs at y miles per hour. Craig walks for two hours and runs for one hour, covering a total distance of 14 miles. His friend Corrina walks for three hours and runs for two hours, covering a total distance of 25 miles. Use this information to find the speed at which the average person walks and runs.

6 In an antiques shop a cup costs £x and a plate costs £y. I buy three cups and three plates and pay £24. On my next visit, to add to my collection I buy seven cups and six plates at a cost of £51. Find the cost of:
   a) a cup.   **b)** a plate.

7 A shop sells CDs and DVDs. Joey buys three CDs and two DVDs for £69 whilst Alex buys two CDs and three DVDs and pays £71. What is the cost of:
   a) one CD?   **b)** one DVD?

8 I think of a number and multiply it by 3. I think of a second number and multiply it by 5. When I add the two results together I get 47. The difference between the two original numbers is 5. Find my two original numbers.
Hint: Let the larger of the two original numbers be x, and the smaller y.

9 LBC Sports advertises special offers in its sale. It has T-shirts selling for £3 each and shorts for £7 each. Harry buys x T-shirts and y pairs of shorts. His sister Lucy buys twice as many T-shirts and the same number of pairs of shorts as Harry.
If Harry spends £27 and Lucy spends £33, how many T-shirts does Lucy buy?

10 A garden centre stocks quantities of bulbs and seeds. In the spring it sells bulbs at £2.50 a pack and seeds at £2 a packet for all varieties. Dougal buys x packs of bulbs and y packets of seeds for his garden and the total cost is £26. His friend buys y packs of bulbs and x packets of seeds and spends £28.
Work out how many packs of bulbs and packets of seeds they each buy.

11 A triangle has sides of lengths $2x$ cm, $3x - y$ cm and $4x - 2y$ cm; its perimeter is 30 cm.
A second triangle has sides of lengths $x + y$ cm, $3x - 4y$ cm and $2x - y$ cm; its perimeter is 16 cm.
Find the values of x and y.

12 A school shop sells pens and pencils to the students. A pen costs 40p and a pencil costs 30p. Jake buys x pens and y pencils, which cost him £2.70. After the school holiday there is a price increase. Pens are increased by 15% and pencils by 10%. For the same number of pens and pencils Jake now has to pay £3.03. How many of each does he buy?

(continued)

  **Worksheet 8.4**   GCSE Maths for Edexcel: Higher Teacher's Resource © Hodder Murray 2006

13 Travel by train costs x pence per mile and travel by bus costs y pence per mile. Darren travels 5 miles by train and 2 miles by bus and his total fare is £1.90. Sophie travels 7 miles by train and 4 miles by bus and pays a total of £2.90.
   a) Write two simultaneous equations to express this information.
   b) Solve your equations, to find the cost of travelling one mile by train and bus.

14 In a local café Jordan spends £4.35 on three cakes and four coffees for her friends. At the next table her friend Peter pays £4.05 for four cakes and three coffees for his friends.
   What is the cost of one cake with one coffee?

15 A rectangle has a length of $2x + y$ cm and a width of $x + 2y$ cm. The perimeter of the rectangle is 48 cm.
   a) Write an equation in x and y for the perimeter of the rectangle.

   A square has a side of length $3x + y$ cm. Its perimeter is 72 cm.
   b) Write an equation in x and y for the perimeter of the square.
   c) Use your equations to find the value of x and y.
   d) What are the dimensions of the rectangle?

Use your algebra skills to rearrange each of these so that it is a single inequality in x.

1 $3x < 12$

2 $\dfrac{x}{2} > 4$

3 $2x + 1 < 4$

4 $5x - 2 < 8$

5 $\dfrac{x}{3} > \dfrac{4}{3}$

6 $7 - 3x > 1$

7 $\dfrac{5x}{2} > \dfrac{x}{2} + 3$

8 $2(x - 1) < 4$

9 $3x - 2 \leqslant 4$

10 $\dfrac{1}{x} < 2$

11 $2x < -5$

12 $\dfrac{x}{3} < \dfrac{1}{2}$

13 $2x + 5 > 3$

14 $3(x - 1) \geqslant 9$

15 $\dfrac{x}{2} > \dfrac{1}{2}$

16 $\dfrac{x}{2} + 1 > \dfrac{1}{2}$

17 $2x + 3 < 4x - 5$

18 $5(2 - x) \leqslant -4x$

19 $1.5(2x + 1) \geqslant x$

20 $3 + 5x < 1 + 3x$

21 $3(2x + 1) < 9$

22 $\dfrac{x}{6} + \dfrac{x}{3} > 1$

23 $4(x - 2) \leqslant 10$

24 $3x + 5 < 2x$

25 $5x + 1 < 3x - 5$

26 $\dfrac{x}{2} - \dfrac{x}{5} \geqslant 3$

27 $2(x + 1) + 3(x - 2) < 6$

28 $\dfrac{3}{x} < 6$

29 $2 - \dfrac{1}{x} < 4$

30 $3(4x - 1) + 3 < 0$

For each of these situations, write an inequality and solve it. Do not use a calculator.
You may find it helpful to draw a number line for some questions.

1 Paul and Cath are playing a guessing game about whole numbers. Paul says his number is less than 12 whilst Cath says her number is more than 7. If they both chose the same number, what are the possible numbers?

2 Tina thinks of a whole number which is more than 5 but less than 9. Represent this as an inequality statement and give the possible numbers.

3 Anthony picks a whole number randomly. It is greater than or equal to 7 but less than 14. Write this as an inequality statement and make a list of the possible numbers.

4 In a 'guess the number' challenge Sonia selects a whole number n from the range given by $3 < n < 10$. What is the largest number she could have selected?

5 Write down all the possible negative integer values in the range given by $-7 < n \leqslant -1$.

6 Tony thinks of a number n, multiplies it by 2 and then adds 1. He says that the result is greater than or equal to 7. Represent this information as an inequality statement and give the range of possible values of n.

7 In a shop a customer advisor believes an article costs in the range £14 to £17 inclusive whilst a supervisor reckons the cost is in the range £15 to £19 inclusive.
   a) Write two inequality statements to represent this information.
   The article costs an exact number of pounds and both the advisor and the supervisor are correct.
   b) What are the possible costs?

8 Steph chooses a number n, subtracts 3 from it and then multiplies the result by 2, to get a final answer greater than 8.
   a) Set up an inequality statement to show this information.
   b) If Steph's number is a whole number, what is the least value it could be?

9 The length of a rectangle is $x + 4$ cm and the width is 5 cm. The perimeter of the rectangle must be more than or equal to 36 cm. Set up an inequality and find the least value of x to satisfy this inequality.

10 The whole number x satisfies the inequality $-2 \leqslant x < 4$ and also satisfies the inequality $-2 < x < 7$. List all the possible values that x could be.

11 A rectangle has sides of length $x - 3$ cm and $x + 3$ cm. If the area of the rectangle has to be more than or equal to 72 cm^2, what are the shortest possible lengths of the sides of the rectangle?

12 I think of a whole number n, multiply it by 4 and add 7 to the result. My final answer is less than 30. Find the greatest possible value of the original number.

13 Romil estimates the temperature (T) as being greater than -7 degrees but less than -3 degrees.
   a) Write this as an inequality.
   Marcus estimates the same temperature (T) as being greater than -5 degrees but less than -2 degrees.
   b) Write this as an inequality.
   c) Romil and Marcus are both correct, and the temperature is a whole number of degrees. Work out what the temperature must be.

14 $2x > 5$ and $3x < 13$. Find two integer values that satisfy both equations.

15 $-6 < 4x < 9$ and $-8 < 3x < 5$. Write this information as a single inequality for x.

Worksheet: Chapter 9

For each of questions **1** to **6**, draw a coordinate grid in which x and y can range from 0 to 7.

1 **a)** Draw the graph of $2x + y = 4$.
   b) Shade the region corresponding to the inequality $2x + y > 4$.

Shade the region corresponding to each of these inequalities.

2 $x + 3y < 6$ **3** $2x + 4y < 8$ **4** $x + 2y > 5$

5 $3x + y \geqslant 6$ **6** $4x + 2y \leqslant 10$

For each of questions **7** to **12**, draw a coordinate grid in which x and y can range from -7 to 7. Shade the region corresponding to the given inequality.

7 $2x - y > 4$ **8** $x - 3y \leqslant 6$ **9** $3y - 2x \geqslant 12$

10 $y - 2x < 7$ **11** $4x - y > 6$ **12** $x - 2y \leqslant -4$

For each of questions **1** to **10**, draw a coordinate grid in which x and y can range from -7 to 7.

1 **a)** On a set of coordinate axes, draw the following straight lines.
   (i) $x > 1$ **(ii)** $x < 3$ **(iii)** $y > 1$ **(iv)** $y < 5$
   b) Shade and label the region R that satisfies the four inequalities $x > 1$, $x < 3$, $y > 1$ and $y < 5$
   c) Hence list the coordinates of the all the points with whole number coordinates which satisfy the four inequalities.

In questions **2** to **10**
a) Draw a diagram to show the region corresponding to the given inequalities.
b) Find the coordinates of all the points with whole number coordinates that satisfy all the coordinates.

2 $x > -3, x < 0, y > 1, y < 5$ **3** $x > 1, x < 4, y > -3, y < 1$

4 $x > -4, x < -2, y > -2, y < 7$ **5** $x > 3, x < 6, \ y > 1, y < 3$

Hint: In questions **6** to **10**, some of the points may be on the boundary lines — look for $\geqslant$ and $\leqslant$.

6 $x \geqslant 1, x < 3, y > 0, y < 3$ **7** $x \geqslant 2, x \leqslant 3, y > 1, y < 4$

8 $x > 1, x < 4, \ y \geqslant 2, y \geqslant 4$ **9** $x > -3, x \leqslant -1, y > 1, y \leqslant 3$

10 $1 < x \leqslant 2, 1 \leqslant y < 2$

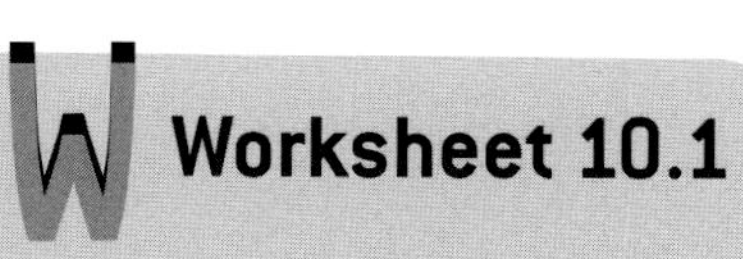

Worksheet 10.1

For questions **1** to **5**, write down the next two terms in the number sequence.

1 10, 13, 16, 19, 22, …

2 10, 13, 17, 22, 28, …

3 80, 77, 74, 71, …

4 3, 6, 12, 24, …

5 2, 6, 12, 20, …

6 A number sequence is defined as follows:
- The first term is 4.
- Each term is 3 more than the previous one.

Use this rule to generate the first six terms of the number sequence.

7 A number sequence is defined as follows:
- The first term is 5.
- Each term is double the previous one.

Use this rule to generate the first six terms of the number sequence.

8 A number sequence is defined as follows:
- The first term is 6.
- Each term is $(10 - n)$, where n is the previous term.

Use this rule to generate the first ten terms of the number sequence.

9 The nth term of a number sequence is $5n - 2$.
  a) Work out the first five terms of the number sequence.
  b) Explain whether this is an arithmetic series or not.

10 The nth term of a number sequence is $n(n + 2)$.
  a) Work out the first five terms of the number sequence.
  b) Explain whether this is an arithmetic series or not.

11 The first five terms in an arithmetic sequence are 10, 12, 14, 16, 18.
  a) Find the value of the 10th term.
  b) Write, in terms of n, an expression for the nth term of this sequence.

12 The first five terms in an arithmetic sequence are 28, 31, 34, 37, 40.
  a) Find the value of the 12th term.
  b) Write, in terms of n, an expression for the nth term of this sequence.

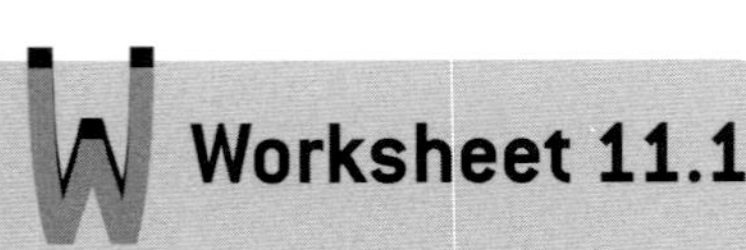

Worksheet 11.1

1. The graph shows the depth of water in a dishwasher over the length of one cycle.
   Choose what you think the dishwasher is doing at each stage.
   Hint: The wash phase lasts longer than rinse.
   Words can be used twice.

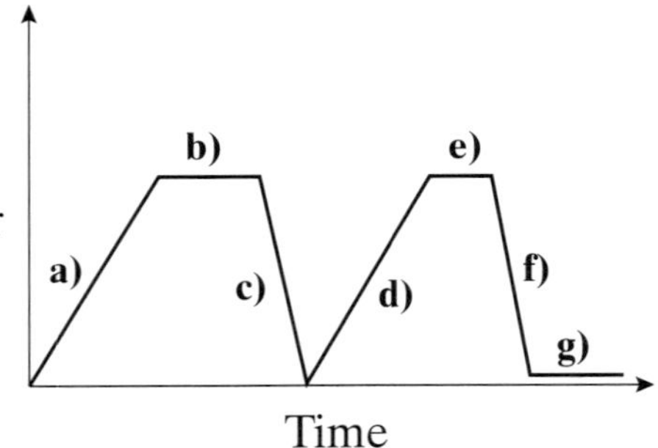

Drying	Draining	Rinsing

Washing	Filling

2. Twin toddlers, Rowan and James, are having their bath. The graph shows the depth of the bath water.

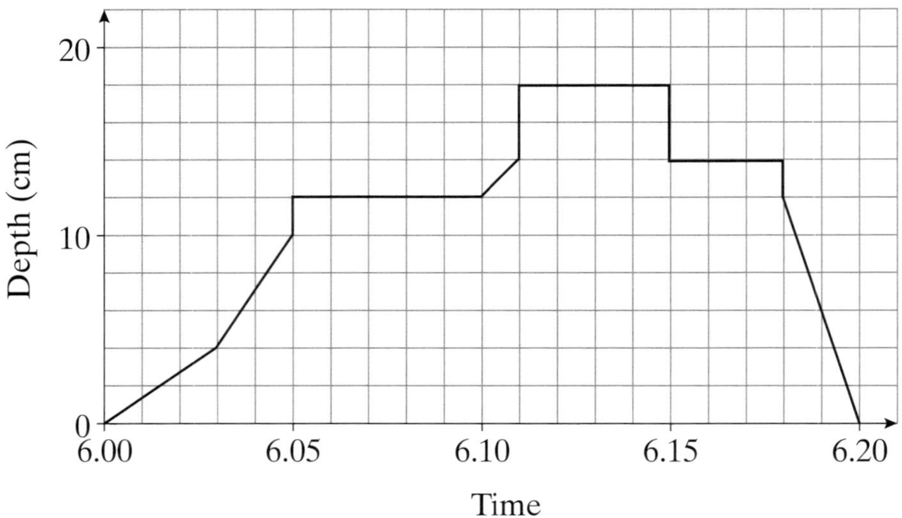

   Use the graph to answer these questions.
   a) Mum runs just the cold tap first. How long is it running before the hot tap is turned on too?
   b) Rowan likes the bath and gets in first. How long is he in the bath on his own?
   c) What does Mum do at 6.10 pm, just before James gets in?
   d) Who do you think is the heavier twin?
   e) James gets out at 6.15 pm. How long has he been in the bath?
   f) What is the maximum depth the water reaches?
   g) What happens to the water at 6.18 pm?

3.
   ## WATTS & STEVENS ELECTRICAL CONTRACTORS

   £40 Call out   **£20 per hour**

   a) Copy and complete the table, which shows the total cost (£T) of calling out an electrician who works for h hours.
   b) Draw a pair of axes with Hours (h) on the horizontal axis and Cost (£T) on the vertical axis. Plot the points from the table and join them with a straight line.

Hours (h)	0	2	4	6	8
Cost (T)	40			160	

   c) Use your graph to answer these questions.
      (i) How much would you have to pay for an electrician to come out and work for $3\frac{1}{2}$ hours?
      (ii) How many hours did the electrician work for if the bill was £140?

(continued)

Worksheet 11.1   GCSE Maths for Edexcel: Higher Teacher's Resource © Hodder Murray 2006

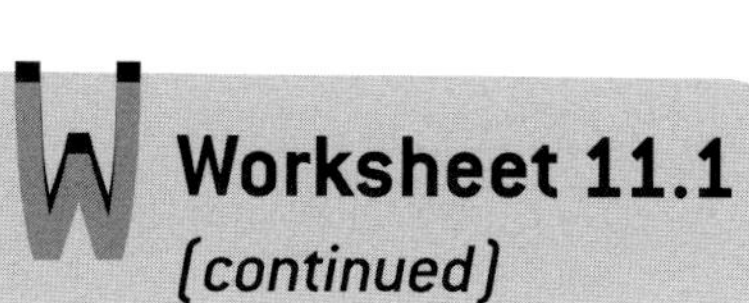

4 Choose the statement which best fits each of these graphs.

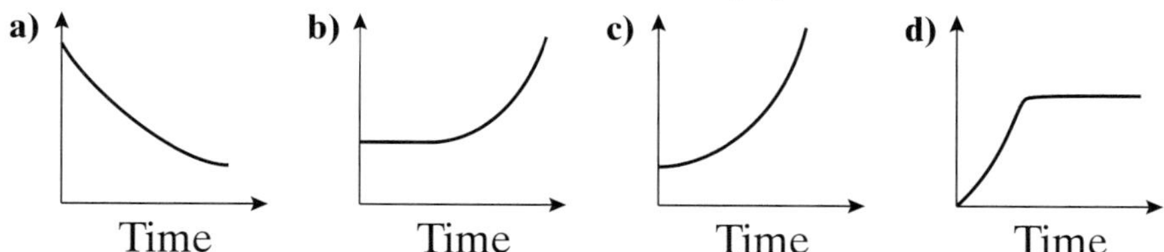

A Temperature of soup that is heated and then left to boil

B Inflation has fallen in the last year, but by less each month

C Unemployment is rising by more each month

D Share prices remained stable last year but have since risen sharply

W Worksheet 11.2

1 The graph shows a bus journey from Luton to London, a distance of 50 km.
The bus stops to pick up passengers, first at Hemel Hempstead, and then in Watford.

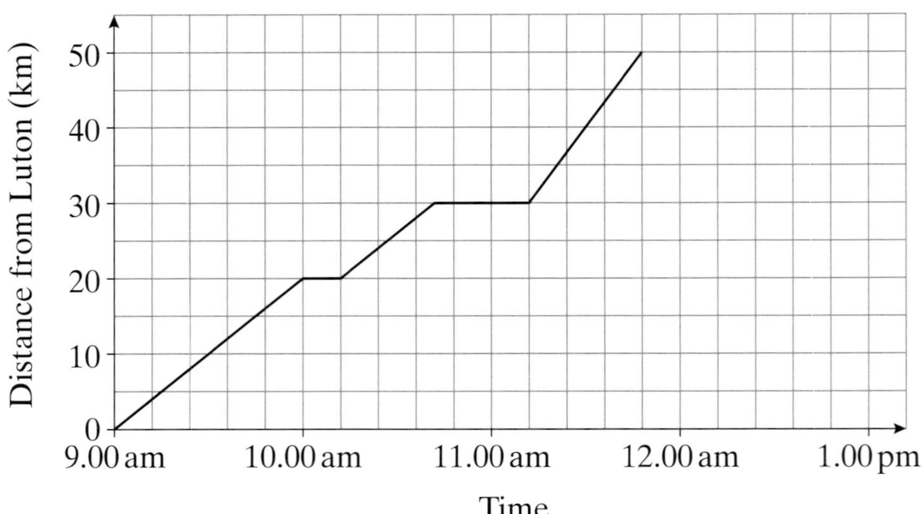

Use the graph to answer these questions.
a) At what time does the bus stop in Hemel Hempstead?
b) How long does it wait for people to get on and off?
c) How far is it from Hemel Hempstead to Watford?
d) At what time does the bus reach London?
e) The driver has a 30 minute break, and then takes 1 hour to return to Luton. Draw this information on the graph.

2 Mr and Mrs Smith are going shopping at Cribbs Causeway. They leave home at 10.00 am and return at 4.20 pm.
The graph shows their journey for the day.

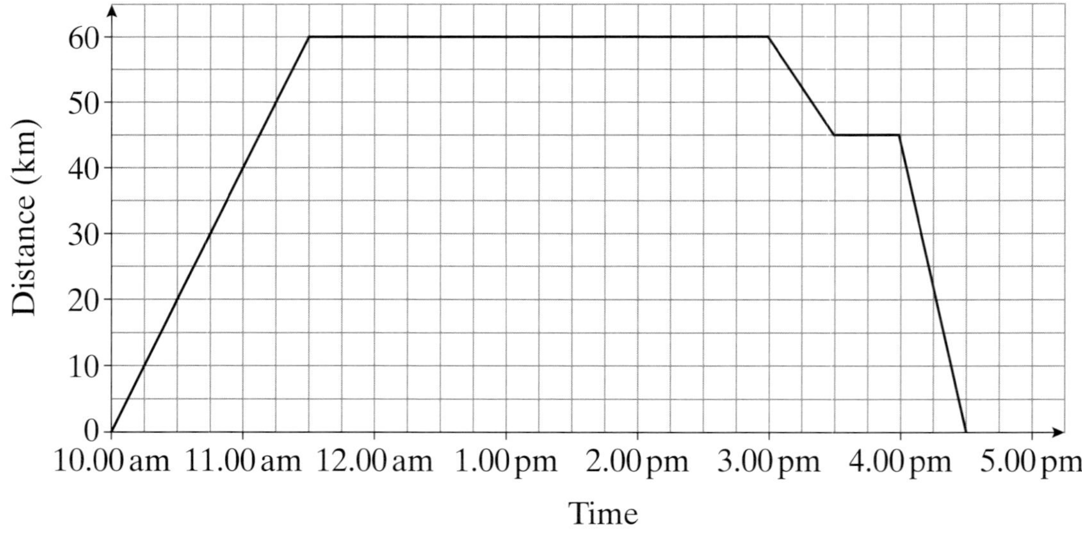

a) How long does it take them to get to Cribbs Causeway?
b) How far is it to Cribbs Causeway?
c) Work out the car's average speed for this part of their journey.

Hint: Speed $= \dfrac{\text{Distance}}{\text{Time}}$

d) How long did they spend shopping?
e) Mr and Mrs Smith stop for petrol and an oil check on the way home. What time do they stop, and for how long?
f) They return home on the motorway. What is their average speed for this final part of their journey?

(continued)

    Worksheet 11.2

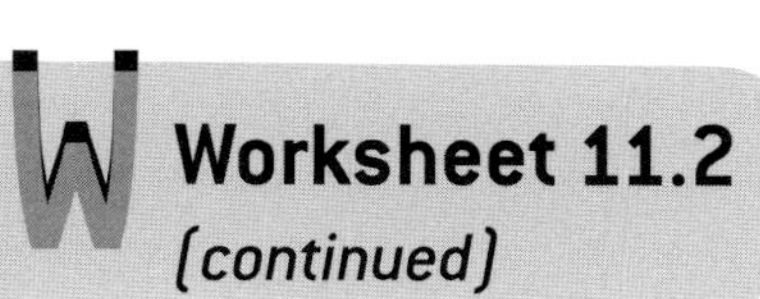

3 Three friends travel from Airton to meet at the beach in Blurham.

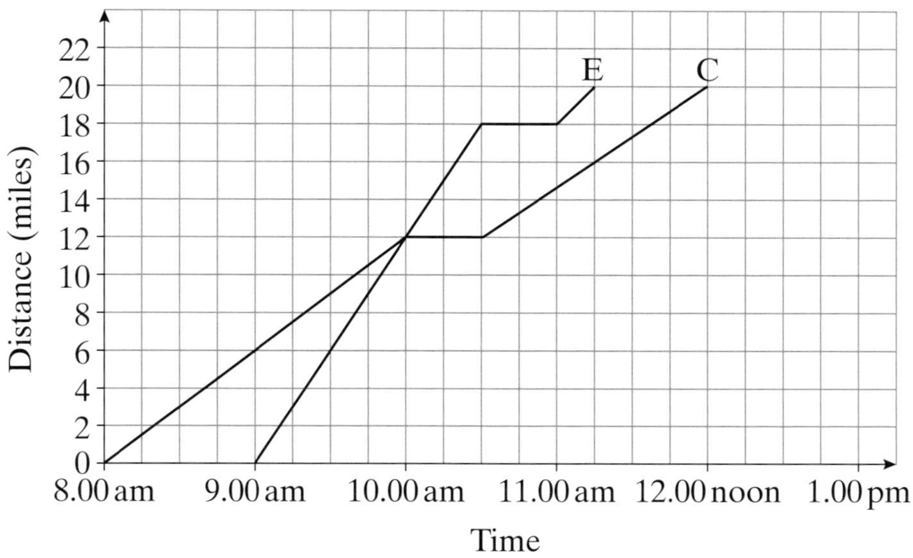

Clive cycles. His journey is shown on graph C.
a) At what time does Clive leave home?
b) What happens at 10.00 am?
c) How far is it from Airton to Blurham?

Derryk goes by car. He leaves at 11.00 am and it takes him 30 minutes to get to Blurham. He drives at a constant speed.
d) Draw his journey on the graph.
e) At approximately what time does Derryk overtake Clive?

Erin takes the bus. Her journey is shown on graph E. She leaves at 09.00 am.
f) At what time does she pass Clive?
g) How long does the bus stop for?
h) Who gets to the beach first?

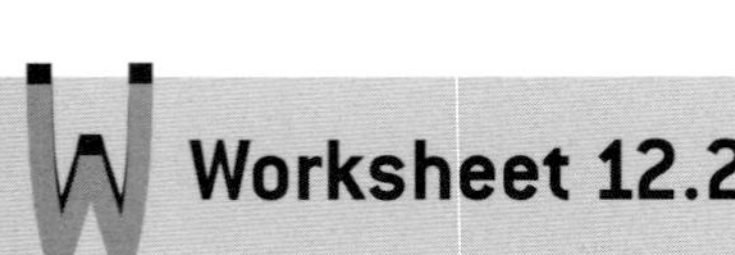

Worksheet 12.2

Find the values of the angles represented by the letters in each question. In each case, give a reason for your answer.

1

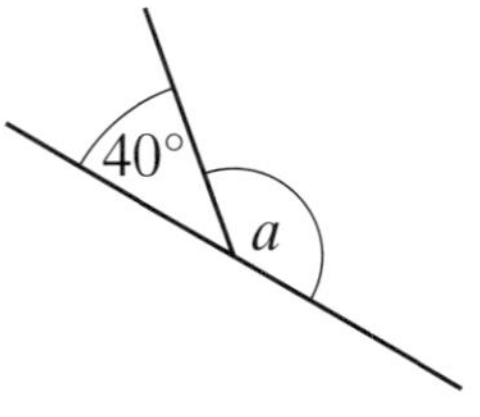

2

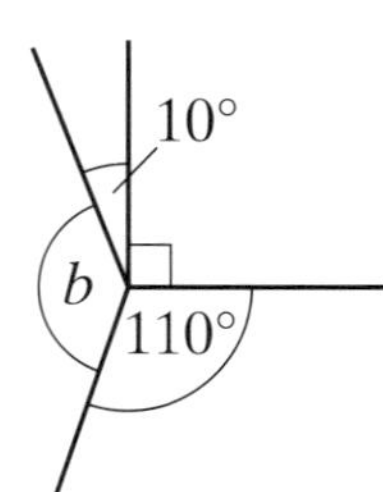

3

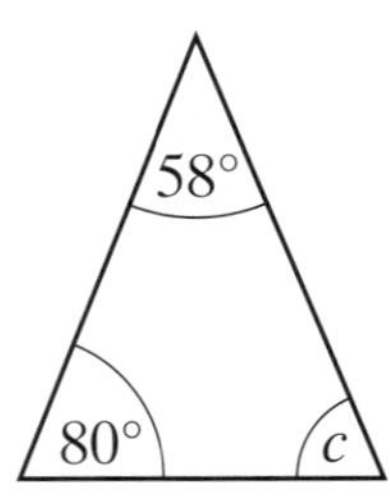

4

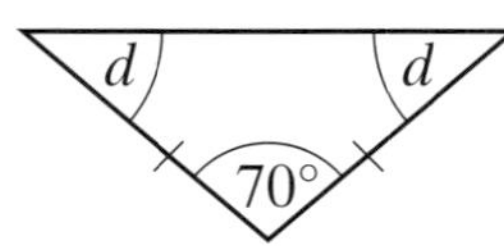

5

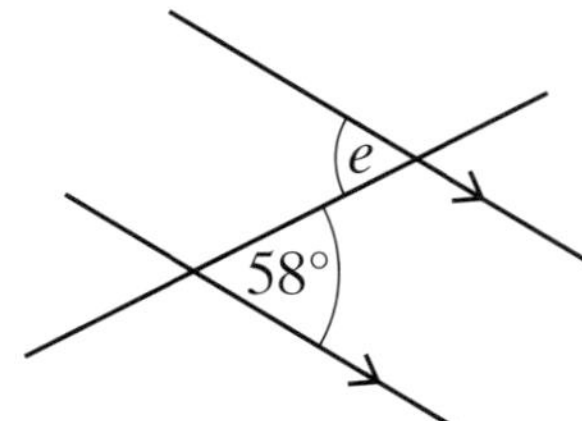

6

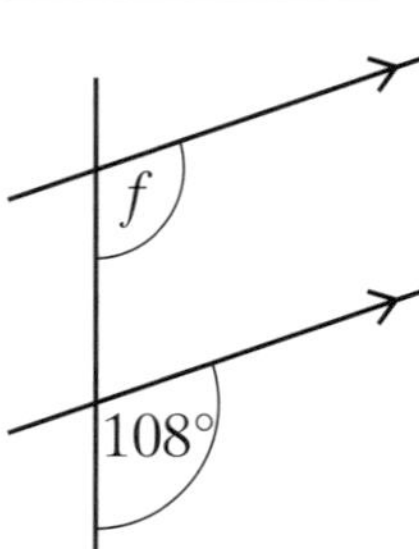

7

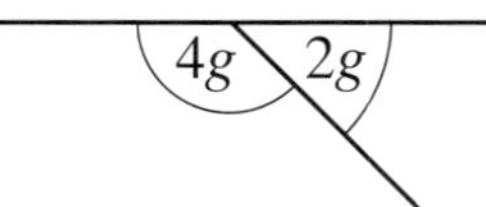

8

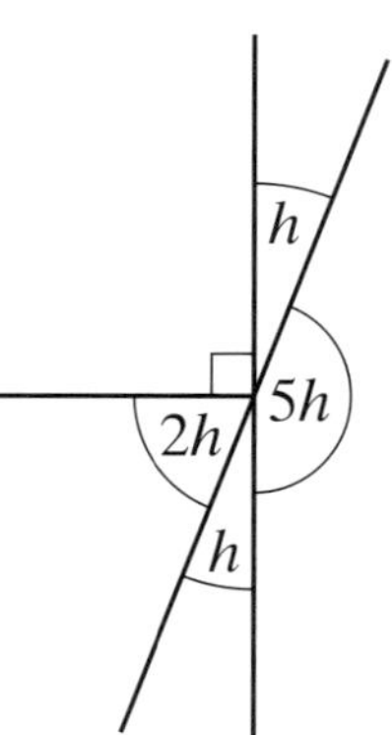

9

10

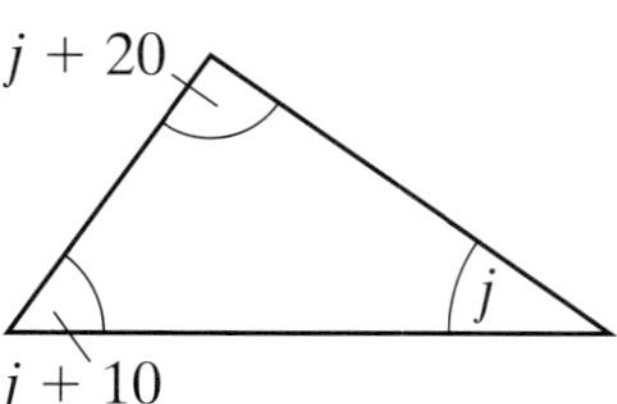

11

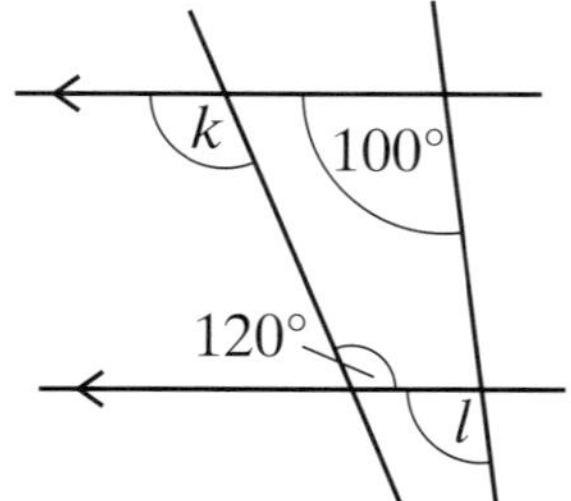

12

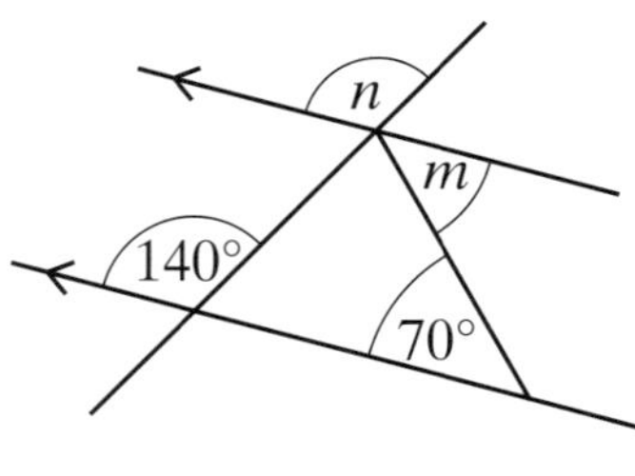

13

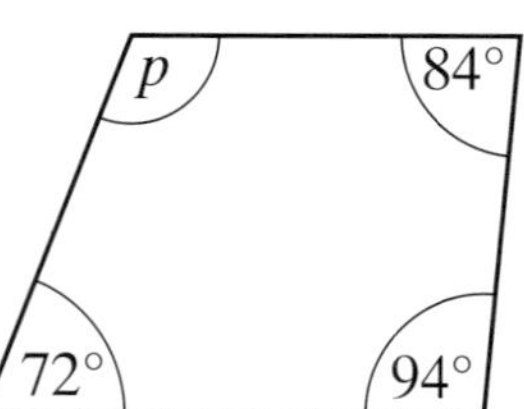

14

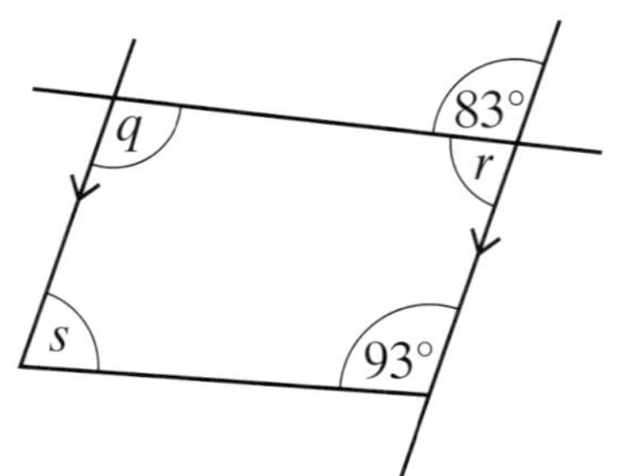

15

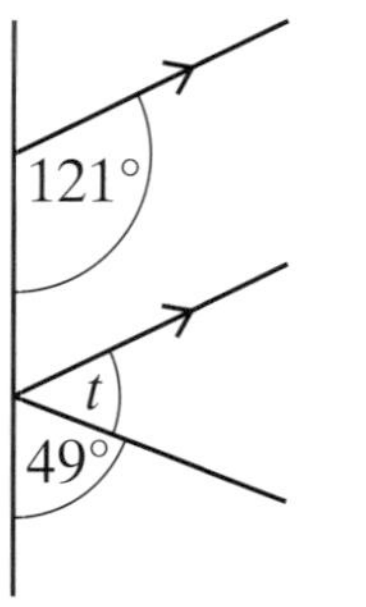

16

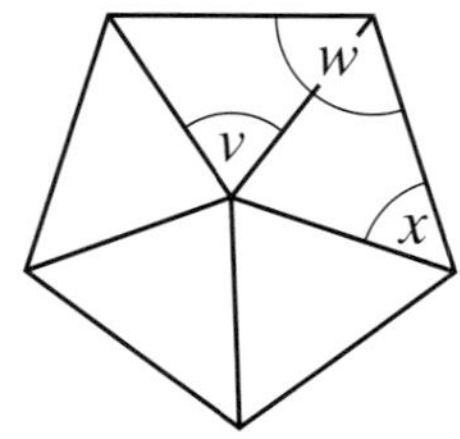

Regular pentagon

17

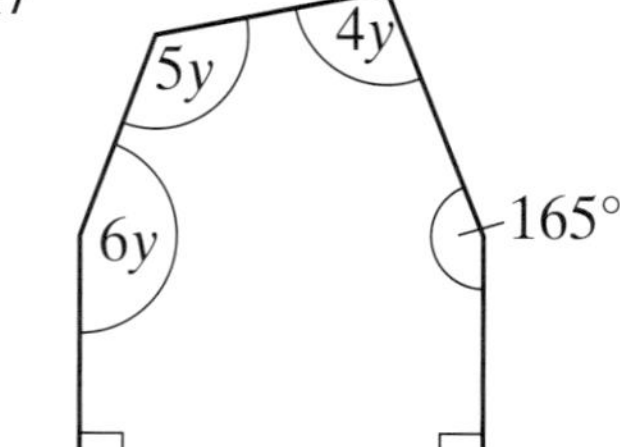

Worksheet 12.4

The drawings are not to scale.

1 Find the area of each shape. Remember to include the units in your answer.

a) 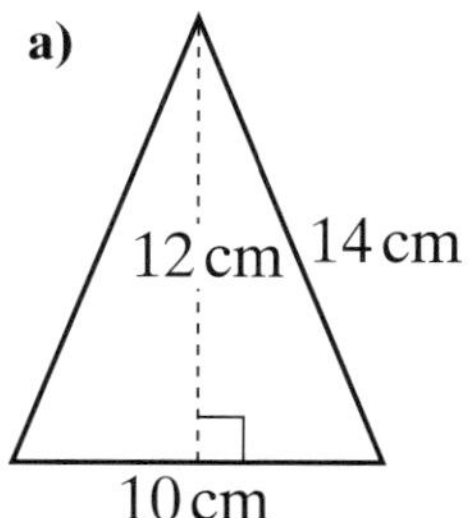

b) 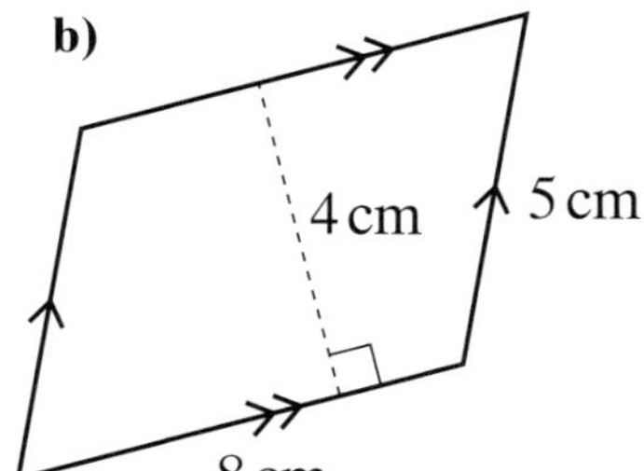

c) 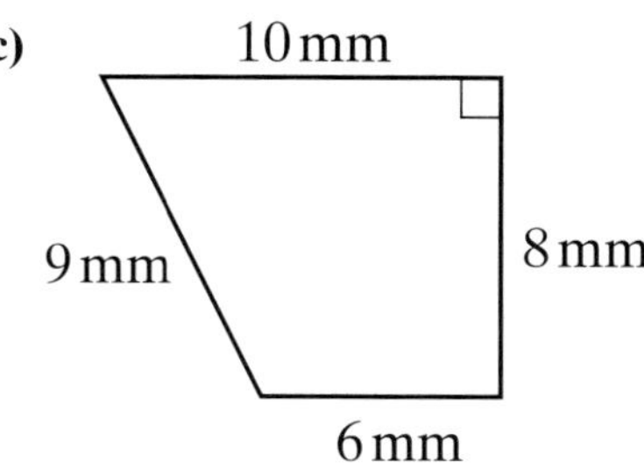

2 a) Find the area of the rectangle on the right.
b) Find the area of the shaded parallelogram.
c) What percentage of the area of the rectangle is taken up by the area of the parallelogram?

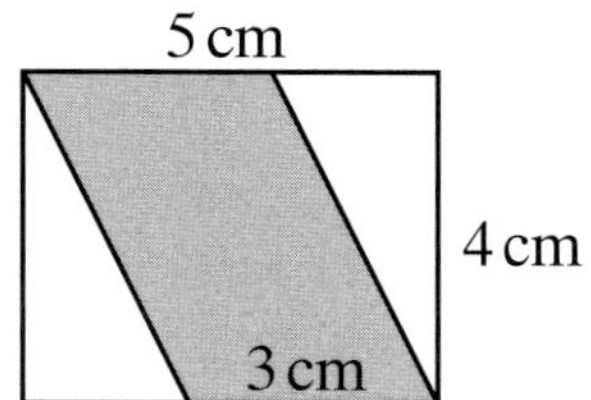

3 Find the area of each shape. Remember to include the units in your answer.

a) 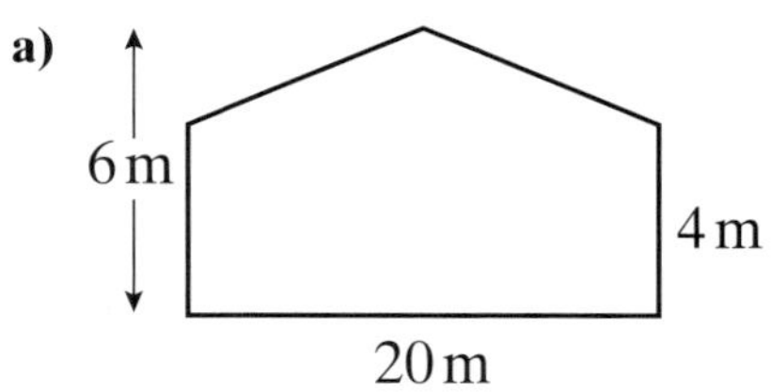

b) 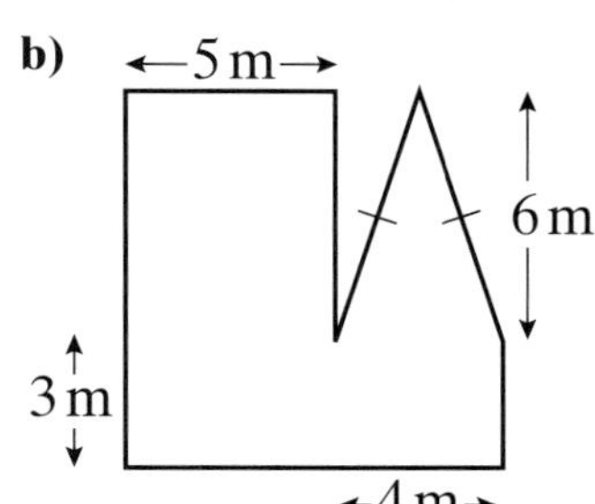

c) 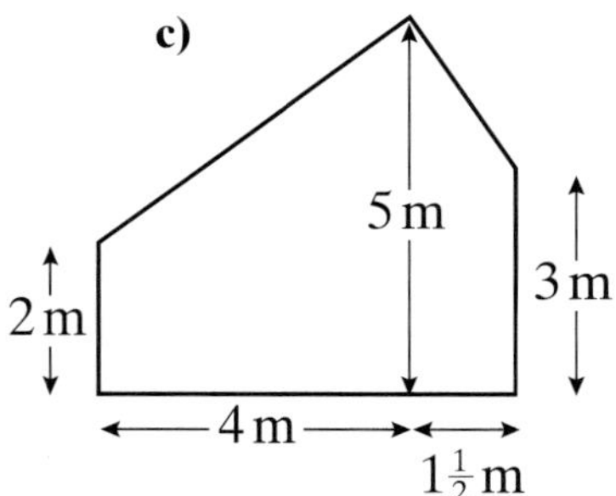

4 The square and triangle have the same area.

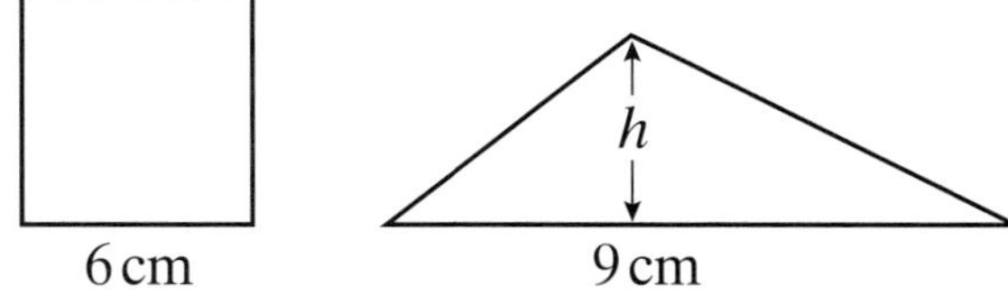

Find the value of h in centimetres.

5 Write an algebraic expression for:
a) the perimeter of the parallelogram.
b) the area of the parallelogram.

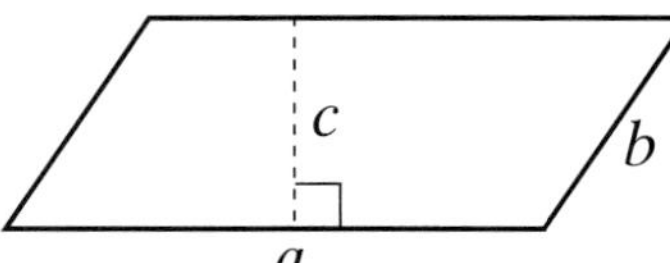

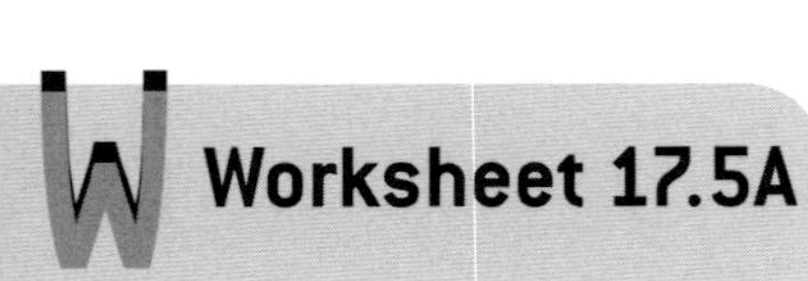

Worksheet 17.5A

Use the sine, cosine or tangent ratio to answer the following questions. You will find it helpful to draw a clear diagram as part of your solution and then decide which trigonometrical ratio to use.

You will need to know the meaning of **angle of elevation** and **angle of depression**:

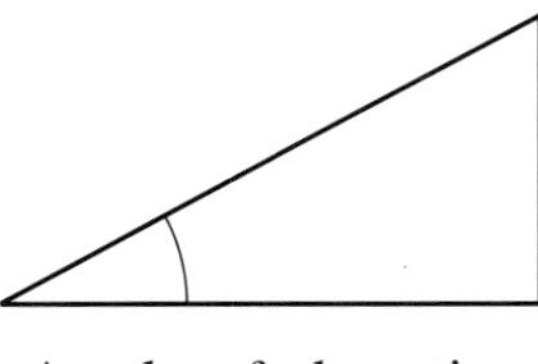 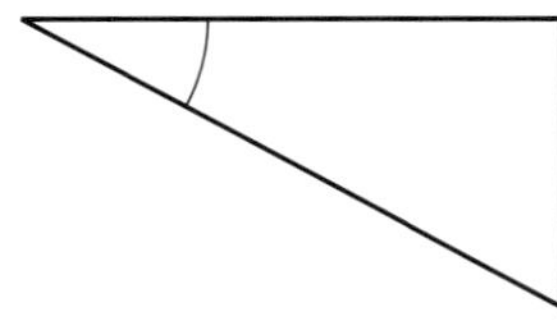

Angle of elevation   Angle of depression

1. In a right-angled triangle the hypotenuse is 12 cm long and the angle between the hypotenuse and the base of the triangle is 33°. Find the length of the base, correct to 1 decimal place.

2. Jasmine walks 2 km on a bearing of 042°. How far North is she from her starting point? Give your answer correct to 3 significant figures.

3. A ruler measures 30 cm and rests against the side of a desk at an angle of 65° to the horizontal. What is the vertical height of the ruler above the desk, correct to 1 decimal place?

4. The roof of a lean-to shed makes an angle of 14° to the horizontal. If the base of the shed is 4 m long, what is the vertical rise in the height of the roof, correct to 1 decimal place?

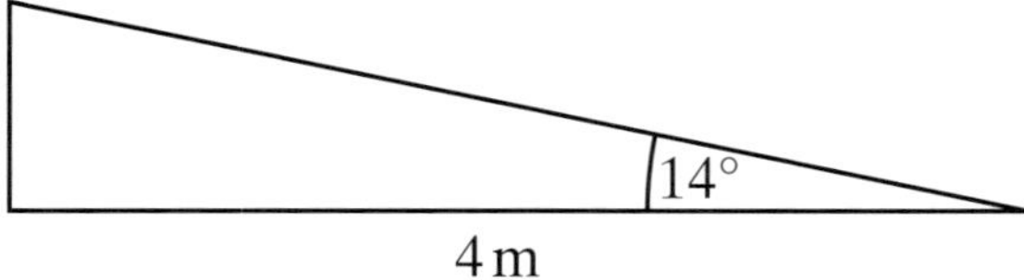

5. From a point on the ground the angle of elevation of the top of a mast is 48°. If the observation point is 20 m from the base of the mast, what is the height of the mast, correct to the nearest metre?

6. The diagonal of a rectangle is 13 cm long and the angle between the diagonal and a side of the rectangle is 64°. Calculate the length of the side of the rectangle. Give your answer correct to 1 decimal place.

7. A right-angled triangle has a hypotenuse of 12 cm and two angles of 45° each. Work out the length of the two remaining sides. Give your answer correct to 3 significant figures.

8. A washing line post is held in position by a supporting stake. The stake is nailed to the top of the post and secured in the ground at a distance of 0.75 m from the base of the post. The stake makes an angle of 74° with the ground. Use this information to find the height of the post, correct to the nearest centimetre.

9. Jamie cycles 9.5 km in an Easterly direction and then turns to cycle in a Southerly direction to the Sports Centre. The bearing of the Sports Centre from his starting point is 120°. How far did Jamie cycle in the Southerly direction? Give your answer correct to 3 significant figures.

10. A bird is hovering at a height of 48 m. The angle of elevation of the bird from an observation point on the ground is 53°. What is the horizontal distance of the bird from the observation point, correct to the nearest metre?

11. A ladder is leaning against a wall and reaches a height of 5.4 m up the wall. If the ladder is inclined at an angle of 72° to the horizontal, what is the length of the ladder correct to 1 decimal place?

12. Lisa is driving in a hilly region and passes along a road inclined at 9° to the horizontal. During the time she drives along the road it rises through a vertical height of 100 m. Calculate the distance she drives along the road. Give your answer correct to the nearest metre.

(continued)

 **Worksheet 17.5A**   GCSE Maths for Edexcel: Higher Teacher's Resource © Hodder Murray 2006

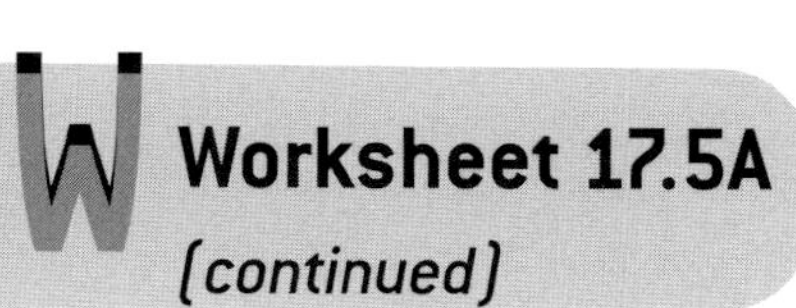

13 Will is flying a kite. He estimates that the string attached to the kite makes an angle of 83° to the ground. He estimates that at this angle the kite is at a horizontal distance of 10 m from where he is standing. Find the length of the string, correct to the nearest metre.

14 A garden plot is in the shape of a rectangle with the shorter side of length 24.8 m. A diagonal path running corner to corner across the plot makes an angle of 67° with this shorter side. What is the length of the path, correct to 3 significant figures?

15 A cone is of vertical height 12.8 cm, and the sloping edge makes an angle of 37° with the base of the cone. Calculate the length of the sloping edge correct to 1 decimal place.

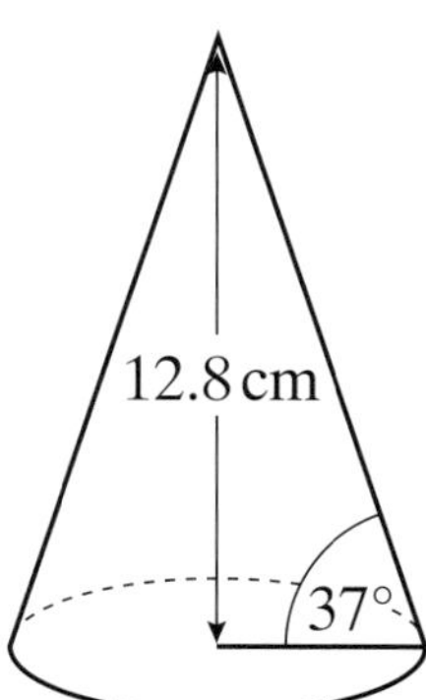

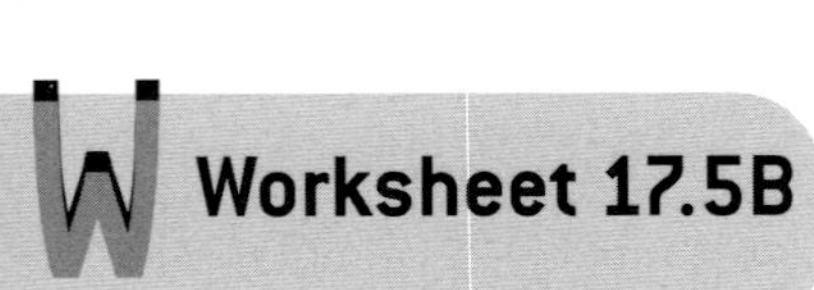

Worksheet 17.5B

Use the sine, cosine or tangent ratio to answer the following questions. Show details of your calculations, including a diagram.

You will need to know the meaning of **angle of elevation** and **angle of depression**:

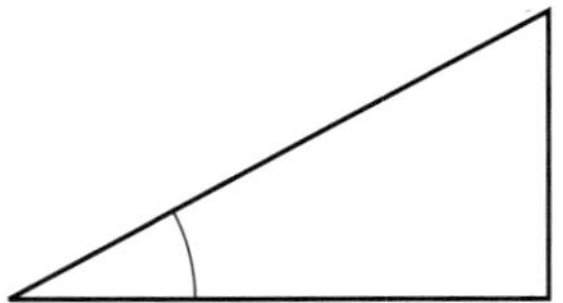
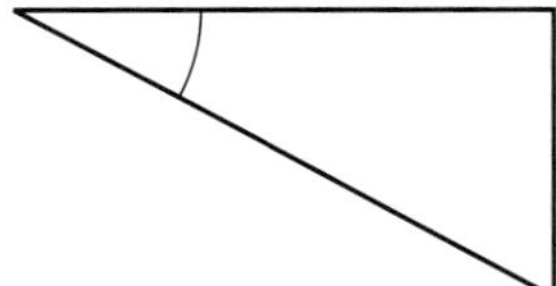

Angle of elevation   Angle of depression

1   A woman walks 3 km in an Easterly direction and then 5 km in a Northerly direction. What is her bearing from the starting point? Give your answer correct to the nearest degree.

2   A telegraph pole is 8 m tall. A wire support is attached to the top of the pole and anchored in the ground at a distance of 3 m from the base of the pole. Calculate the angle the wire makes with the horizontal ground, correct to 1 decimal place.

3   The diagonal of a rectangle is of length 15 cm. The longest side of the rectangle is 11 cm. Find the angle between the diagonal and the longest side correct to 3 significant figures.

4   A ladder is 7 m long and rests against a wall, reaching up to a height of 6.2 m. What angle does the ladder make with the ground, correct to 3 significant figures?

5   An isosceles triangle has two sides of length 8.5 cm and the third side of length 6.4 cm. Find the angle, correct to 1 decimal place, between a long side and the shorter side of the triangle.
Hint: You will need to create a right-angled triangle inside the isosceles triangle.

6   From a point on the ground, the angle of elevation of the top of a tree is used to calculate the height of the tree. If the point of observation is 19 m from the base of the tree and the height of the tree is worked out to be 16.5 m, what was the angle of elevation, to the nearest degree?

7   The length of the hypotenuse of a right-angled triangle is 8.45 cm. Find the angle between the hypotenuse and a side of the triangle which measures 4.8 cm. Give your answer correct to 3 significant figures.

8   Jason travels 4 km in a Northerly direction and then 6 km in an Easterly direction. Calculate his angle from the starting point. Give your answer as a three figure bearing correct to the nearest degree.

9   A plank of wood 1.87 m long rests against a wall. The base of the plank is 0.45 m from the wall. Calculate the angle the plank makes with the ground, correct to 2 decimal places.

10   A boat is observed at a distance of 247 m from the base of a cliff of height 30.8 m. What is the angle of depression of the boat from the observation point on the cliff top? Give your answer correct to 2 decimal places.

11   In a right-angled triangle the two sides which make up the right angle are 17 cm and 19 cm long. Find the size of each of the angles in the triangle, each correct to 1 decimal place.

12   A tent pole is supported by a rope attached to the top of the pole and the ground. If the tent pole is 2.4 m high and the rope is 3.5 m long, what is the angle, correct to 1 decimal place, between the rope and the ground?

13   A mini-football tournament has goals which are 1.5 m high. Scott shoots at the goal from a distance of 12.4 m and the ball hits the cross-bar. Calculate the angle of the shot, correct to the nearest degree.

14   The pages in a book measure 200 mm by 295 mm. A line is drawn diagonally across the page corner to corner. What angles does this line make with the sides of the page? Give your answers correct to the nearest degree.

   **Worksheet 17.5B**   GCSE Maths for Edexcel: Higher Teacher's Resource © Hodder Murray 2006

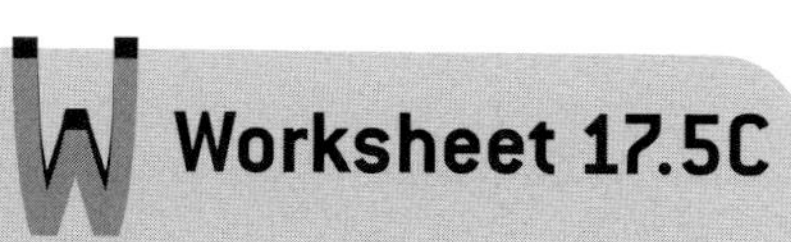

Worksheet 17.5C

In each of questions **1** to **10**, use the sine, cosine or tangent ratio or Pythagoras' theorem to find the missing side or angle. Give all lengths correct to 2 decimal places. Give angles correct to 0.2°

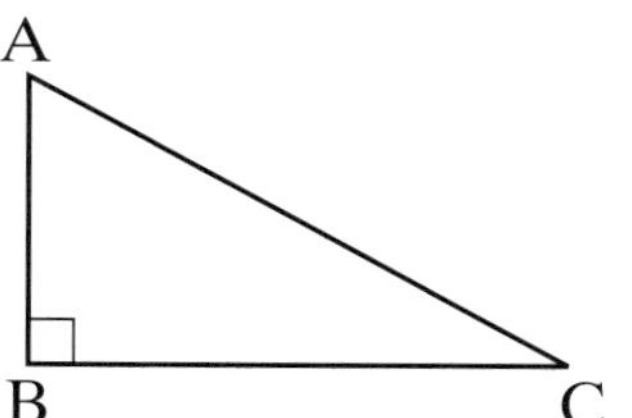

  1 Side AC = 4 cm, angle C = 47°. Find side AB.

  2 Side BC = 5.5 cm, angle C = 38°. Find side AB.

  3 Side AC = 9 cm, angle A = 63°. Find side BC.

  4 Side AC = 12.4 cm, angle A = 29°. Find side AB.

  5 Side AB = 8.2 cm, side BC = 14 cm. Find angle C.

  6 Side AC = 13 cm, side AB = 9 cm. Find angle A.

  7 Side BC = 8.3 cm, side AC = 20.4 cm. Find angle A.

  8 Side AB = 7.5 cm, side BC = 8 cm. Find side AC.

  9 Angle A = 35°, side AC = 27 cm. Find side BC.

10 Angle A = 74°, side AB = 9.8 cm. Find side BC.

In each of questions **11** to **20**, use the sine, cosine or tangent ratio or Pythagoras' theorem to find the missing side or angle. Give all lengths correct to 2 decimal places, and angles correct to 0.1°.

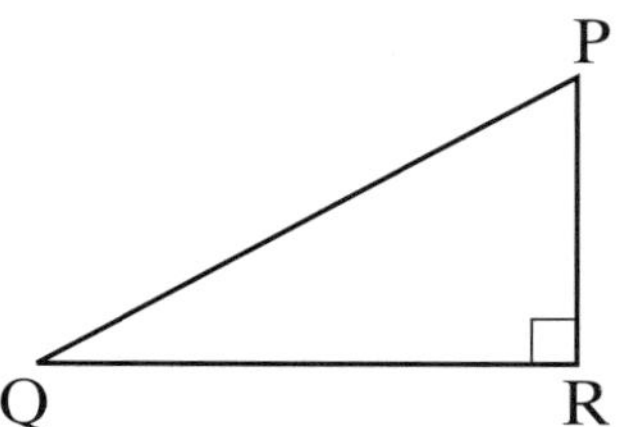

11 Angle Q = 35°, side PR = 7 cm. Find side PQ.

12 Side PR = 4.8 cm, angle P = 72°. Find side PQ.

13 Side PR = 9.4 cm, side PQ = 13.2 cm. Find side QR.

14 Angle P = 49°, side QR = 14 cm. Find side PR.

15 Side QR = 32 cm, angle Q = 32°. Find side PQ.

16 Side QP = 8.4 cm, side RP = 5.9 cm. Find angle P.

17 Angle P = 75°, side QR = 18 cm. Find side PQ.

18 Angle P = 28.5°, side PR = 5.8 cm. Find side QR.

19 Side PQ = 15 cm, side QR = 7.8 cm. Find side PR.

20 Angle Q = 19°, side PR = 4.85 cm. Find side PQ.

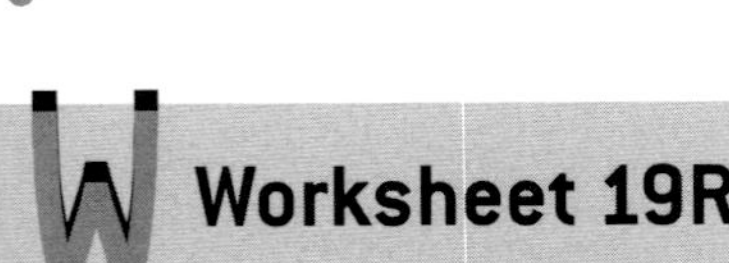

Worksheet 19R

Find the value of each angle marked with a letter. (A dot marks the centre of each circle.)

1

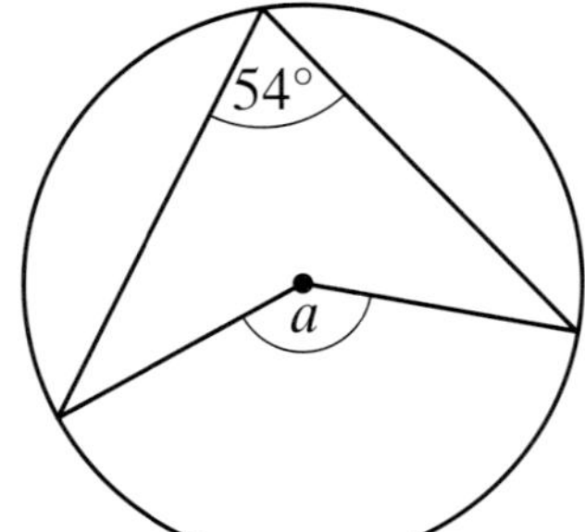

2

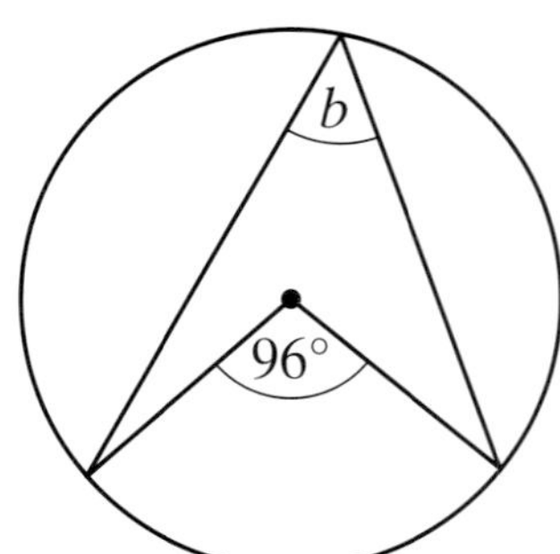

3

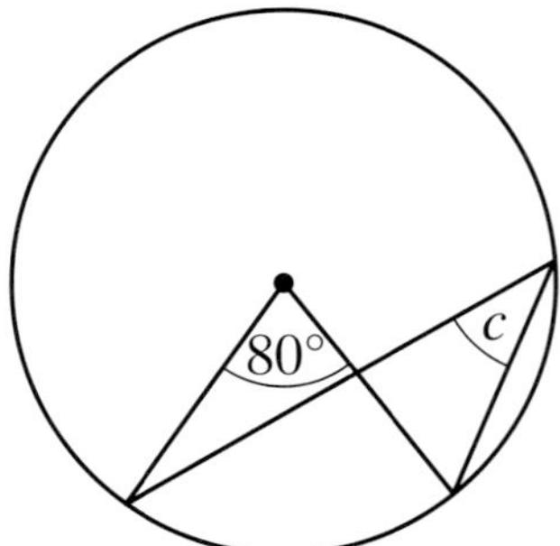

4

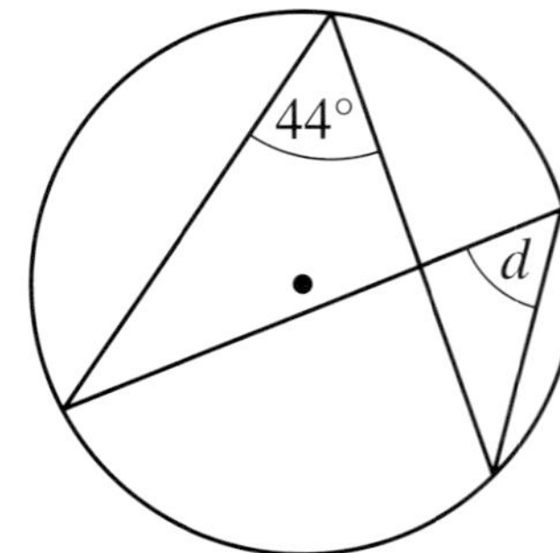

5

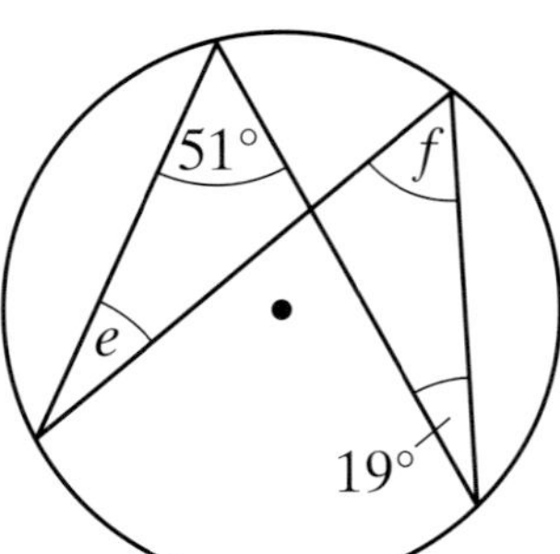

6

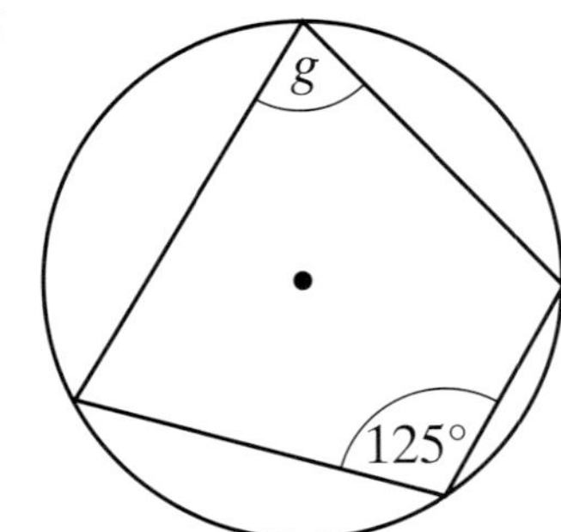

7

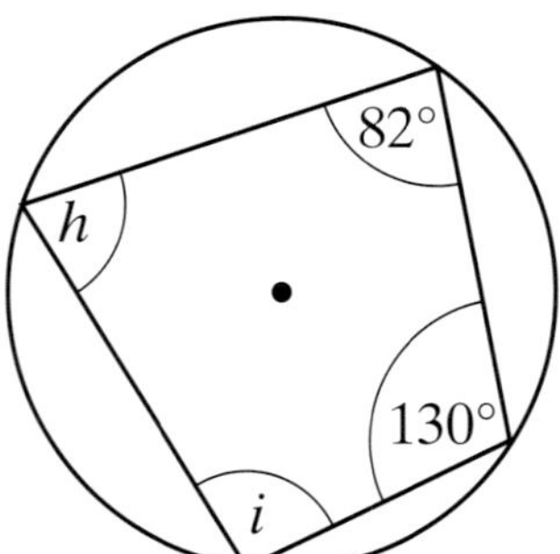

8

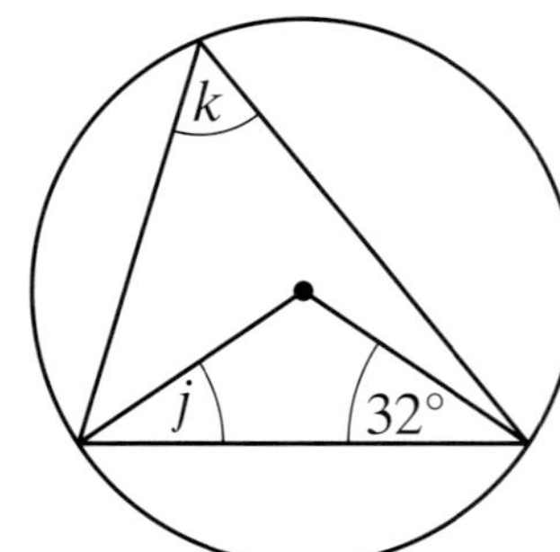

9

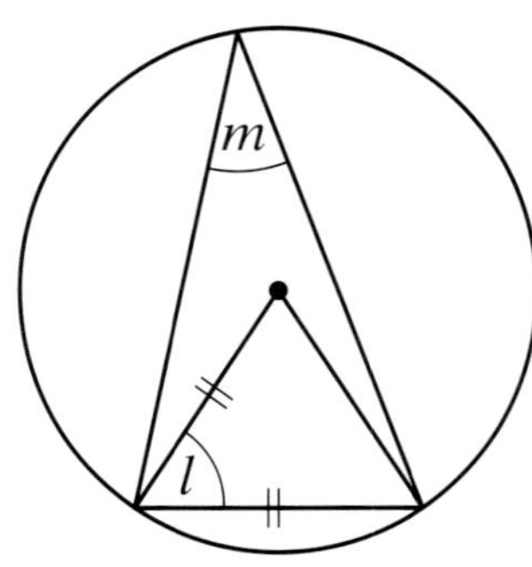

10

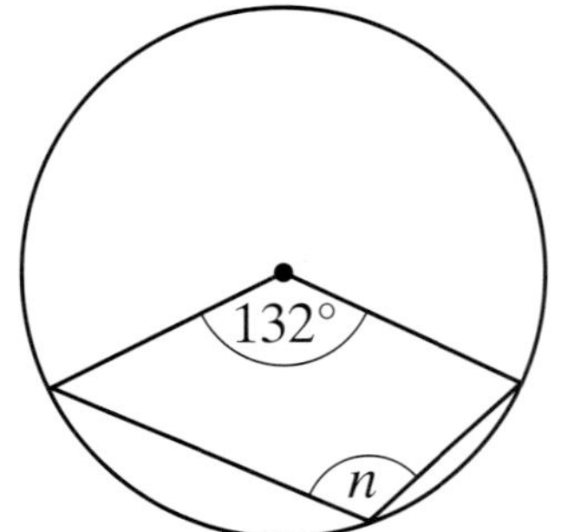

(continued)

 Worksheet 19R

11 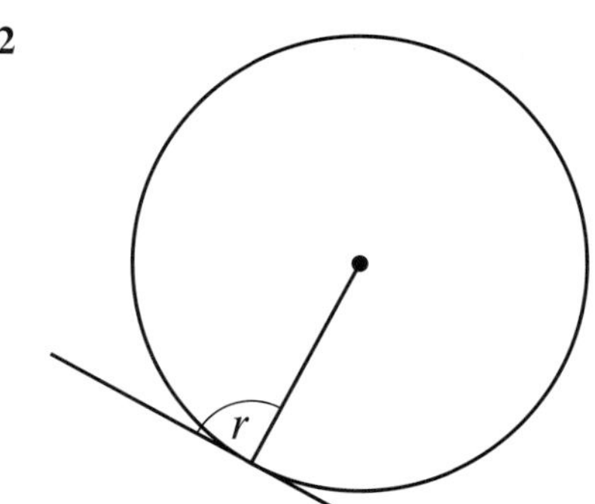

12 

13 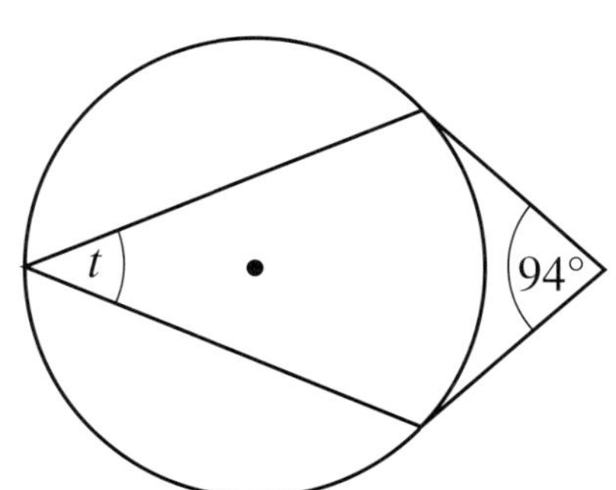

14

15 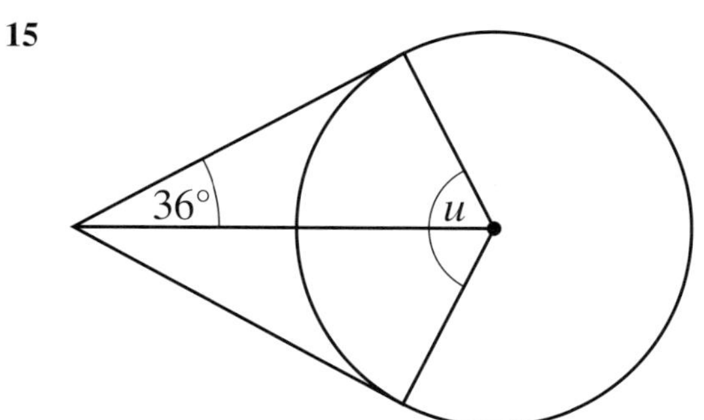

16 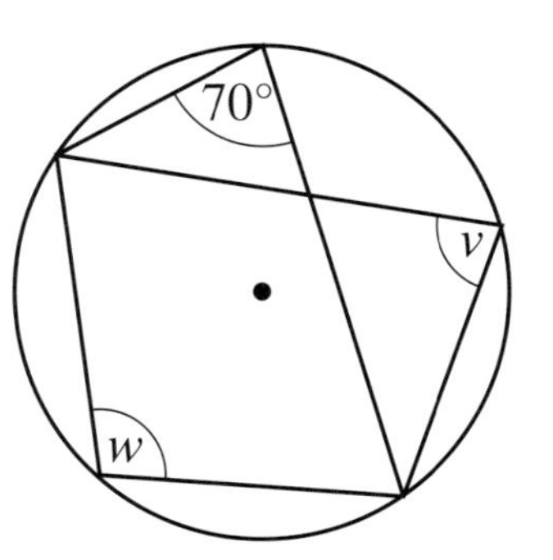

17 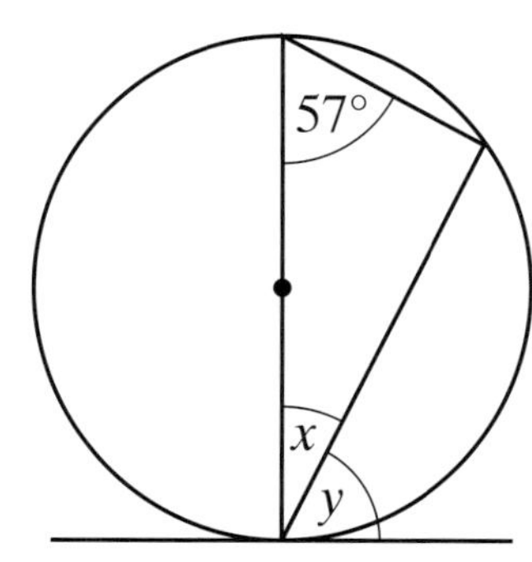

18 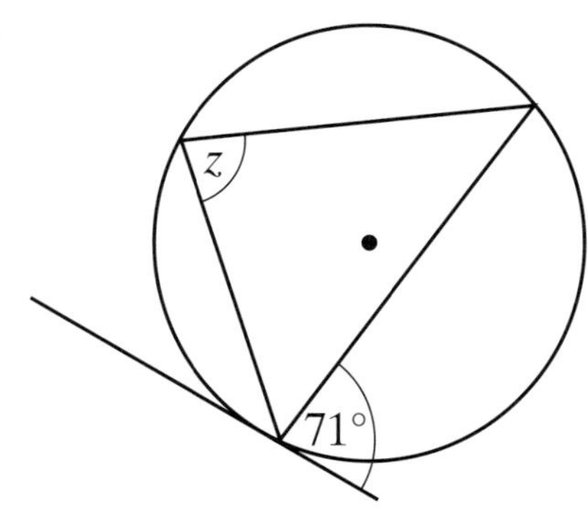

19 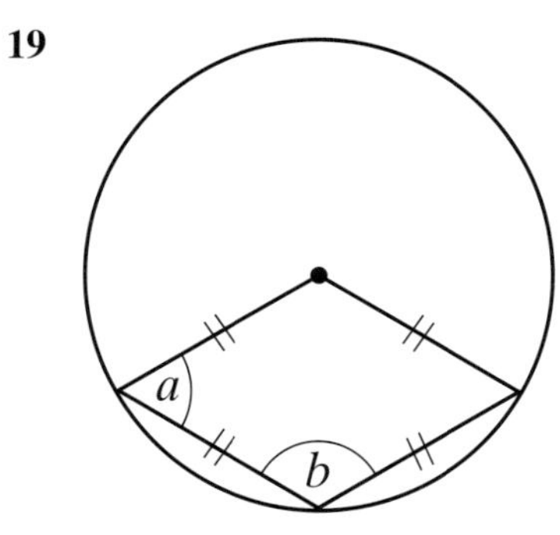

20 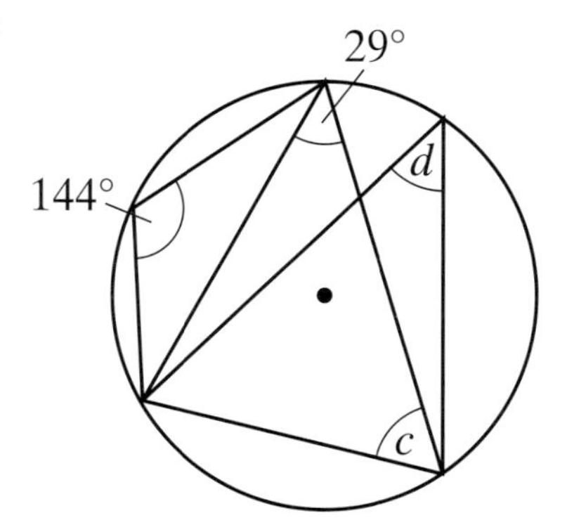

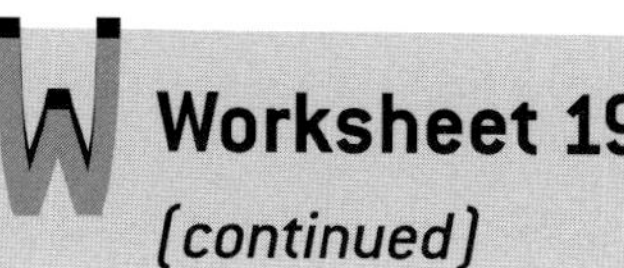

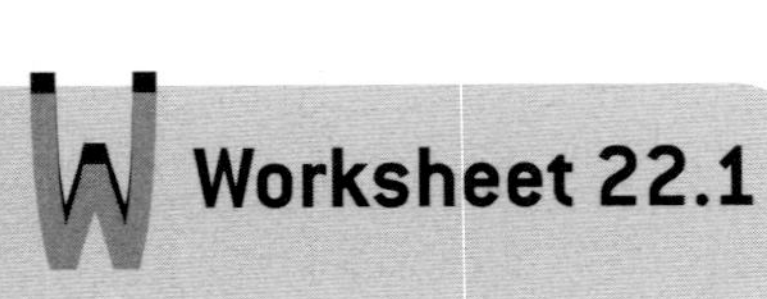

Worksheet 22.1

1 A bag contains three red counters and four blue counters. A counter is drawn at random.
Find the probability that it is:

a) red.   **b)** blue.   **c)** green.   **d)** red or blue.

2 In a Hockey Club raffle, 30 red tickets and 90 white tickets are sold. The tickets are put into a container and shaken up. A single raffle ticket is drawn to decide the prizewinner. Find the probability that the winning ticket is:

a) red.   **b)** white.
Give your answers as fractions in their simplest form.

3 A fair six-sided dice is thrown and the number recorded. Calculate the probability that the number thrown is:

a) 5.   **b)** less than 3.   **c)** a multiple of 2.
d) a square number.   **e)** a factor of 12.

4 Luke has made some quadrilateral shapes out of card. He has cut out a square, a rectangle, a parallelogram, an isosceles trapezium and a rhombus. He randomly picks out one of the shapes. Calculate the probability that it has:

a) two pairs of parallel sides.   **b)** four equal sides.
c) four equal angles.   **d)** only one pair of parallel sides.
e) diagonals of equal lengths.

5 A factory production line manufactures plastic ducks. The ducks are inspected, and those which are not up to standard are rejected. The probability of a duck being rejected is $\frac{2}{15}$.

a) Find the probability that a duck is not rejected.

During a production run 300 plastic ducks are produced.

b) Estimate how many of these ducks are rejected.

6 A football team has played 12 matches of which 5 were wins and 3 were draws.

a) How many matches did the team lose?
b) Using the information given, find the probability that the football team will win its next match.

7 Melissa is revising for a maths exam. She predicts that there is a probability of 0.45 that she will get full marks for any particular question.

a) Write down the probability that she does not get full marks for a particular question.
b) There are 60 questions on the paper. Estimate the number of questions for which Melissa expects to get full marks.

8 The whole numbers from 1 to 20 are written on pieces of paper, put into a container and shaken up. A number is drawn out at random. Find the probability that it is:

a) an odd number.   **b)** a prime number.
c) a cube number.   **d)** a multiple of 6.
e) a factor of 30.

9 A letter is chosen at random from the word **ISOSCELES**. Calculate the probability that the letter chosen is:

a) S.   **b)** E.   **c)** C.   **d)** N.

10 Lisa works in a shoe shop during the summer holiday. The probability of her being early is 0.35 whilst the probability of being on time is 0.45.

a) Write down the probability of her being late.
b) Lisa works a five-day week for a total of four weeks. Estimate the number of days she will be:
     (i) on time.   **(ii)** late.

Answer these questions without using a calculator.

1 A six-sided dice has a bias such that the probability of getting a score of five is 0.25. Work out the probability of not getting a score of five.

2 Five red counters, two blue counters and three green counters are placed in a bag. The bag is shaken to mix them up. A counter is pulled at random from the bag. Calculate the probability that the counter is:
a) green. **b)** blue. **c)** green or blue. **d)** not blue.

3 Three classes, 10X, 10Y and 10Z, compete against each other in a netball tournament. The probability of winning is shown in the table.

Class	10X	10Y	10Z
Probability of winning	0.3	0.35	p

a) Calculate the value of p.
b) Calculate the probability of either 10X or 10Y winning.
c) The three classes compete against each other in a series of ten tournaments. Estimate the number of times that either 10Y or 10Z would be expected to win.

4 At the end of term, a class of students measured the length of their pencils. The number of pencils of different lengths are shown in the table.

Length (cm)	$x \leqslant 5$	$5 < x \leqslant 10$	$10 < x \leqslant 15$	$x > 15$
Frequency	3	14	7	6

A pencil is taken at random from the box. Calculate the probability that the length of the pencil is:
a) less than or equal to 10 cm.
b) in the range $5 < x \leqslant 15$ cm.
c) less than or equal to 5 cm or greater than 15 cm.

5 Mr Baker can take three routes to work: the motorway, the A-road or the B-road. The probability that he chooses to take the motorway is 0.18. The probability that he chooses to take the A-road is 0.36.
a) Calculate the probability that Mr Baker chooses to take the B-road.

Due to road works, the B-road is closed for a period of time.
b) Assuming the probabilities of Mr Baker taking the motorway and the A-road remain in the same ratio, what is now the probability that he will choose to take the A-road?
c) During the closure of the B-road Mr Baker makes 60 journeys. Estimate the number of these journeys that he would make using the motorway.

6 A five-sided spinner has the numbers 1, 2, 3, 4 and 5 written on it. Due to the shape of the spinner there are different probabilities for each number. These are shown in the table.

Number	1	2	3	4	5
Probability	0.2	0.14	0.32	p	0.28

a) Calculate the probability, p, of getting a 4.
b) Find the probability of getting a 1 or a 3.
c) The probability of getting number A or number B is 0.2. Work out what the two numbers must be.
d) Work out the probability of an odd number coming up on the spinner.

(continued)

Worksheet 22.2
(continued)

7 When Bobby plays chess against his computer he either wins, draws or loses. For any particular game the probability that Bobby wins is 0.07. The probability of a draw is 0.18.
Bobby plays one game against the computer. Find the probability that:
a) Bobby does not win.
b) Bobby does not lose.
After going on a chess coaching course, Bobby finds that the probability of winning against the computer is doubled; so too is the probability of a draw.
c) Work out the probability that Bobby loses a game against the computer after the course.

8 Last year Jackie entered a number of swimming competitions and came first in five of them. The probability of her coming first in a competition is estimated to be 0.3125. How many swimming competitions did she enter?

9 A pack of 52 playing cards has equal numbers of hearts (red), diamonds (red), spades (black) and clubs (black). Each card has an equal chance of being selected. Find the probability of selecting:
a) a red card.   **b)** a black club.
c) a red heart or a black spade.   **d)** not a black club.
e) neither a black spade nor a red diamond.

10 A bag contains coloured discs. Each disc is either red, white, blue, green or yellow. The bag does not contain equal numbers of each colour of disc. When a disc is chosen at random from the bag, the probability that it is of a certain colour is given by the table.

Colour	Red	White	Blue	Green	Yellow
Probability	$\frac{1}{6}$	$\frac{1}{4}$	$\frac{1}{5}$	$\frac{3}{10}$	$\frac{1}{12}$

a) Use the table to find the probability that a randomly chosen disc is:
    (i) red or white.
    (ii) blue or yellow.
    (iii) not red.
There are 36 green discs in the bag.
b) Work out the total number of discs in the bag.

 **Worksheet 22.2**   GCSE Maths for Edexcel: Higher Teacher's Resource © Hodder Murray 2006

In order to answer some of these questions, you will need to draw a probability tree.

1. A bag contains four red counters and six blue counters. A counter is taken from the bag at random and the colour noted. It is then replaced in the bag and the bag is shaken. A second counter is then taken out at random. Find the probability that both counters are of the same colour.

2. On Mrs Green's way to school there are two sets of traffic lights which she has to drive through. The probability that the first set of traffic lights shows red when Mrs Green reaches them is 0.3. The probability that the second set of traffic lights shows red when she reaches them is 0.6.
   a) Illustrate this situation with a probability tree diagram.
   b) Use the diagram to calculate the probability of both sets of traffic lights showing red.
   c) Use the diagram to calculate the probability that Mrs Green will have to stop at at least one set of traffic lights.

3. Ten green counters and ten red counters are placed in a bag. One counter is chosen at random, then replaced in the bag. A second counter is then chosen at random.
   a) Find the probability that both counters are green.
   b) Find the probability that both counters are red.

4. Matthew has been taking driving lessons. He believes he is now ready to take his driving test. His instructor reckons that his probability of passing the test on his first attempt is 0.4. If he fails the first time the instructor reckons he has a probability of 0.7 of passing on his second attempt. Calculate the probability of Matthew failing his driving test on both attempts.

5. A car is taken for an MOT test. The owner estimates that the probability of it passing is 0.25. If it fails the owner will carry out minor repairs and he estimates that this will increase the probability of passing the re-test to 0.65. Calculate the probability that the car fails the MOT test but then passes the re-test.

6. A bag contains five red balls, four green balls and three blue balls. One ball is taken from the bag at random and its colour noted before it is put back in the bag. Another ball is then taken from the bag and its colour recorded. Work out the probability that the two balls are:
   a) red.
   b) green.
   c) blue.

7. A bag contains 10 cards, each showing the picture of a mathematical shape. Four cards show pictures of triangles and the other six show rectangles. One card is taken from the bag and examined; it is then replaced. A second card is then taken from the bag. Find the probability that the cards show:
   a) a triangle followed by a rectangle.
   b) not a triangle on either draw.
   c) two shapes the same.

8. In a raffle 10 red tickets and 20 white tickets are sold. Each ticket has an equal chance of being chosen. There are two prizes in the raffle. A ticket is chosen at random, and wins the first prize. The ticket is then returned to the bag, and a ticket is again chosen at random to determine the winner of the second prize.
   Calculate the probability that:
   a) two red tickets are drawn.
   b) a white ticket is drawn, followed by a red ticket.
   c) two white tickets are drawn.
   d) two tickets of the same colour are drawn.

(continued)

9 Inside a container there are seven white squares and five black squares. A square is taken at random from the container.
   a) Work out the probability that the square drawn is:
       (i) white.   **(ii)** black.

   The square is returned to the container. A second square is taken out and the colour noted.
   b) Work out the probability that the two squares drawn out are:
       (i) both white.   **(iv)** both black.

10 In a school canteen there are two main courses: burger or salad. The probability that a student will choose a burger is 0.65. There are also two desserts: yoghurt or treacle pudding. The probability that a student will choose treacle pudding is 0.45.
   Each student chooses a main course and a dessert. The choice of dessert is made independently of the choice of main course.
   Work out the probability that a particular student will choose:
   a) salad.   b) yoghurt.   c) salad followed by yoghurt.

        **Worksheet 22.4**   GCSE Maths for Edexcel: Higher Teacher's Resource © Hodder Murray 2006

 **Worksheet 25.1**

Solve these quadratic equations, using the factorising method.
The first one has been partly done to start you off.

1
$$x^2 - 6x + 8 = 0$$
$$(x - 4)(x\ldots\ldots\ldots\ldots) = 0$$
$$x - 4 = 0 \text{ or } \ldots\ldots\ldots$$
$$x = 4 \text{ or } x = \ldots\ldots\ldots$$

2 $x^2 + 5x + 6 = 0$

3 $x^2 - 4x - 5 = 0$

4 $x^2 - 5x + 6 = 0$

5 $x^2 + x - 56 = 0$

6 $x^2 + 2x - 35 = 0$

7 $x^2 + 8x + 15 = 0$

8 $x^2 + 3x - 40 = 0$

9 $x^2 - x - 2 = 0$

10 $x^2 - 9x + 20 = 0$

11 $2x^2 + 5x + 3 = 0$

12 $2x^2 - 11x + 5 = 0$

13 $3x^2 + x - 4 = 0$

14 $6x^2 + x - 1 = 0$

15 $5x^2 - 13x - 6 = 0$

16 $6x^2 + 5x - 6 = 0$

17 $10x^2 - 30x = 0$

18 $x^2 - 36 = 0$

19 $12x^2 + 18x = 0$

20 $4x^2 - 100 = 0$

▶▶ STARTER 2

3

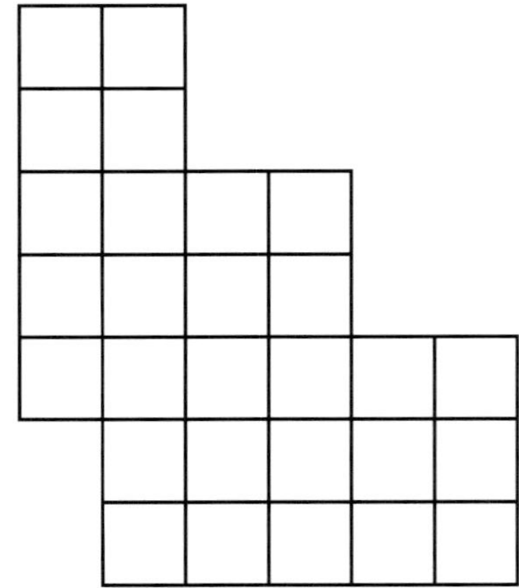

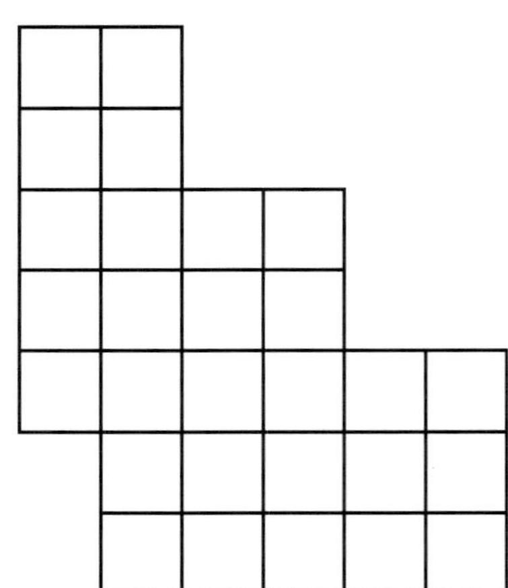

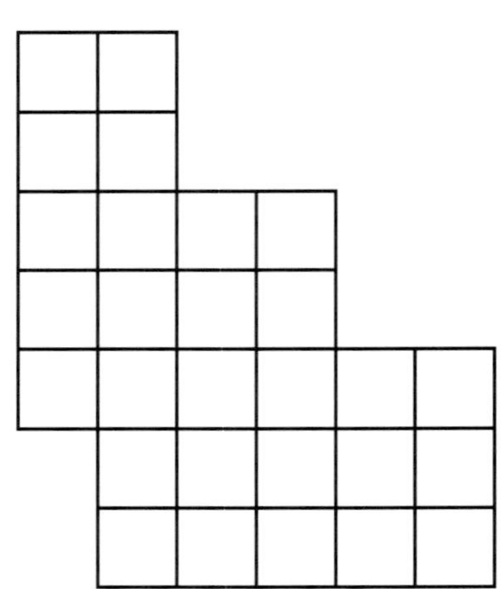

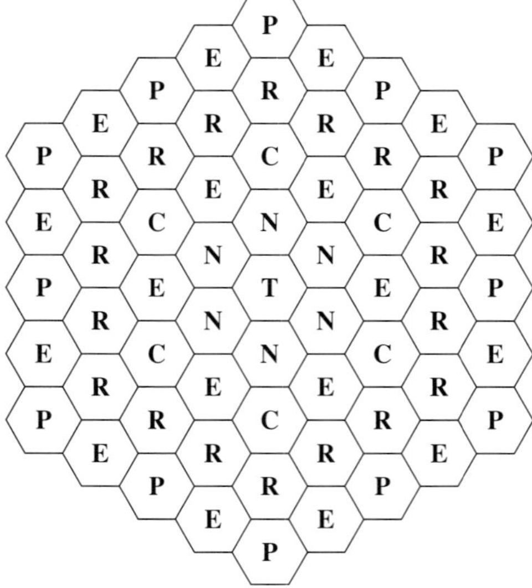

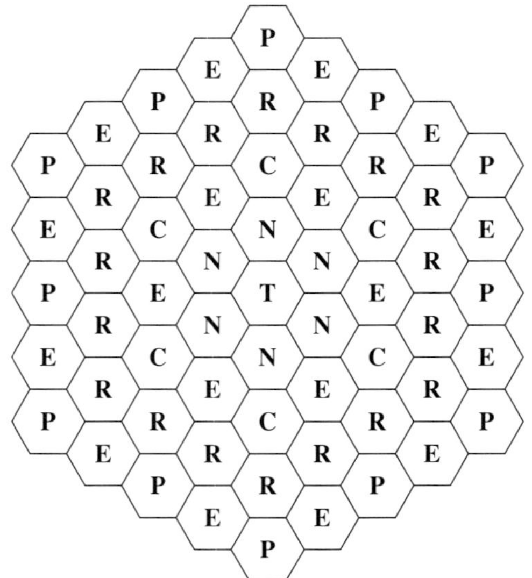

▶▶ STARTER 3

Speed-up sheet

GCSE Maths for Edexcel: Higher Teacher's Resource © Hodder Murray 2006

1   Astronomers have calculated that the mass of the Sun is about .... kg.

2   The Sun is thought to have formed about .... years ago.

3   Each second the Sun's mass decreases by about .... tonnes.

4   The surface temperature of the Sun has been measured to be about .... degrees C.

5   It takes our Solar System about .... years to make one revolution around the Milky Way galaxy.

6   Light travels through space at a speed of .... metres per second.

7   Visible light has a wavelength of about .... metres.

8   X-rays can have wavelengths as short as .... metres.

9   The Andromeda galaxy is so remote that light from it takes .... years to reach us.

10   It is thought that the Universe contains about .... individual galaxies.

2.998×10^8	10^{-9}	6×10^3	10^{11}	2×10^{30}
2.25×10^8	4×10^6	2.8×10^6	5×10^9	5.5×10^{-7}

▶▶ INTERNET CHALLENGE 5

T	E	Q	U	A	T	I	O	N	N	D	P	O	R	I
V	X	S	P	G	F	E	N	Y	M	Z	S	E	R	Y
B	P	O	W	E	R	F	T	L	A	C	I	D	A	R
Q	R	X	O	Y	U	O	M	K	P	P	M	S	T	L
U	E	S	N	E	O	A	N	T	P	N	P	C	I	H
O	S	L	D	R	U	S	A	O	I	S	L	D	O	S
T	S	A	O	D	E	R	L	U	N	F	I	V	N	J
I	I	N	D	E	X	Y	E	A	G	I	F	A	A	Y
E	O	J	E	L	N	C	D	U	H	C	Y	R	L	T
N	N	D	N	O	I	T	C	N	U	F	E	I	Q	I
T	E	R	M	G	E	O	W	L	A	L	H	A	F	T
A	R	I	L	T	C	U	D	O	R	P	N	B	A	N
V	A	D	J	H	Y	T	E	D	W	T	X	L	B	E
L	Y	T	S	F	A	C	T	O	R	I	S	E	L	D
C	I	T	A	R	D	A	U	Q	F	V	H	I	Y	I

EQUATION	EXPAND	EXPRESSION	FACTORISE	FUNCTION
IDENTITY	INDEX	MAPPING	POLYNOMIAL	POWER
PRODUCT	QUADRATIC	QUOTIENT	RADICAL	RATIONAL
ROOT	SIMPLIFY	SURD	TERM	VARIABLE

Speed-up sheet: Chapter 5

3

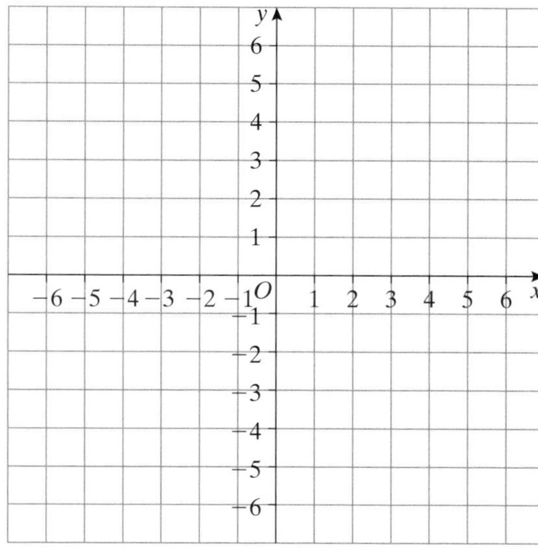

4

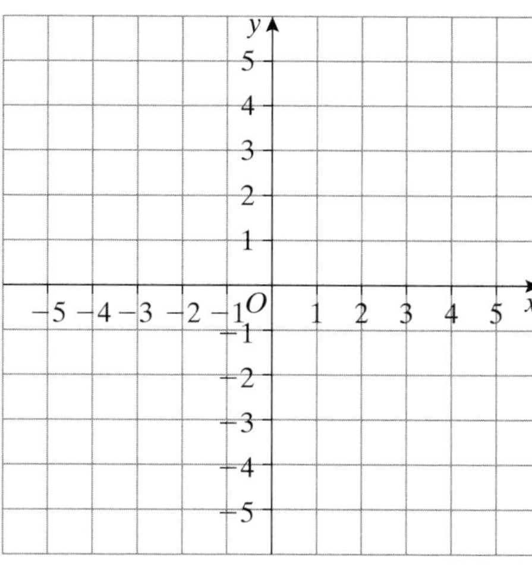

GCSE Maths for Edexcel: Higher Teacher's Resource © Hodder Murray 2006　　　　**Speed-up sheet**　　67

▶▶ EXERCISE 7.2

1

x	-4	0	4
y	-7	1	

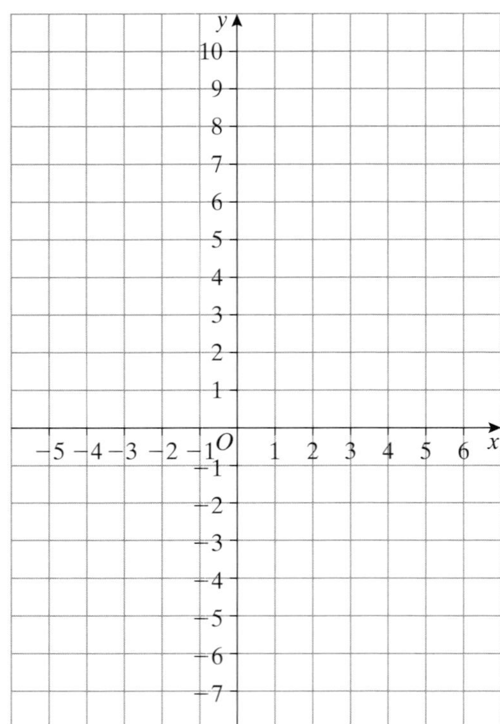

2

x	-5	0	5
y	-1		

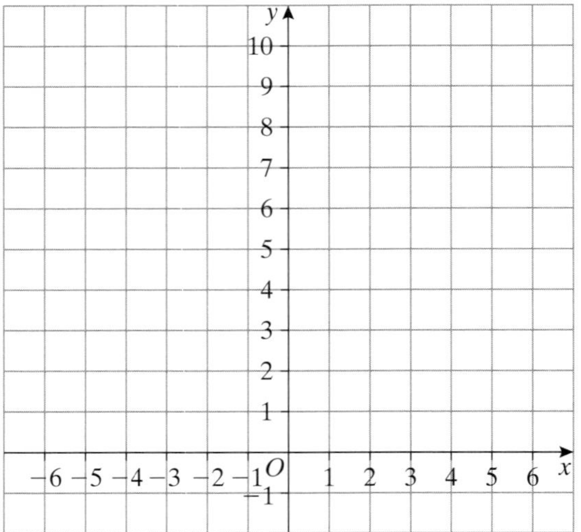

4

x	-2	0	4
y			

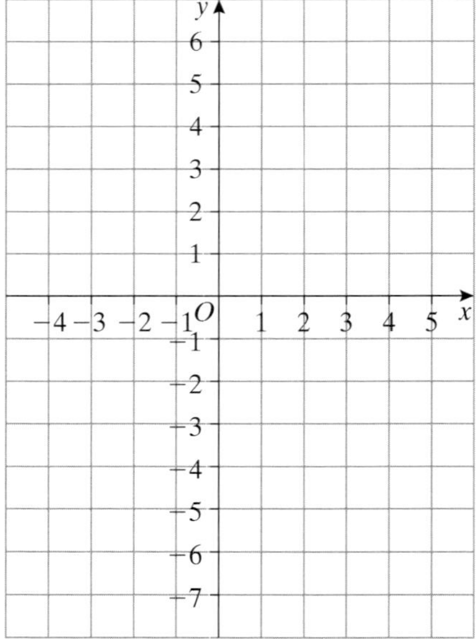

3

x	-4	0	5
y			

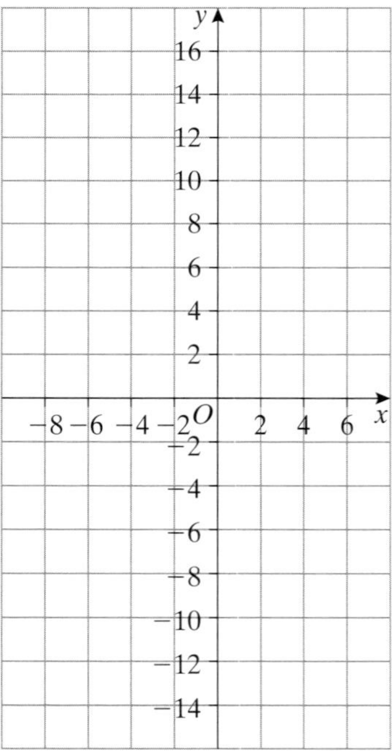

5

x	-6	0	4
y			

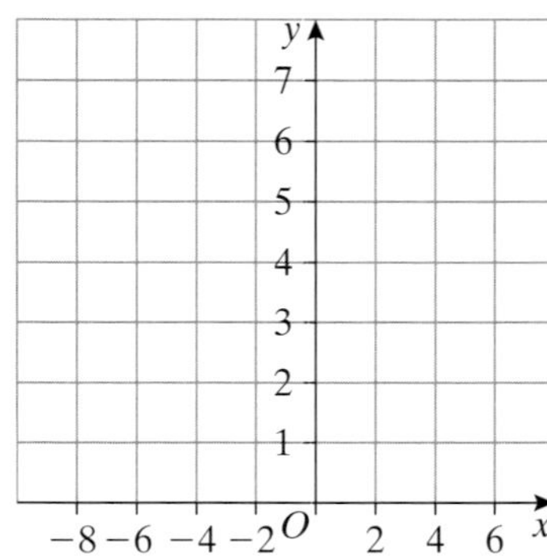

(continued)

GCSE Maths for Edexcel: Higher Teacher's Resource © Hodder Murray 2006

6

x	-5	0	5
y			

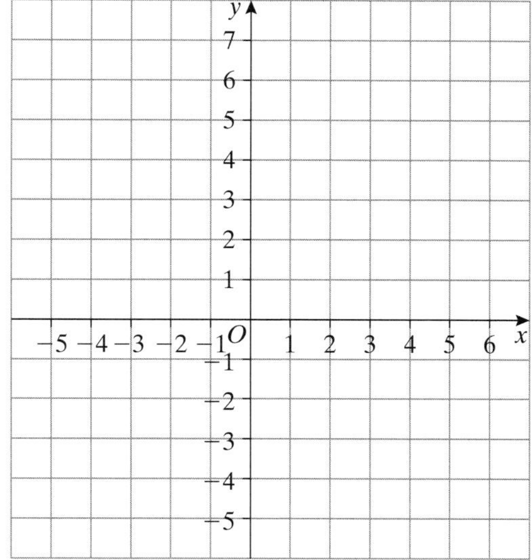

7

x	0	5	10
y			

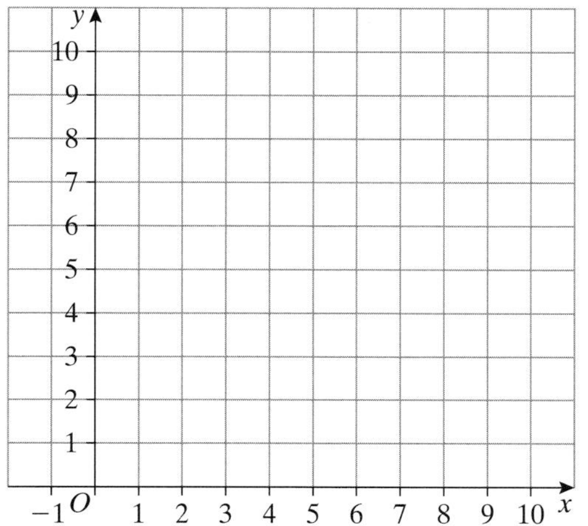

8

x			
y			

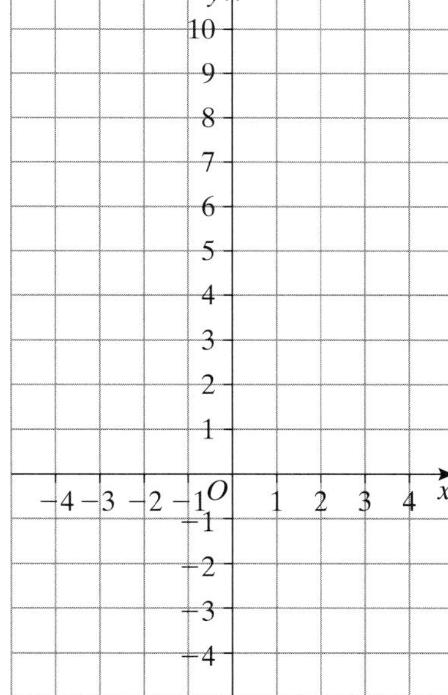

9

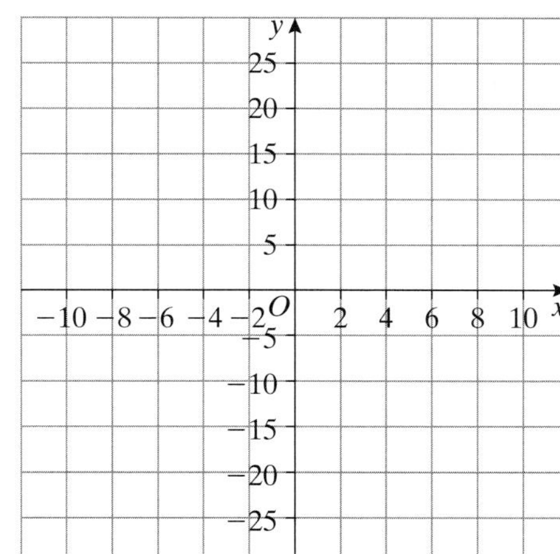

10

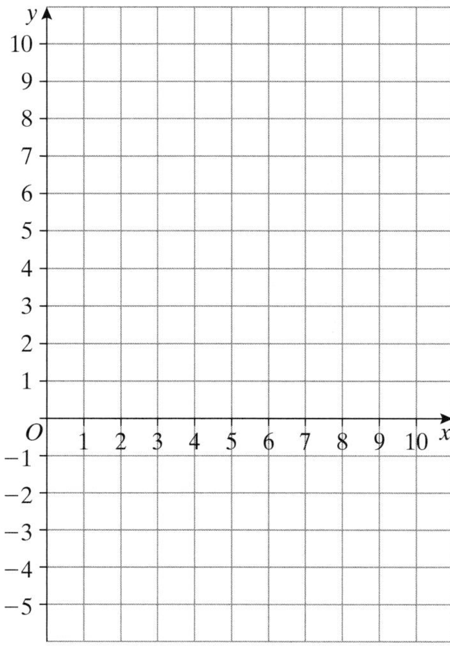

▶▶ REVIEW EXERCISE 7

1

x	-5	0	2
y	-1		6

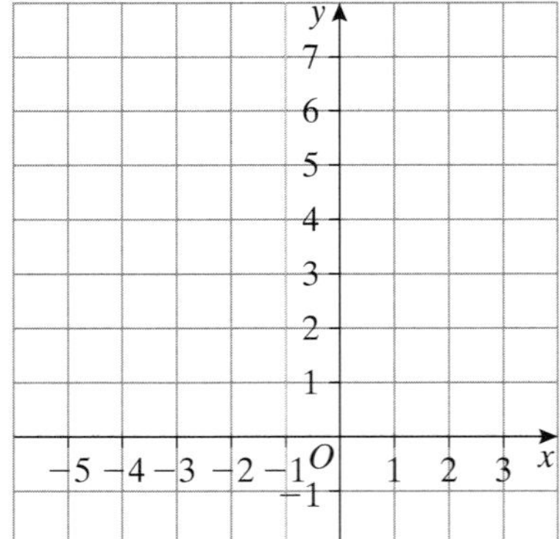

2

x	-2	0	6
y	0		

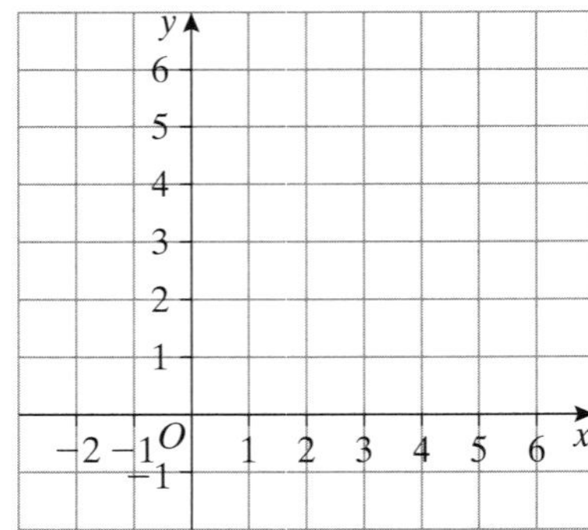

3

x	-6	0	6
y			

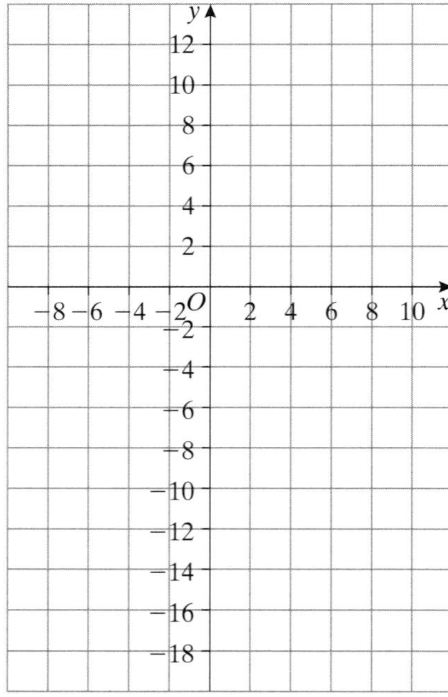

4

x	0	8	20
y			

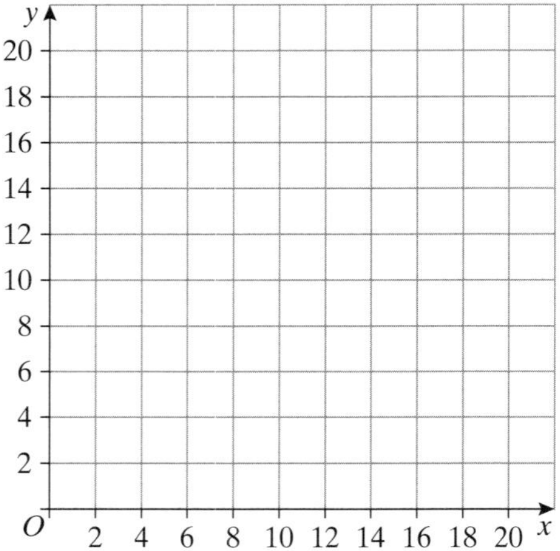

Speed-up sheet

GCSE Maths for Edexcel: Higher Teacher's Resource © Hodder Murray 2006

▶▶ EXERCISE 8.3

1
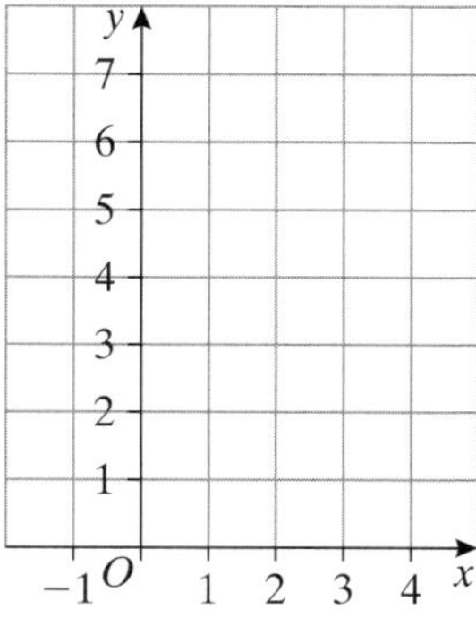

2
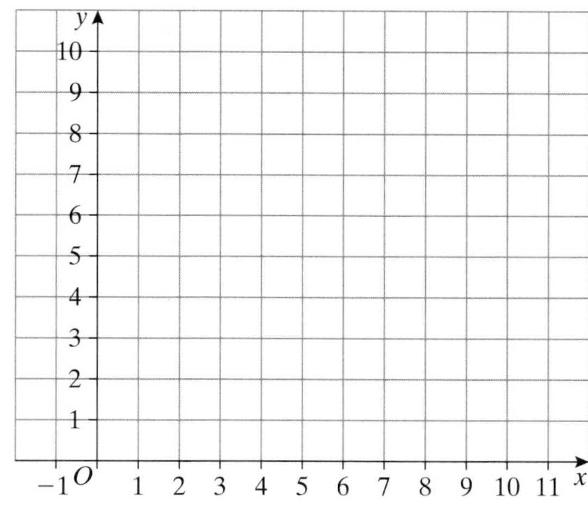

3
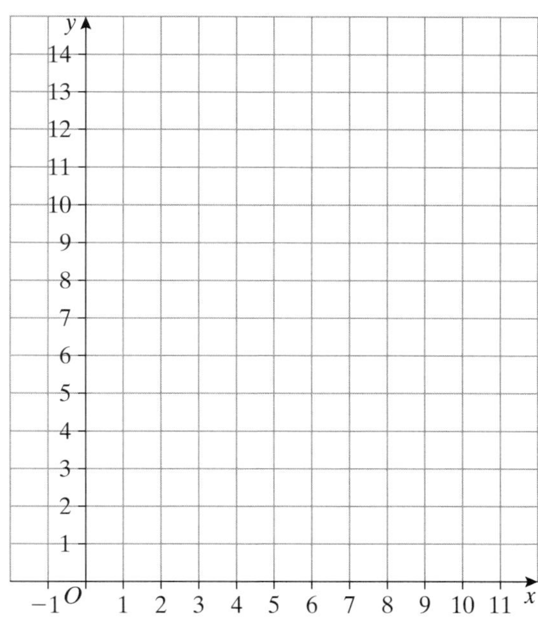

4
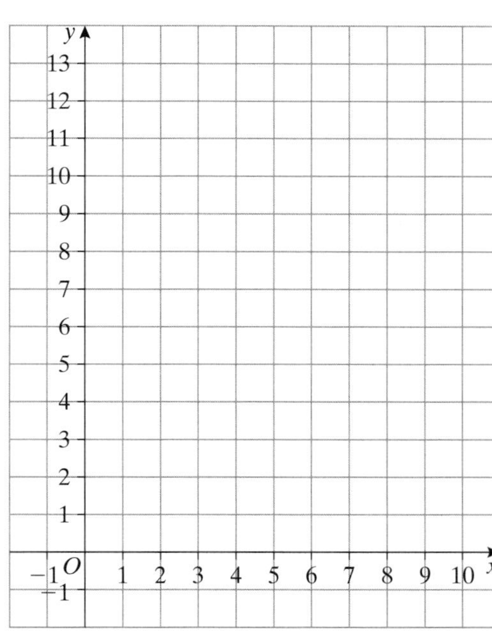

5
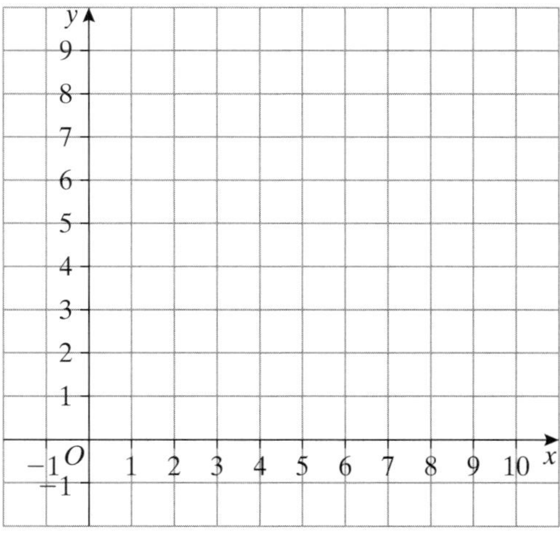

6
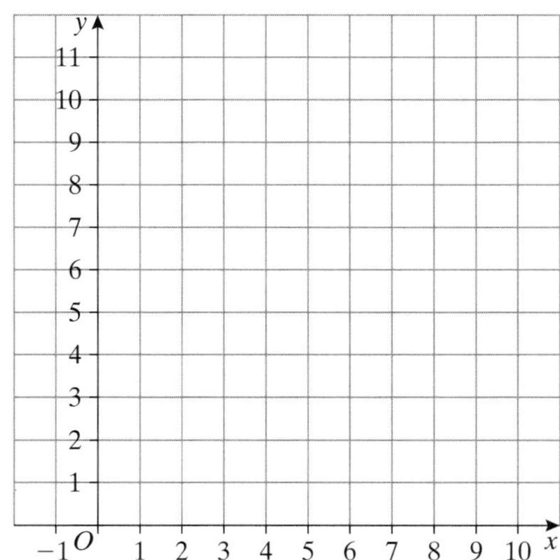

(continued)

7

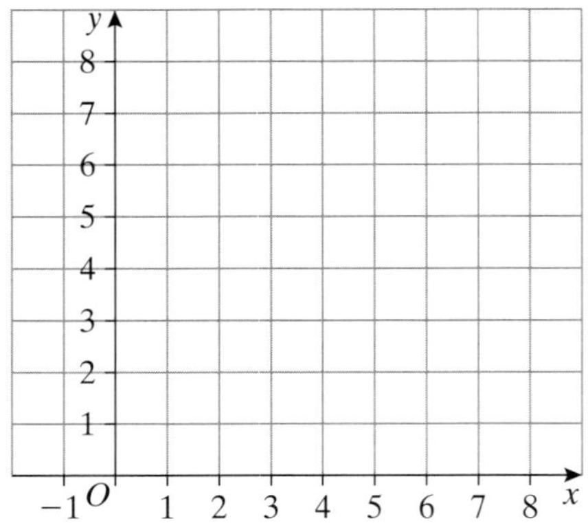

8

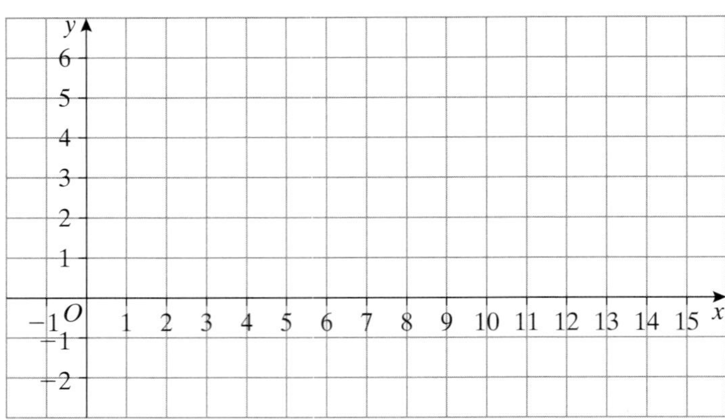

▶▶ **WORKSHEET 8.3**

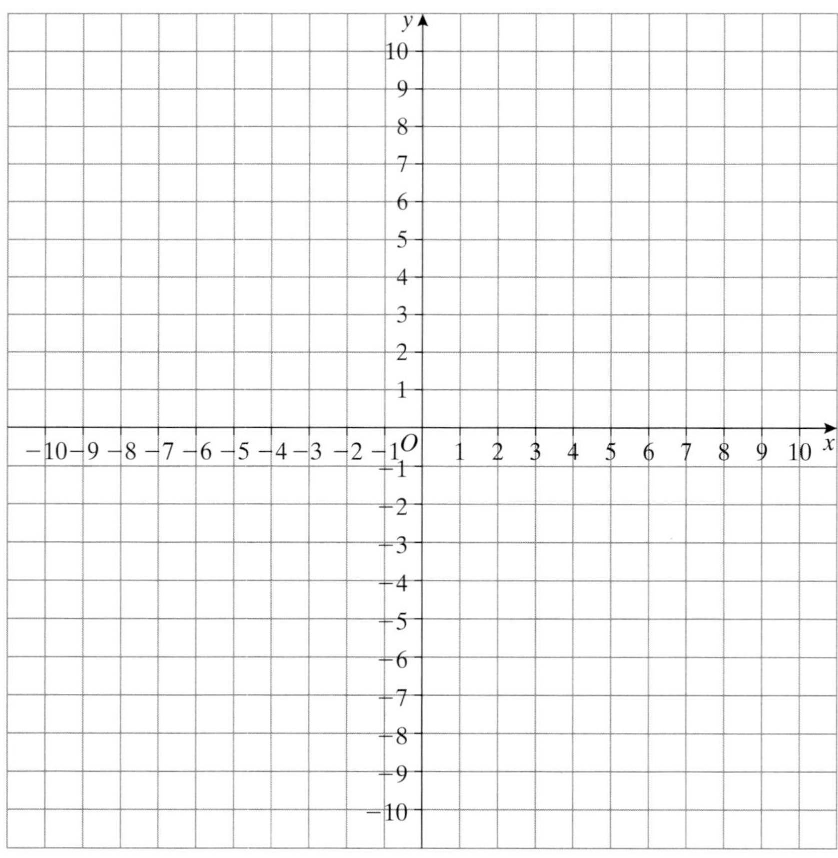

13

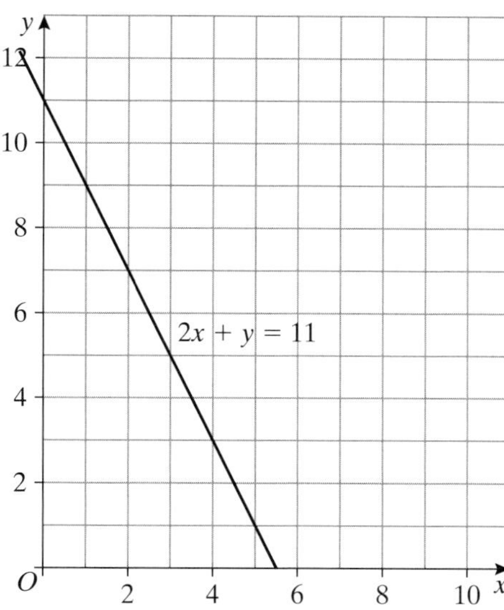

y
12
10
8
6
4
2
O
2   4   6   8   10   x
2x + y = 11

y
6
4
2
O
2   4   6   8   10   x
2x + 3y = 18

y
11
10
9
8
7
6
5
4
3
2
1
−1   O   1   2   3   4   5   6   x

Speed-up sheet: Chapter 8

14

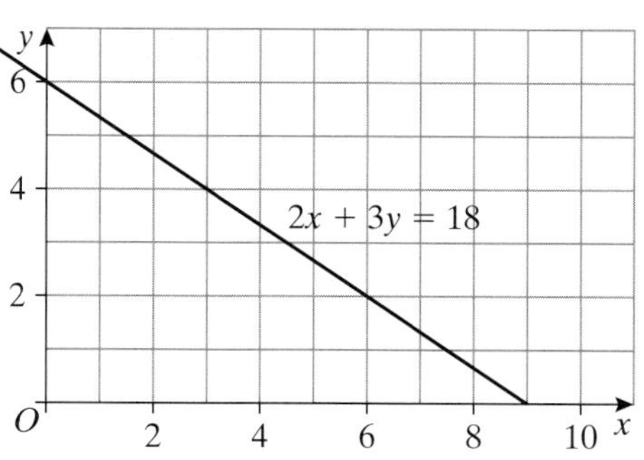

15

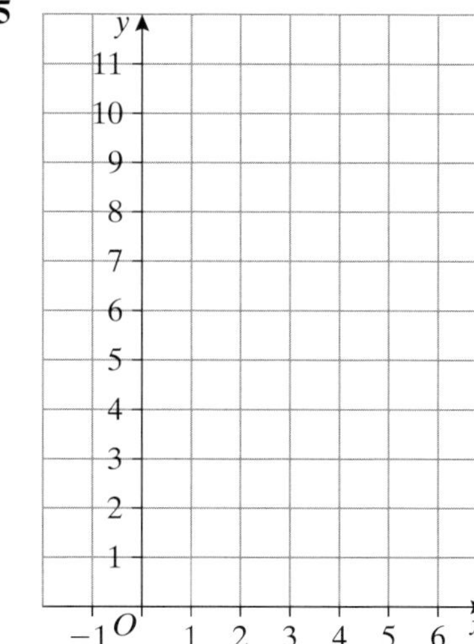

▶▶ EXERCISE 9.4 and WORKSHEET 9.4A

1

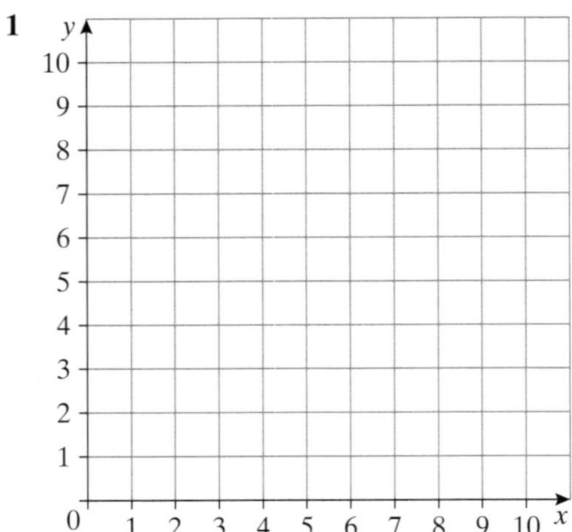

2

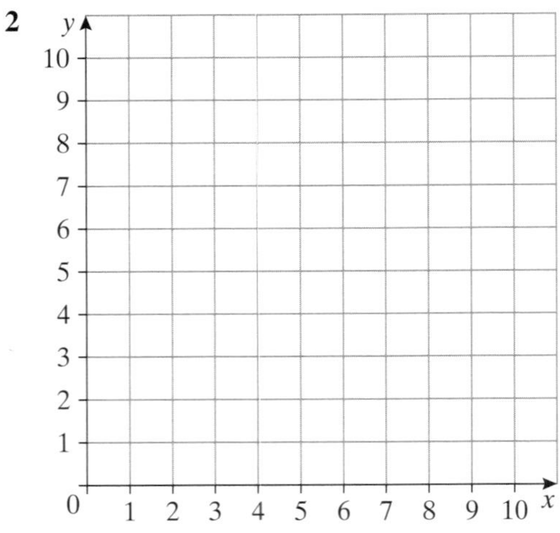

3

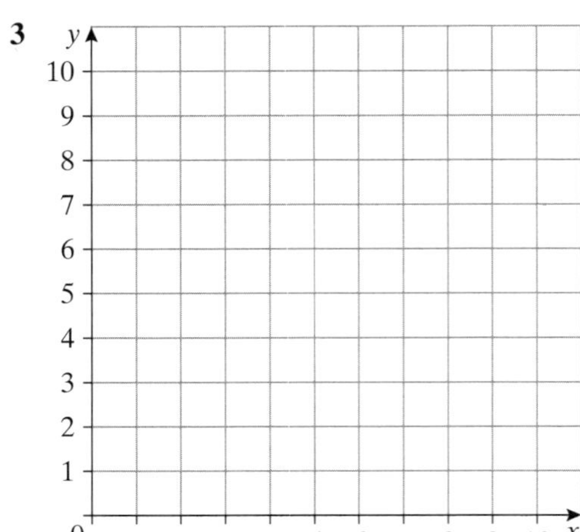

4

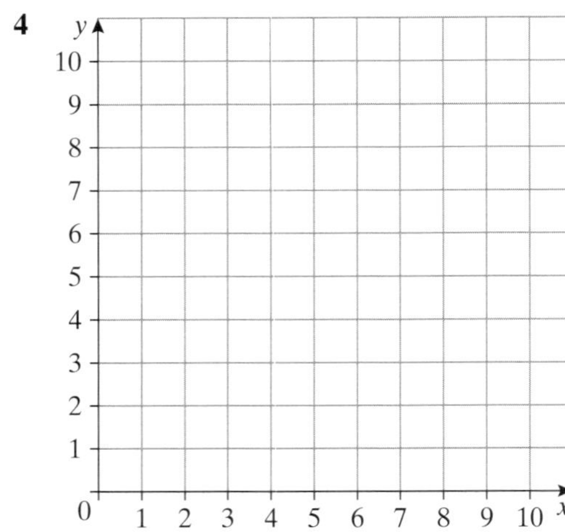

5

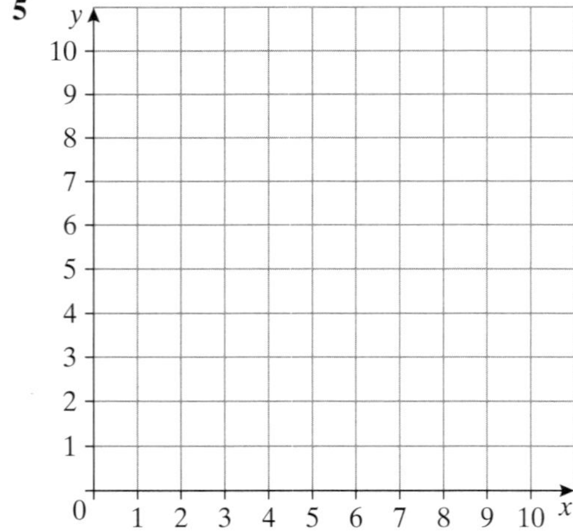

6

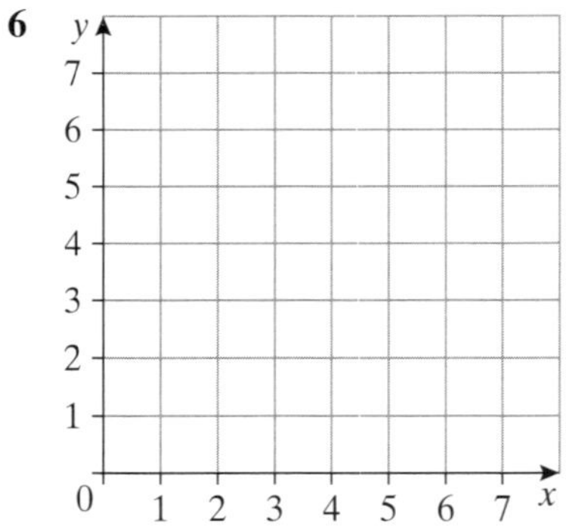

(continued)

 Speed-up sheet GCSE Maths for Edexcel: Higher Teacher's Resource © Hodder Murray 2006

7

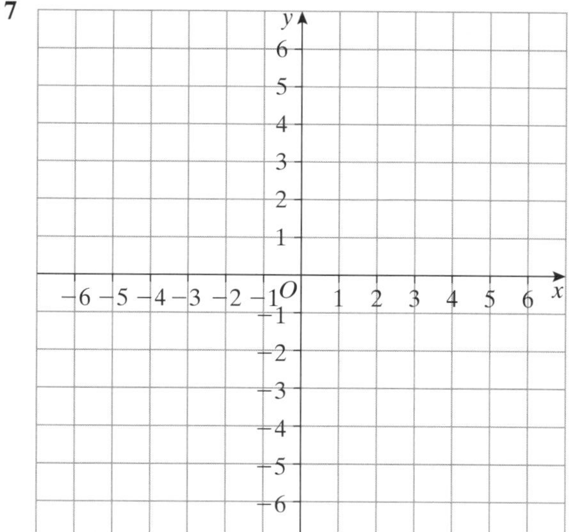

8

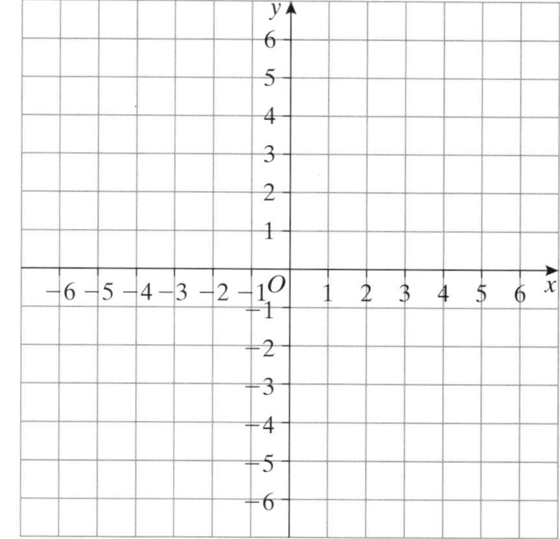

9

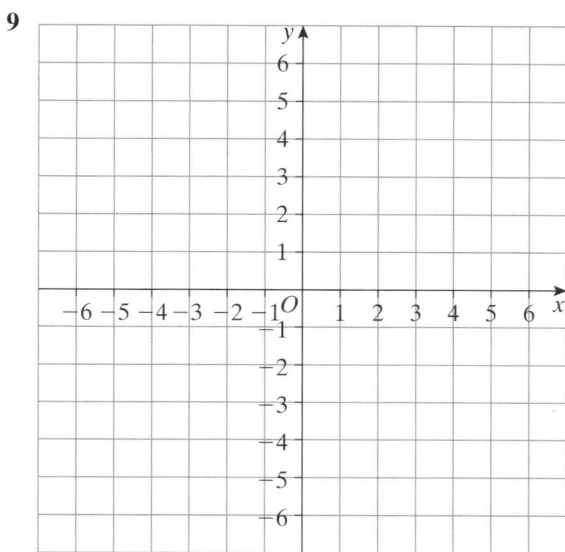

10

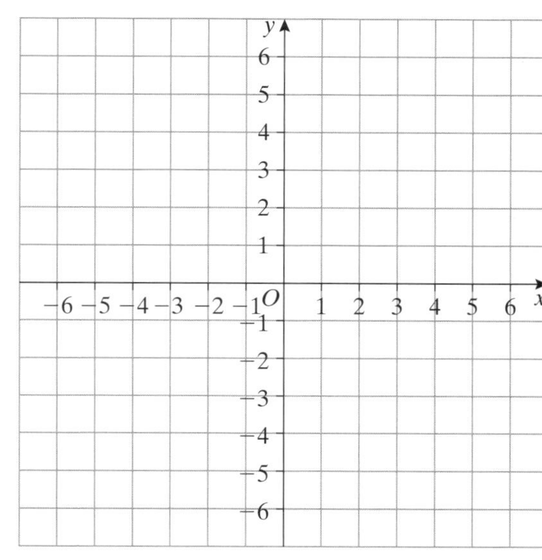

11

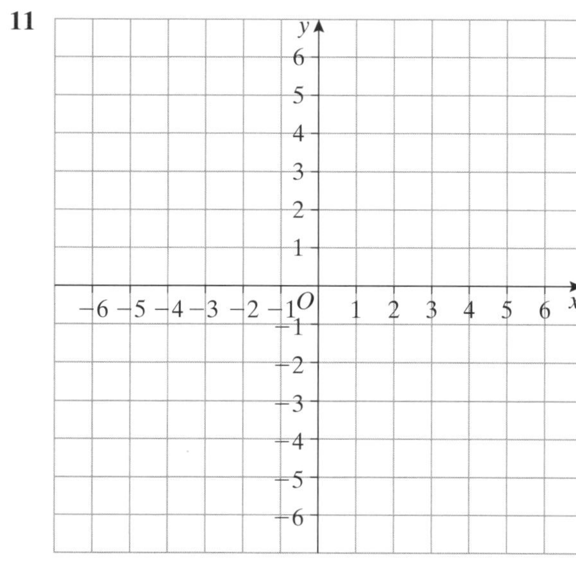

12

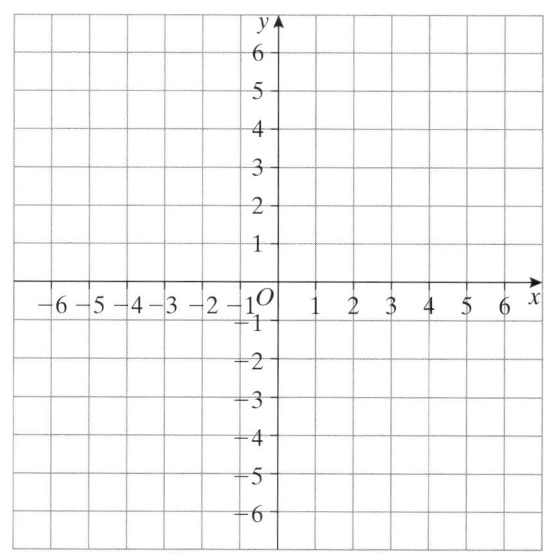

▶▶ WORKSHEET 9.4B

1

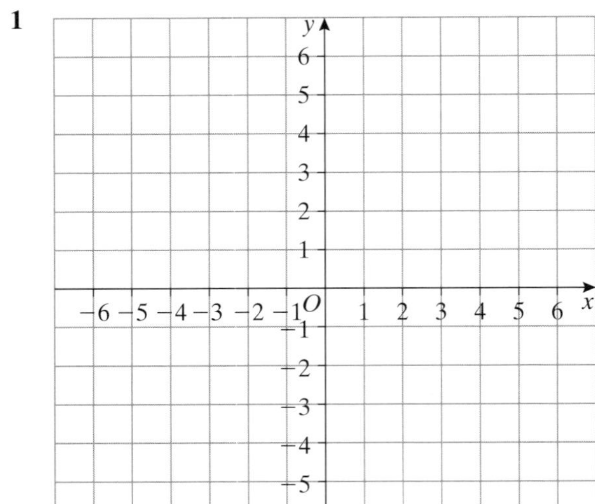

2

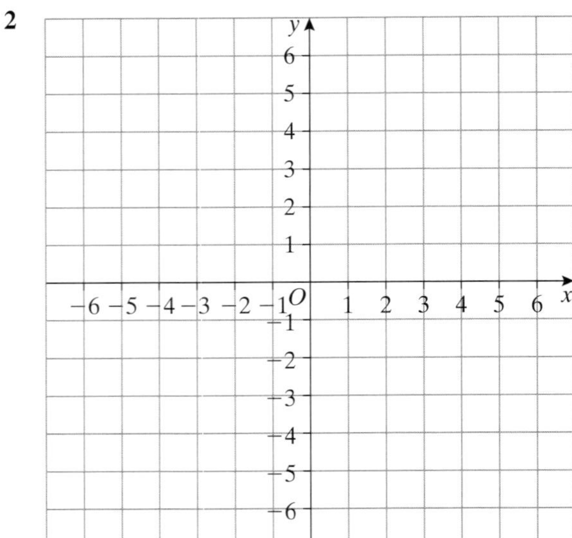

3

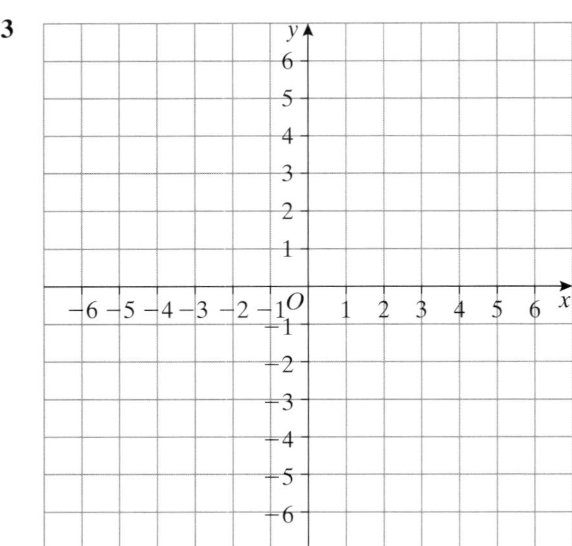

4

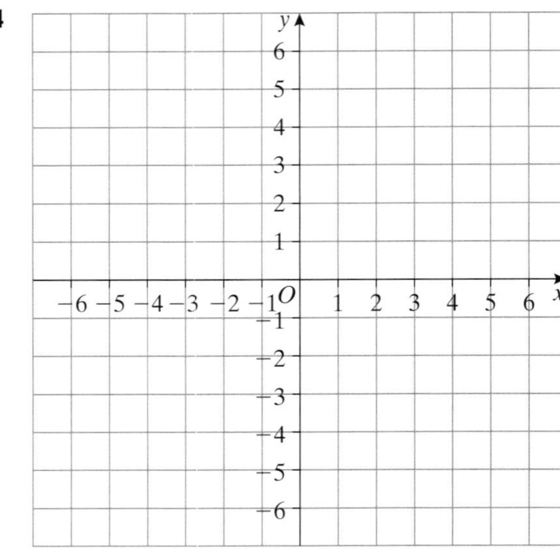

5

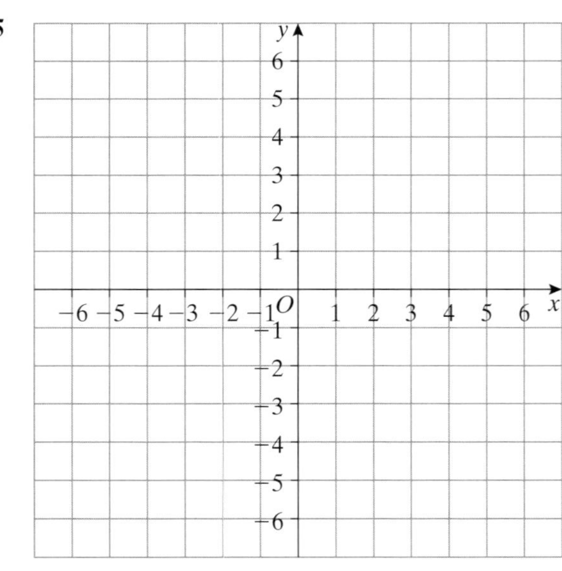

6

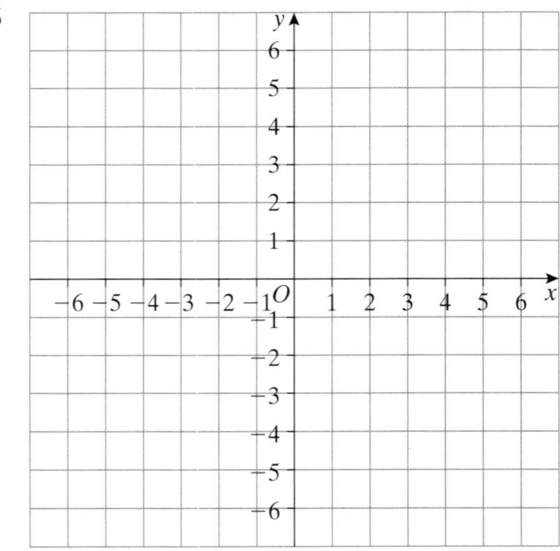

(continued)

Speed-up sheet

7

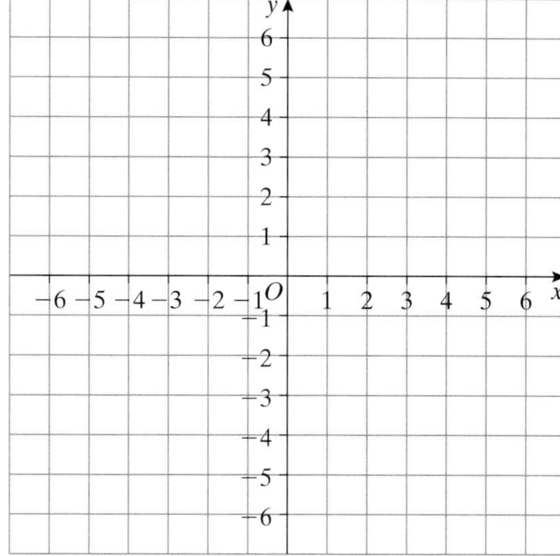

8

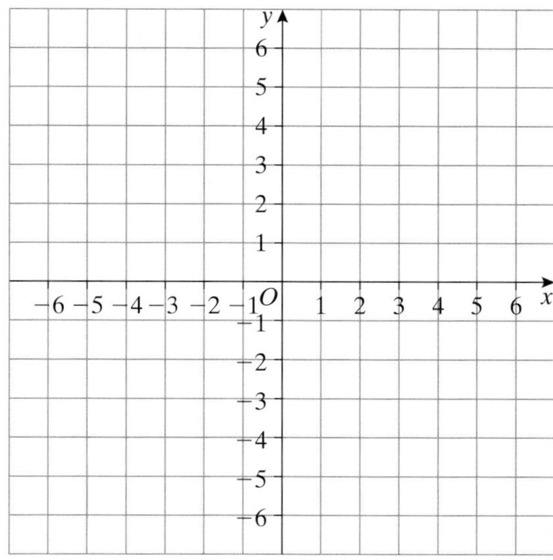

9

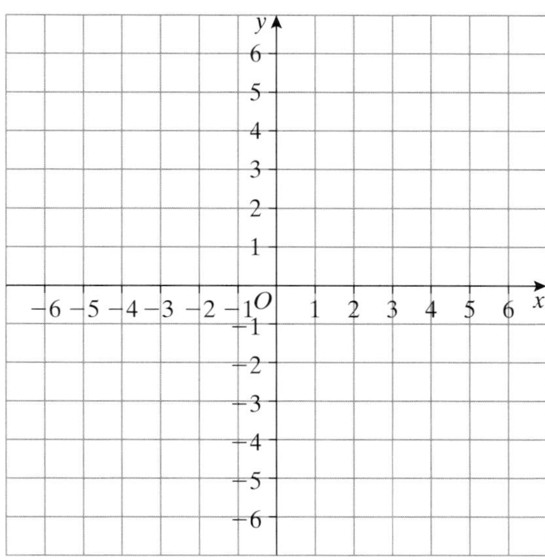

10

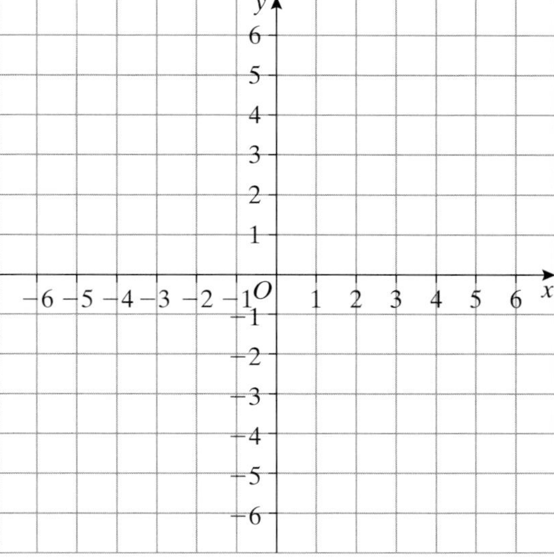

Speed-up sheet

21

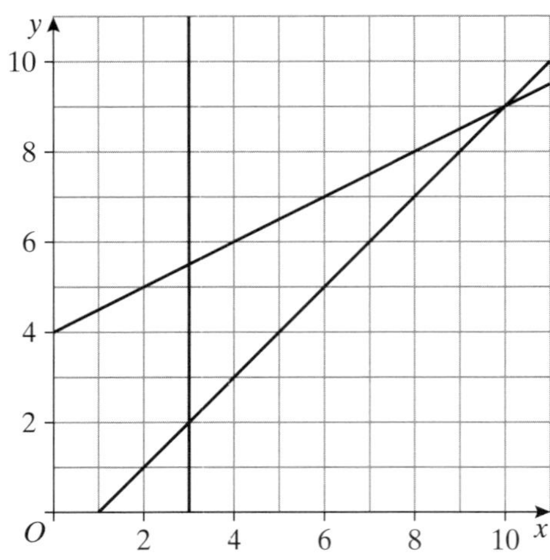

22 b)

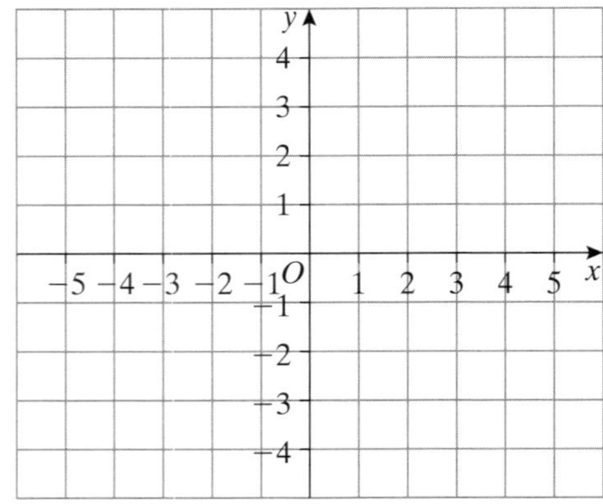

[Edexcel]

25 c)

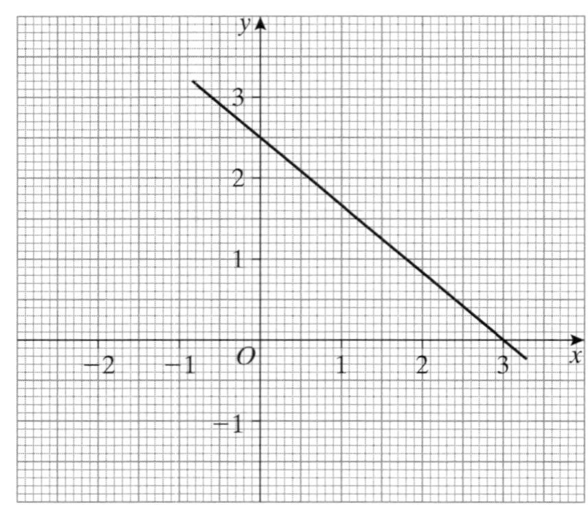

[Edexcel]

▶▶ **INTERNET CHALLENGE 9**

The Golden Ratio.	∞
The eighteenth letter of the Greek alphabet, denotes 'the sum of'.	<
This 17th century symbol was formerly used in Europe to indicate subtraction.	√
A sculpture of this symbol, by Marta Pan, stands on the A6 roadside in France.	φ
The (not real) square root of minus one.	θ
First used in Harriot's *Artis Analyticae Praxis* in 1631.	÷
This originated from Hindu mathematics, where it was known as *sunya*.	=
Invented by Robert Recorde in 1557.	i
The eighth letter of the Greek alphabet, used to denote an unknown angle.	Σ
This 16th century symbol may be a corrupted abbreviation for *radix*.	0

Speed-up sheet   GCSE Maths for Edexcel: Higher Teacher's Resource © Hodder Murray 2006

EXERCISE 11.1

1

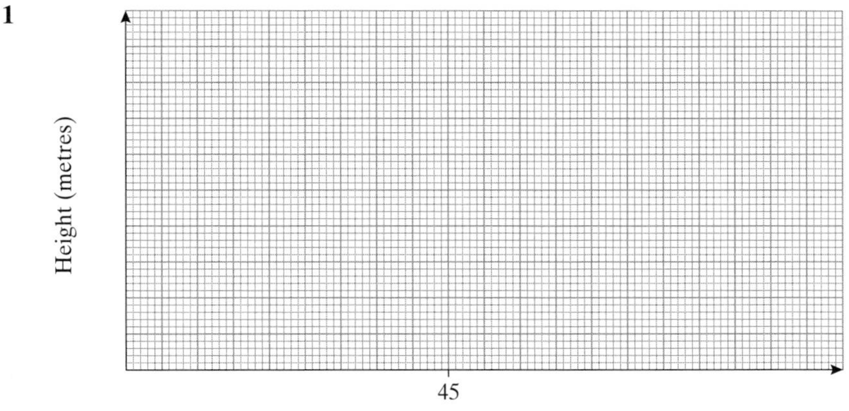

3

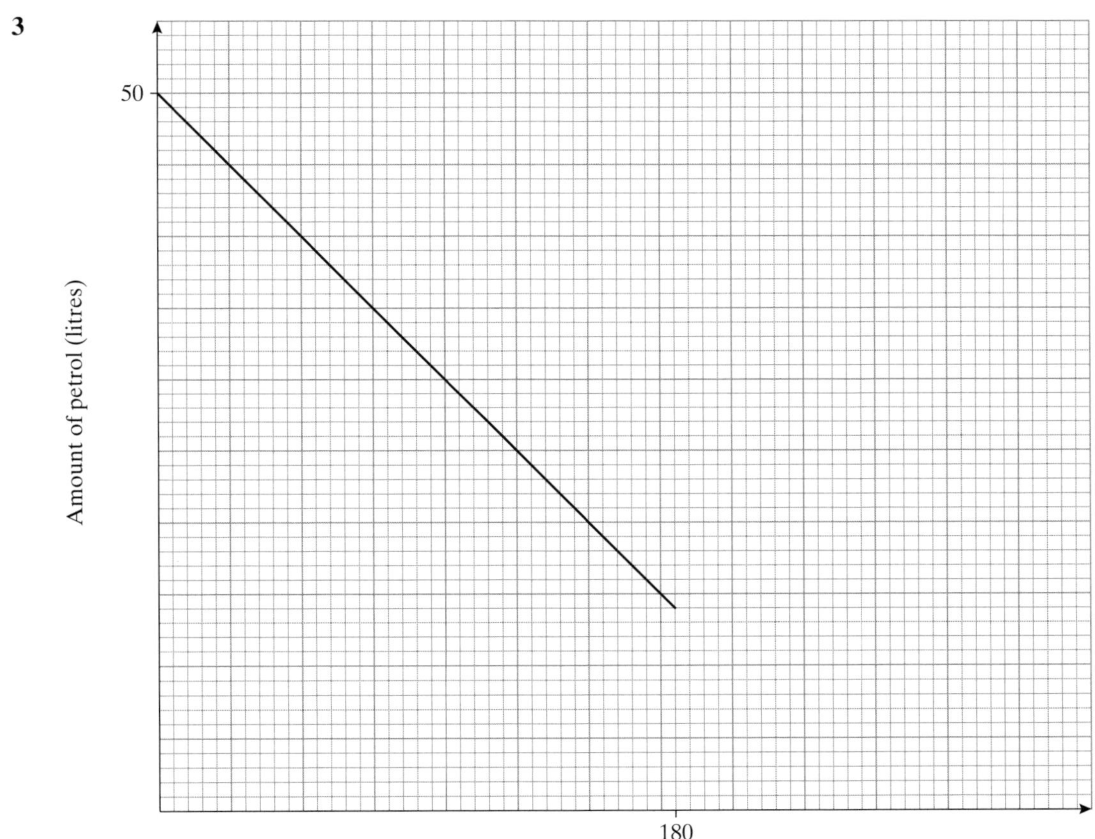

4

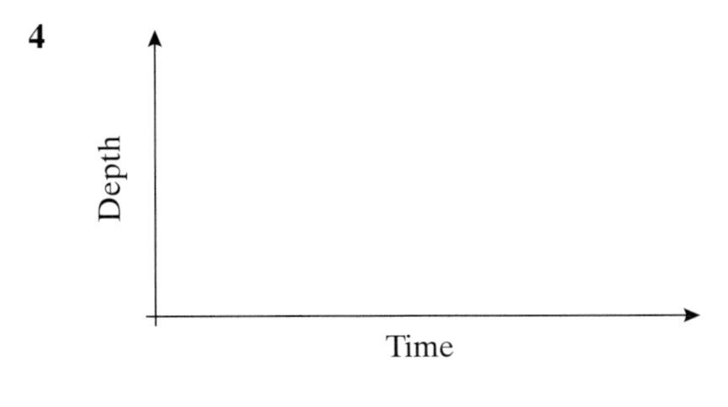

5

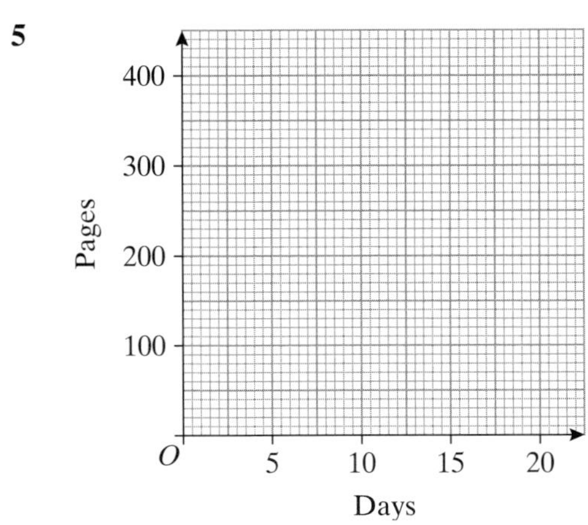

EXERCISE 11.2

80
Beesville 70
Distance (miles)
60
50
40
30
20
10
Ayton 0
1200 1230 1300 1330 1400 1430
Time of day

180
160
140
120
Distance (km)
100
80
60
40
20
O 2 4 6 8 10
Days

32
28
24
20
Distance (km)
16
12
8
4
Base 0
0900 1000 1100 1200 1300 1400 1500 1600 1700
Time

2

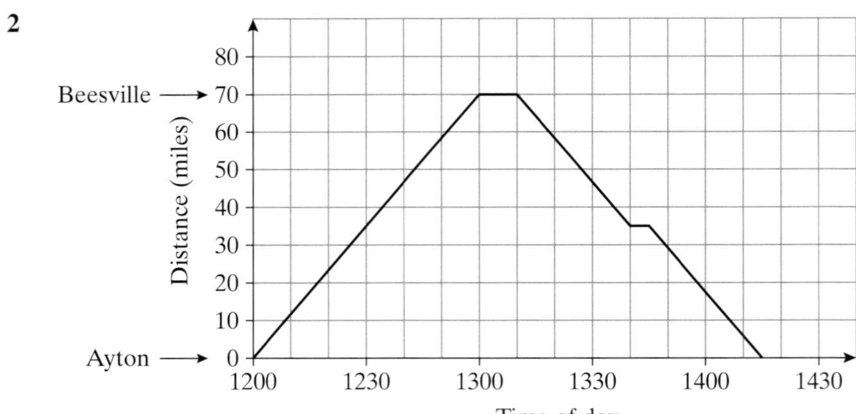

3

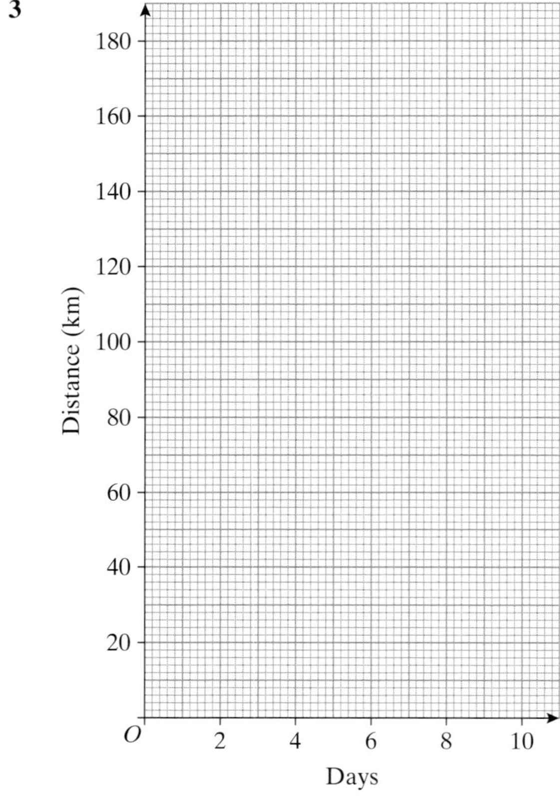

4

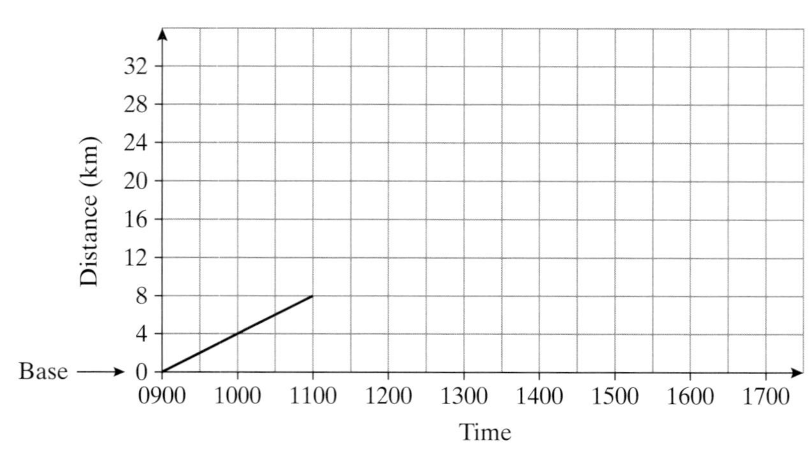

► EXERCISE 11.3

4

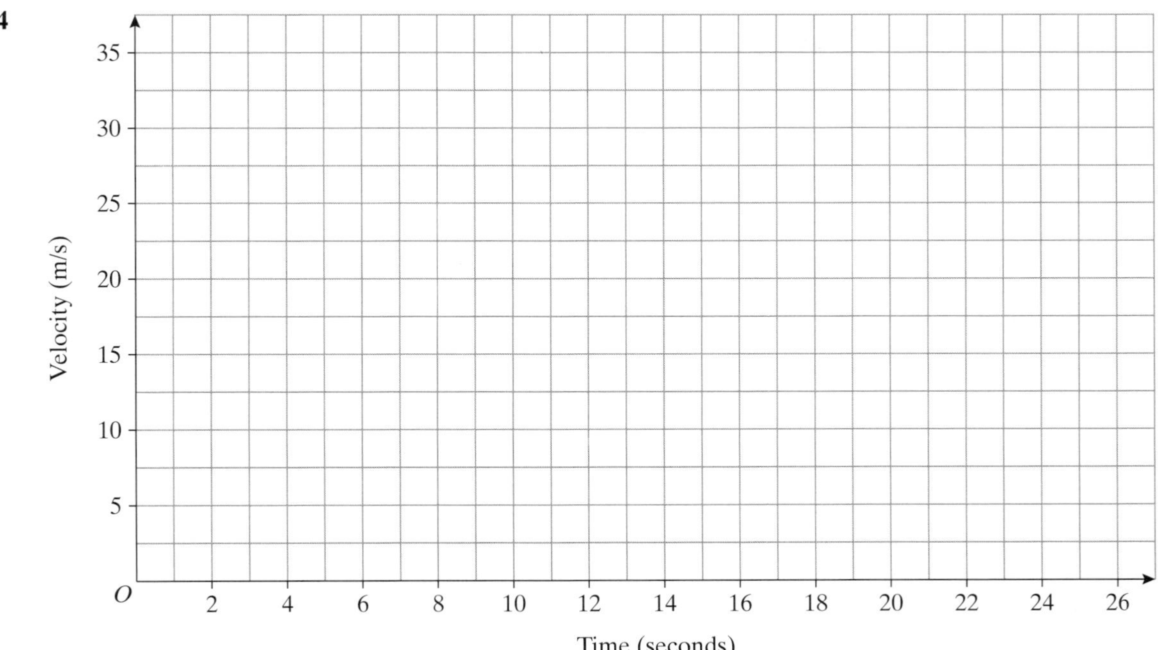

2

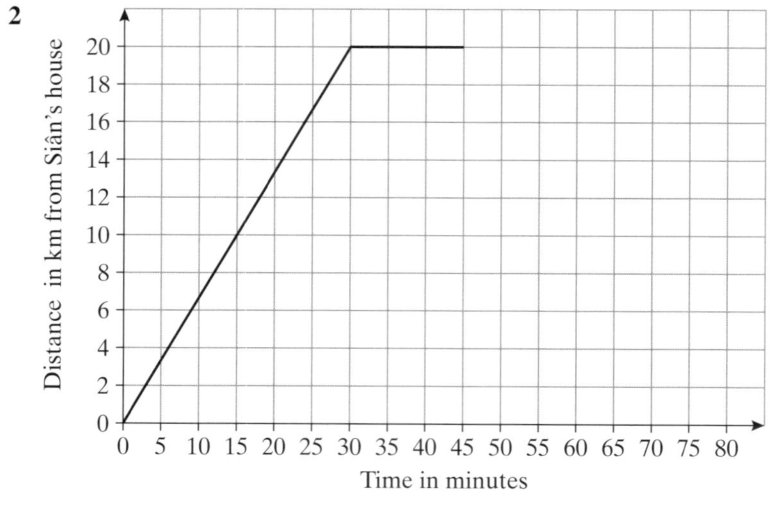

[Edexcel]

3

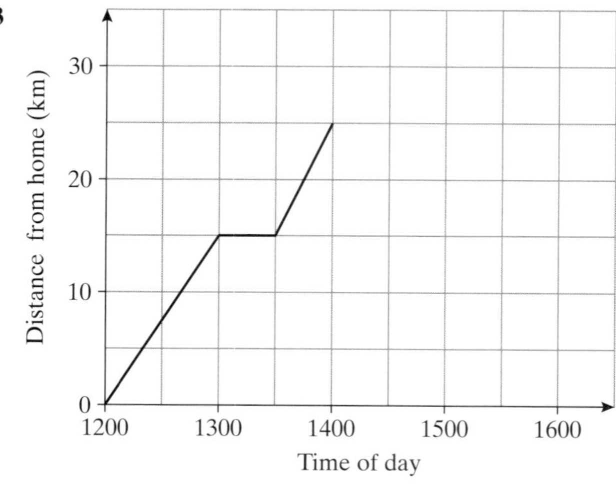

[Edexcel]

4

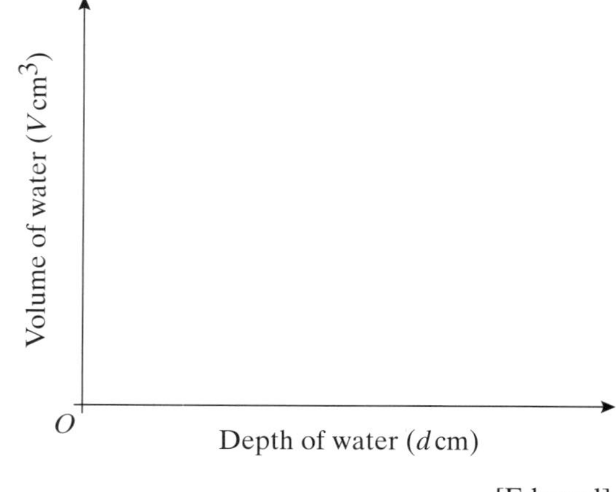

[Edexcel]

6

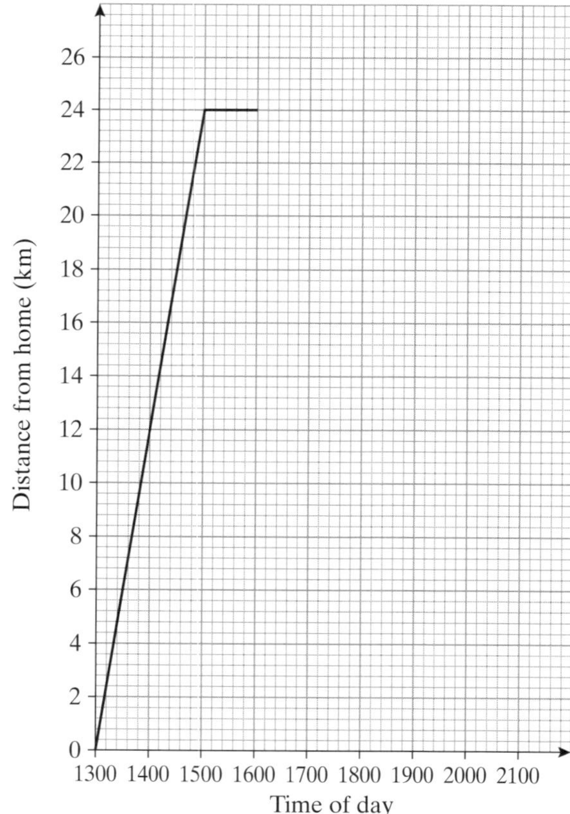

[Edexcel]

 **Speed-up sheet**

GCSE Maths for Edexcel: Higher Teacher's Resource © Hodder Murray 2006

7

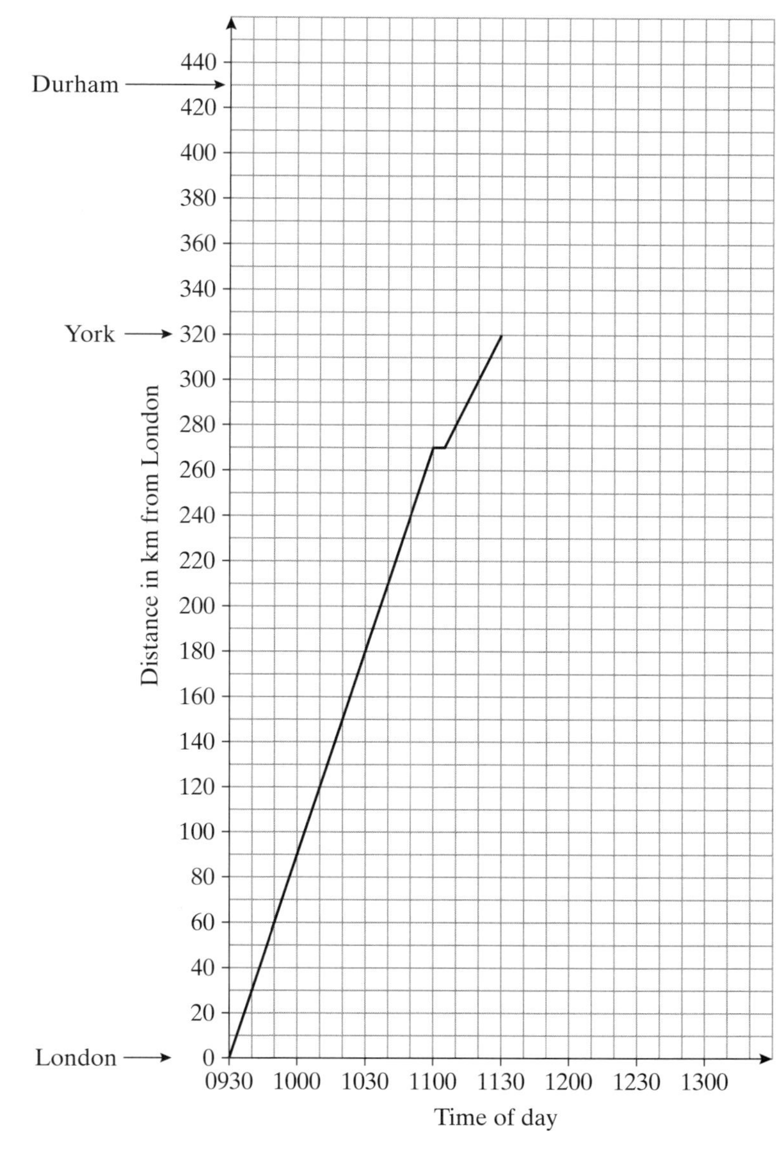

Durham
440
420
400
380
360
340
York
320
300
Distance in km from London
280
260
240
220
200
180
160
140
120
100
80
60
40
20
London
0
0930 1000 1030 1100 1130 1200 1230 1300
Time of day

R
30
28
Q
26
24
22
Distance from P (miles)
20
18
16
14
12
10
8
6
4
2
P
0
1200 1230 1300 1330 1400
Time of day

[Edexcel]

9

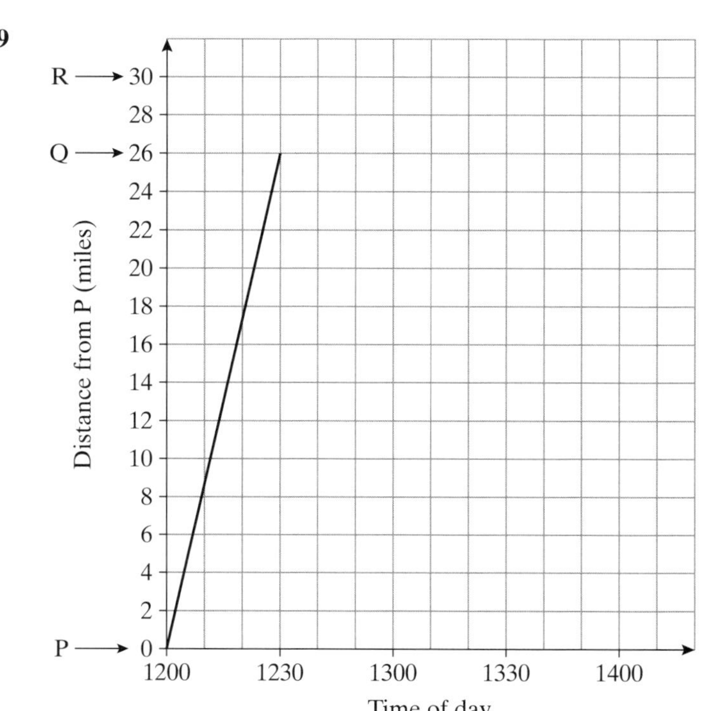

▶▶ EXERCISE 14.2

1

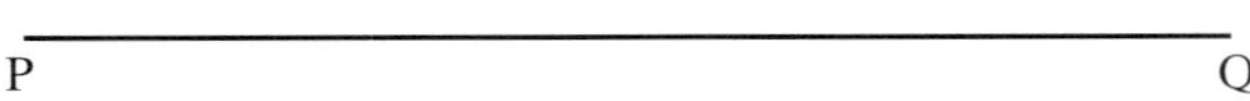

P   Q

2

A

B   C

3

P ×

A   B

(continued)

4

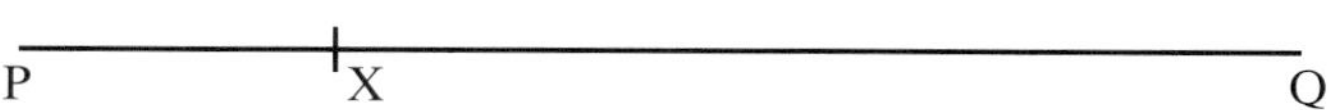

5

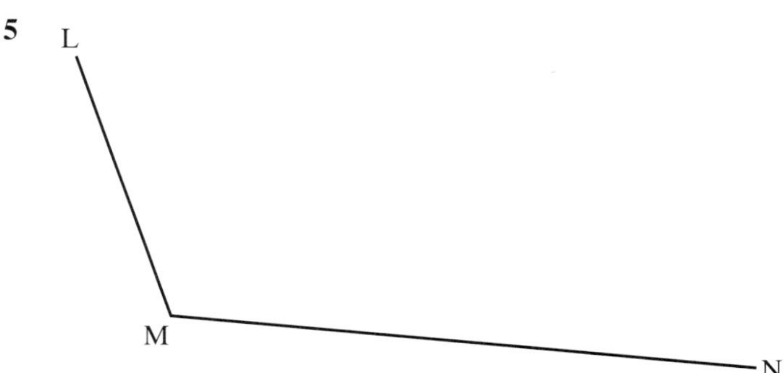

6

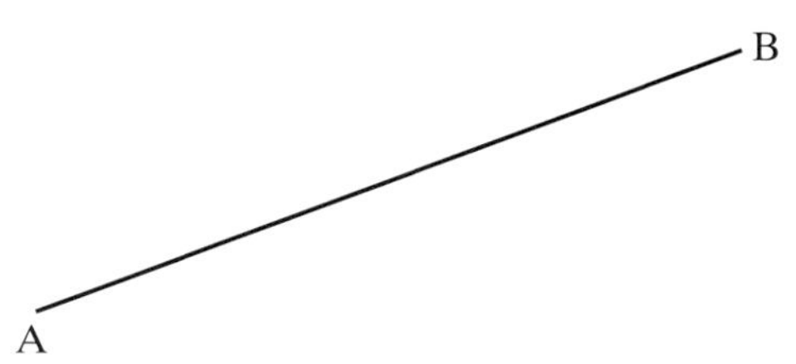

Speed-up sheet

(continued)

▶▶ EXERCISE 14.2 (continued)

7

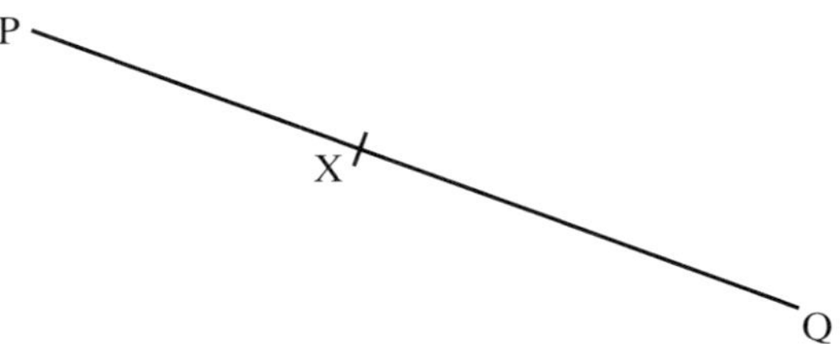

8

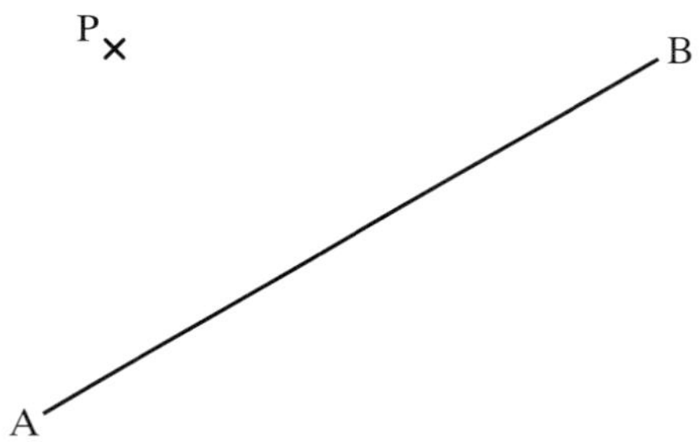

Speed-up sheet

GCSE Maths for Edexcel: Higher Teacher's Resource © Hodder Murray 2006

▶▶ EXERCISE 14.3

1

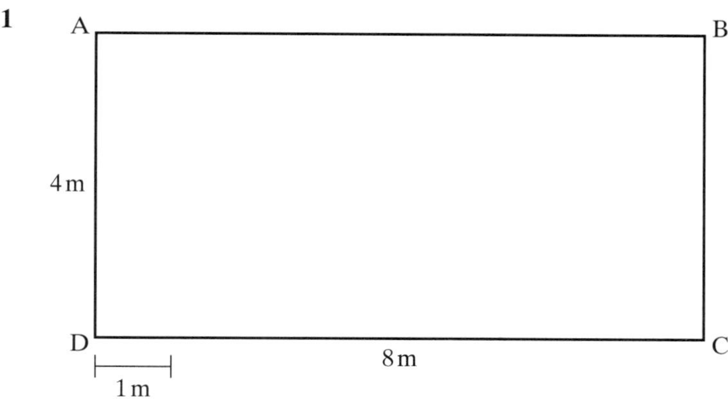

2

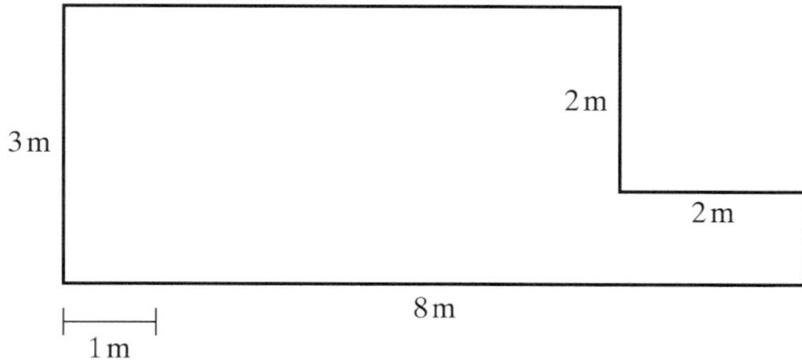

3

(continued)

Speed-up sheet

4

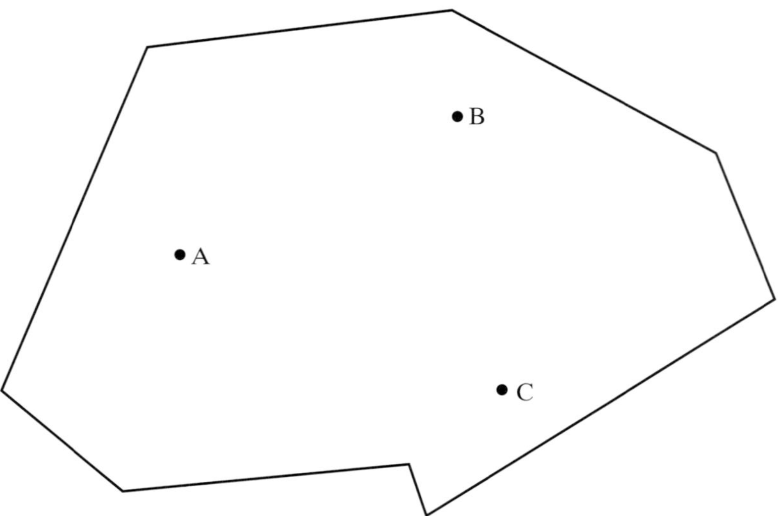

5

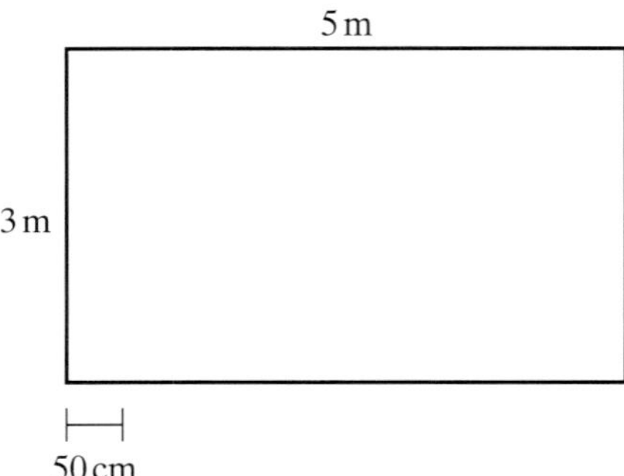

Speed-up sheet   GCSE Maths for Edexcel: Higher Teacher's Resource © Hodder Murray 2006

▶▶ REVIEW EXERCISE 14

1

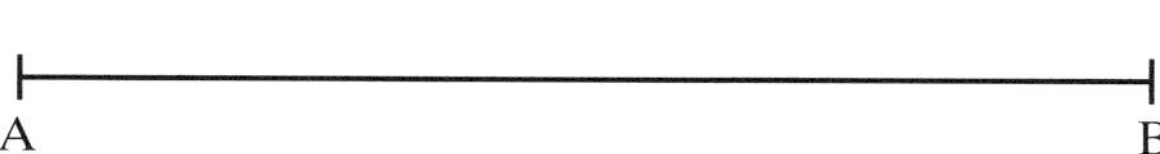

A   B

[Edexcel]

2

[Edexcel]

(continued)

3

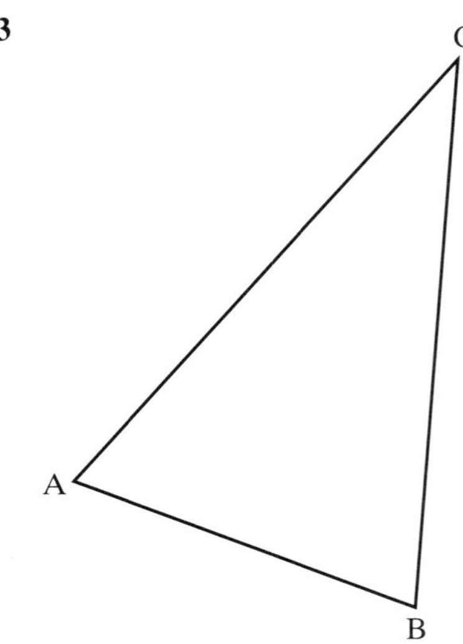

[Edexcel]

6

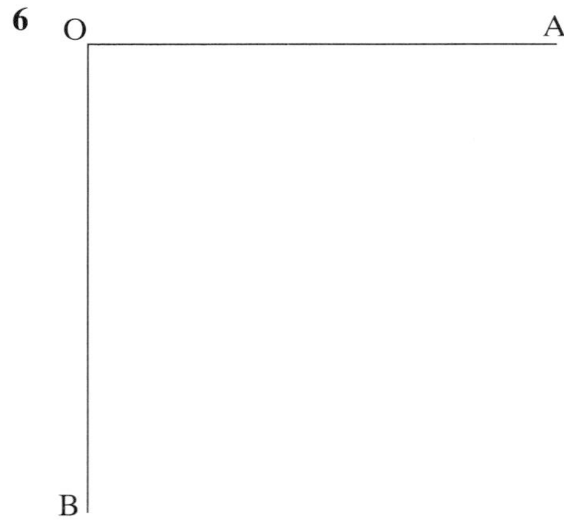

[Edexcel]

8

[Edexcel]

(continued)

Speed-up sheet

GCSE Maths for Edexcel: Higher Teacher's Resource © Hodder Murray 2006

Speed-up sheet: Chapter 14

10

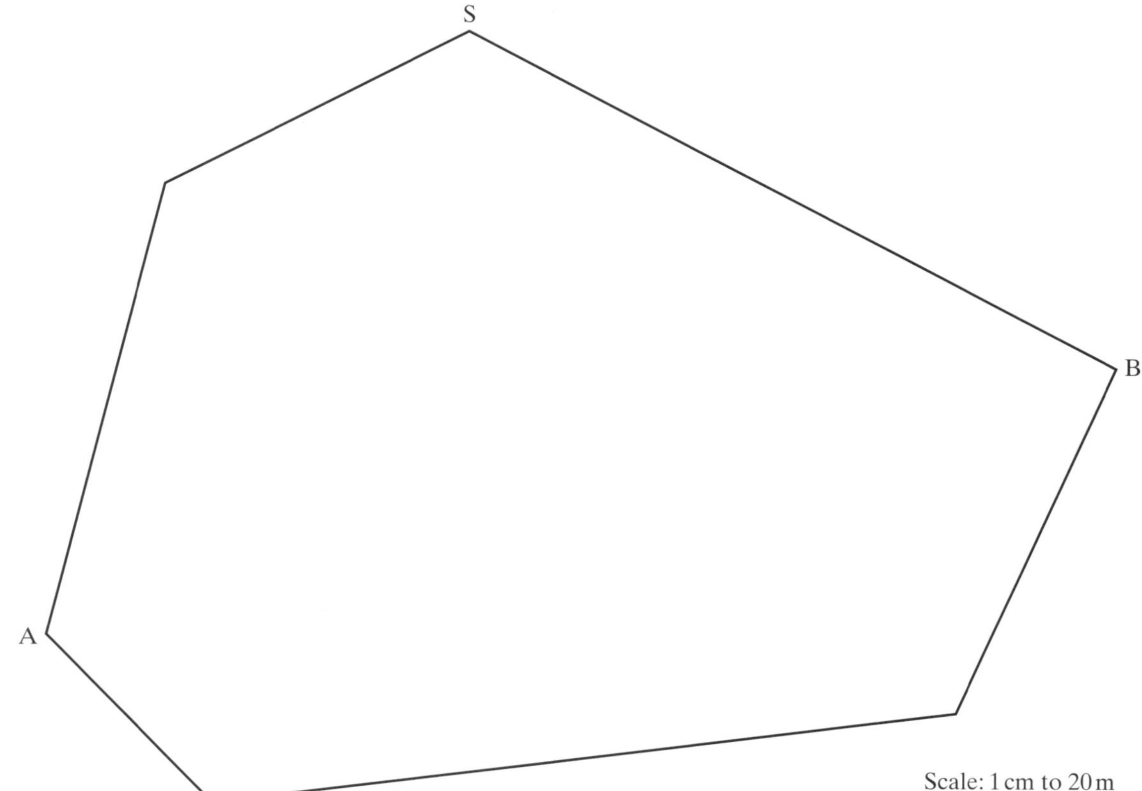

[Edexcel]

11

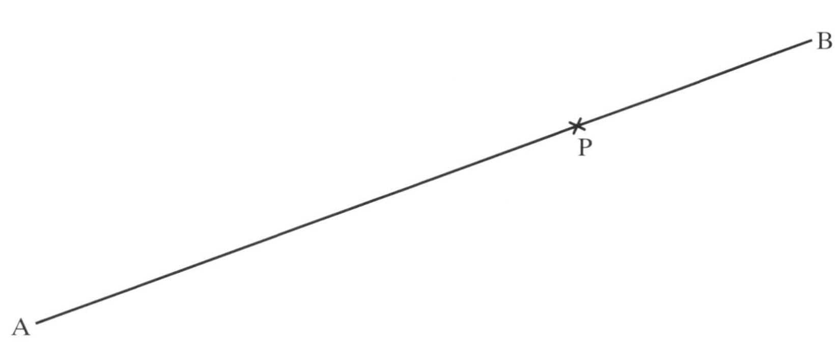

[Edexcel]

13

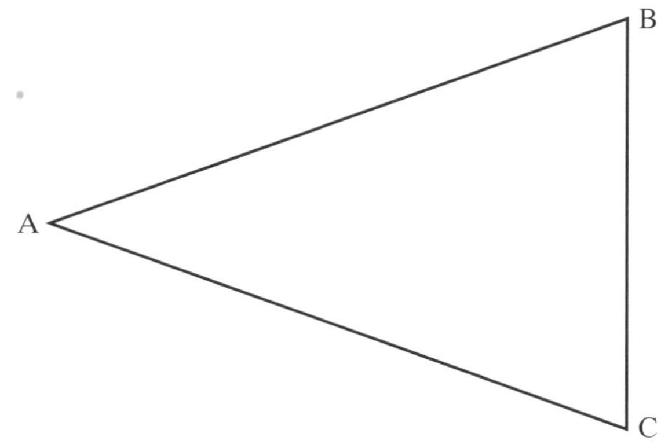

[Edexcel]

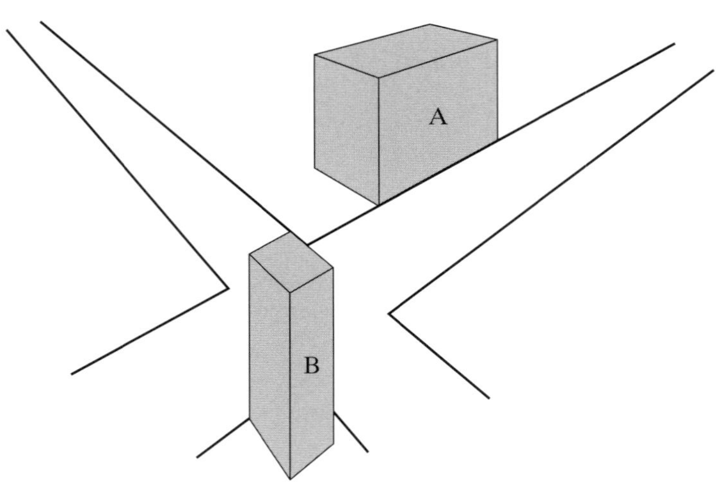

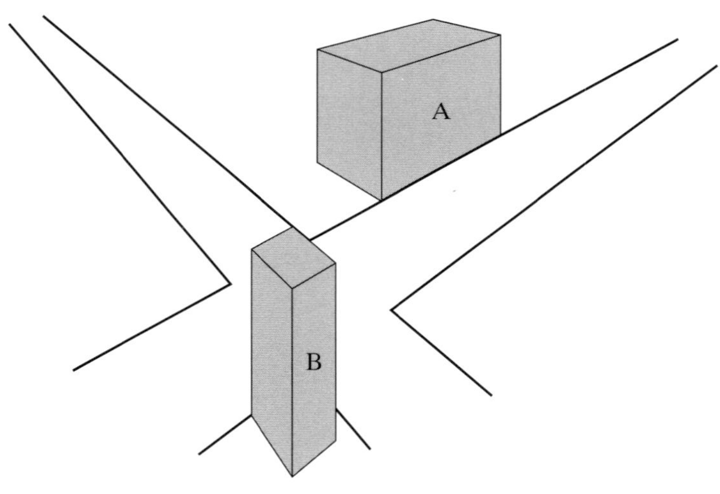

▶▶ EXERCISE 15.1

1 a)

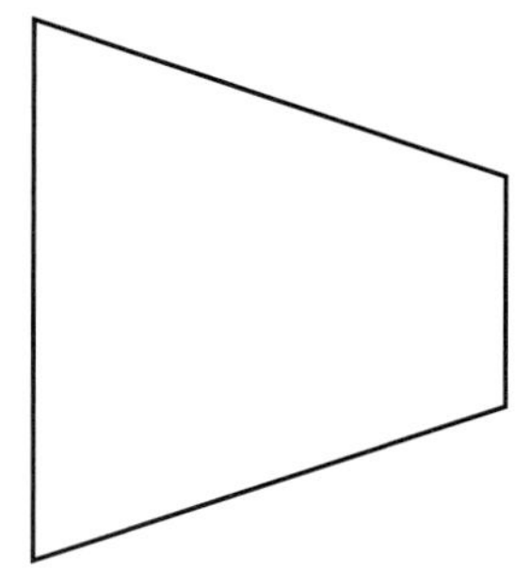

b)

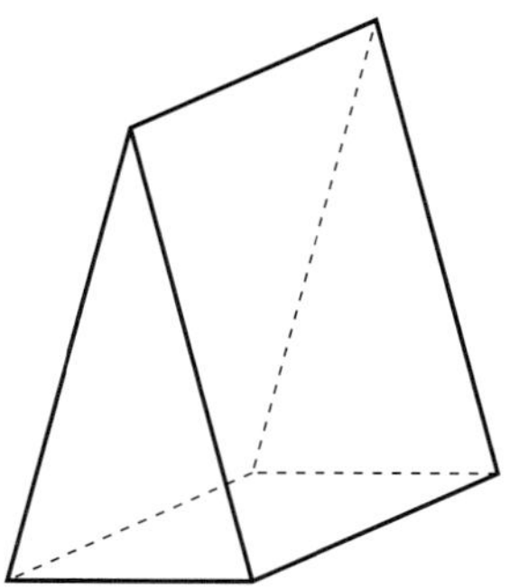

2

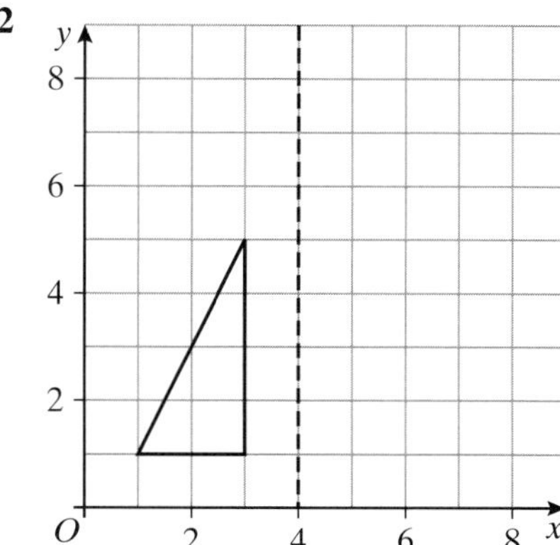

3

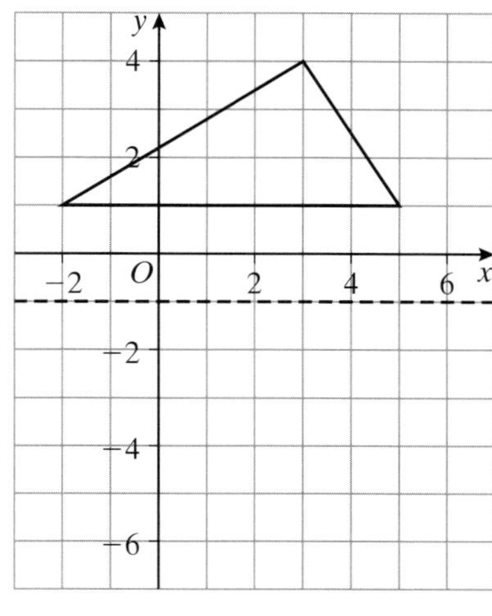

4

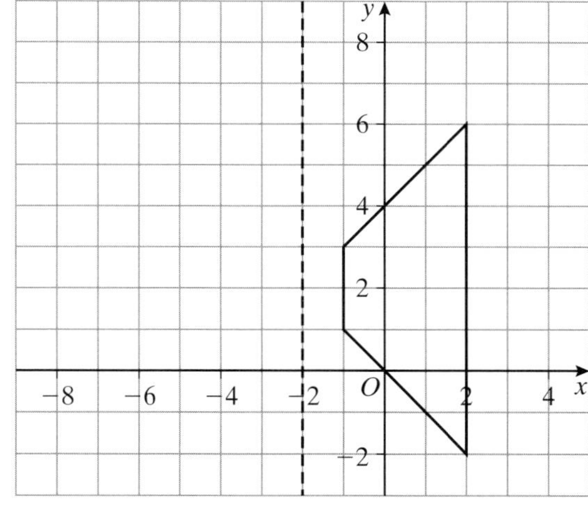

5

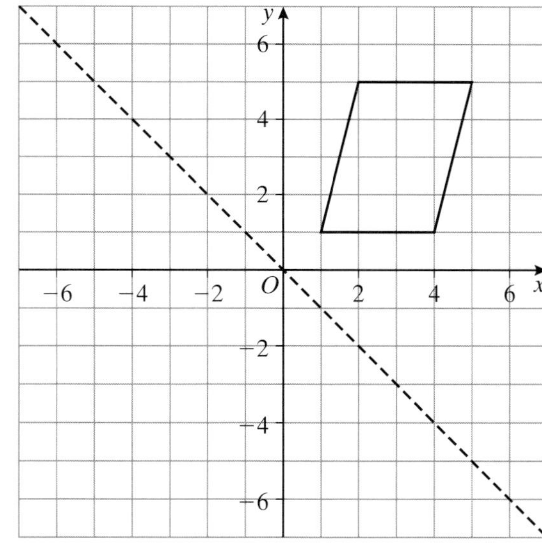

6

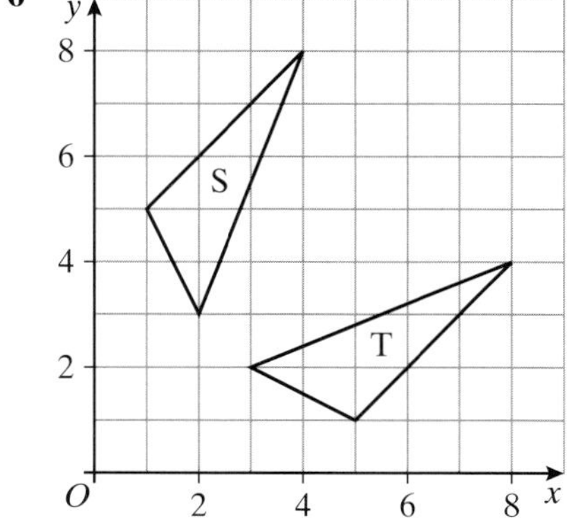

(continued)

7

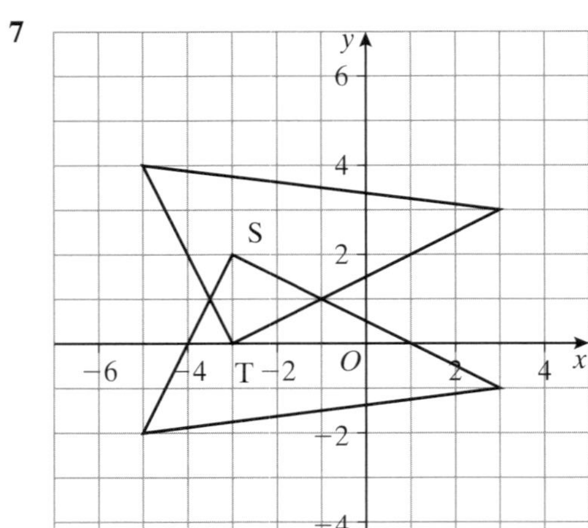

8

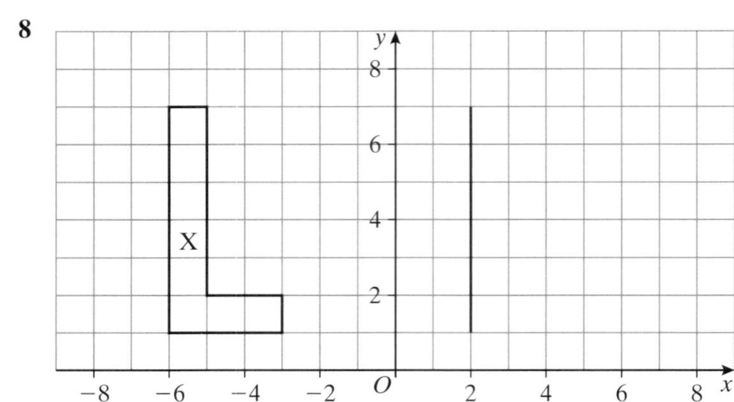

1

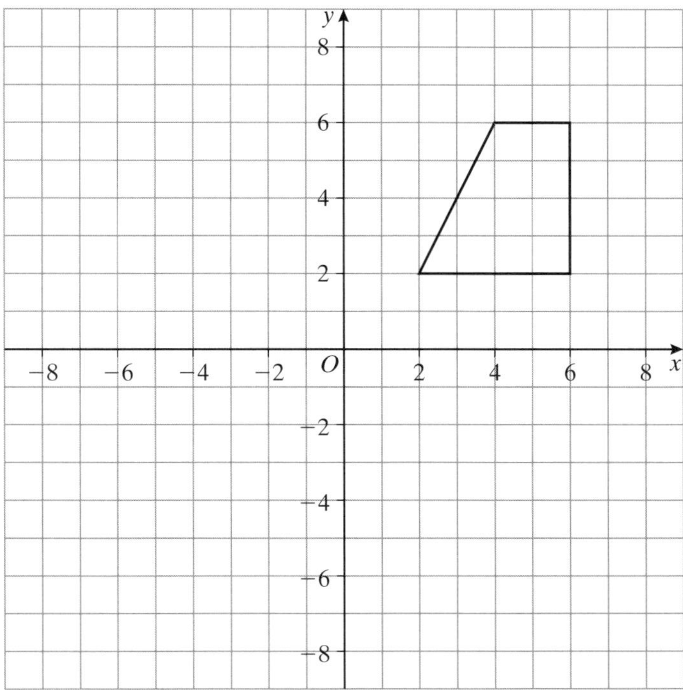

2

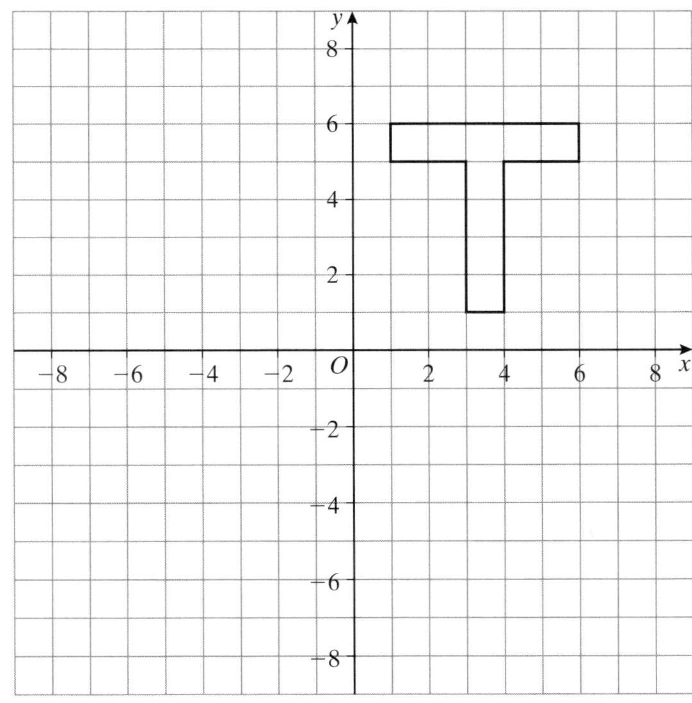

(continued)

3

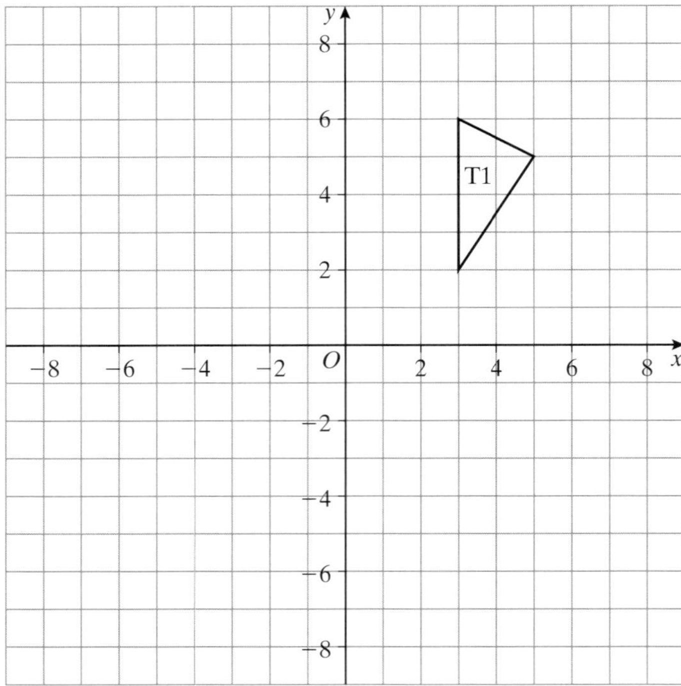

4

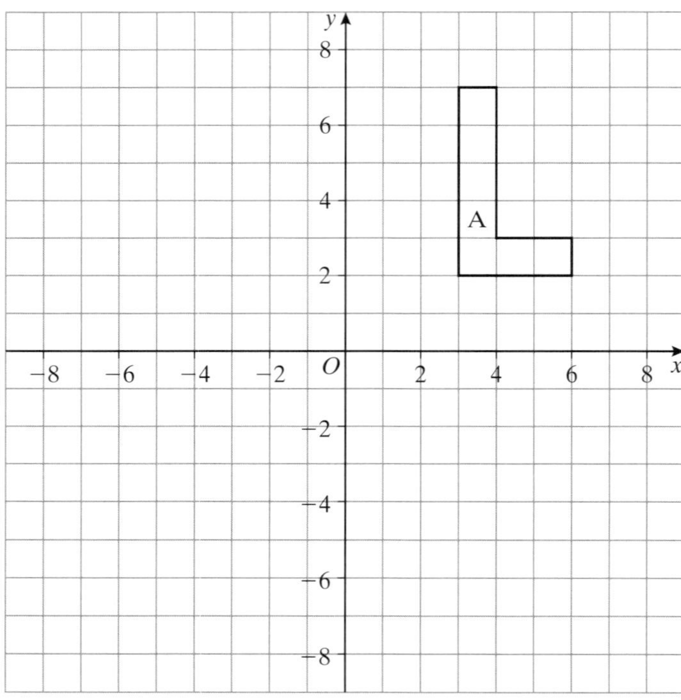

(continued)

Speed-up sheet

GCSE Maths for Edexcel: Higher Teacher's Resource © Hodder Murray 2006

5

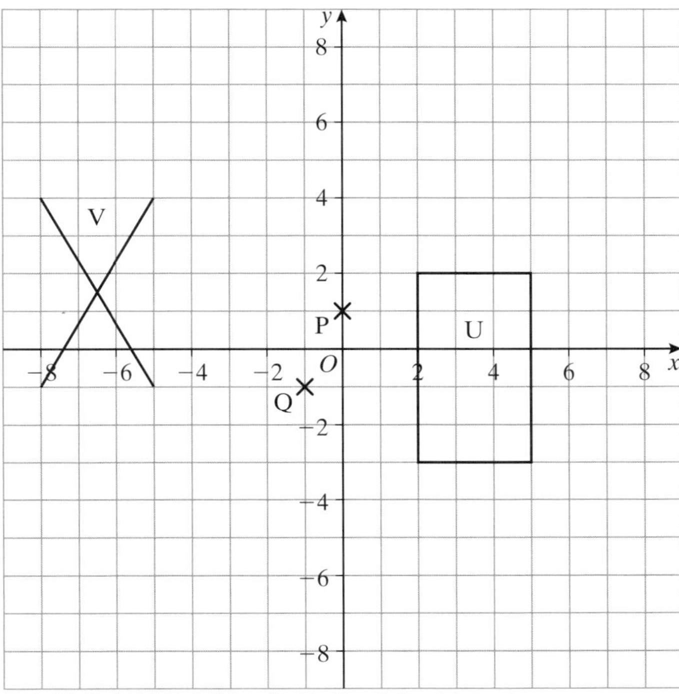

6

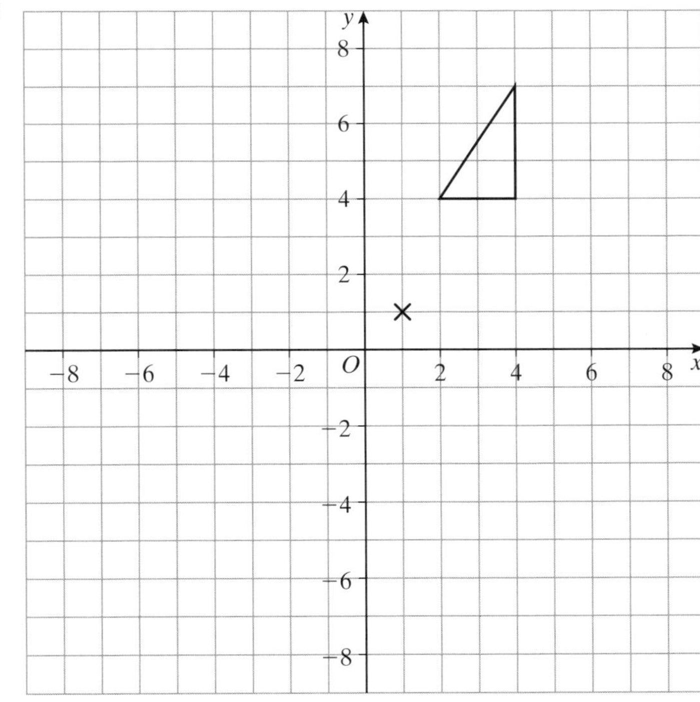

▶▶ EXERCISE 15.3

1

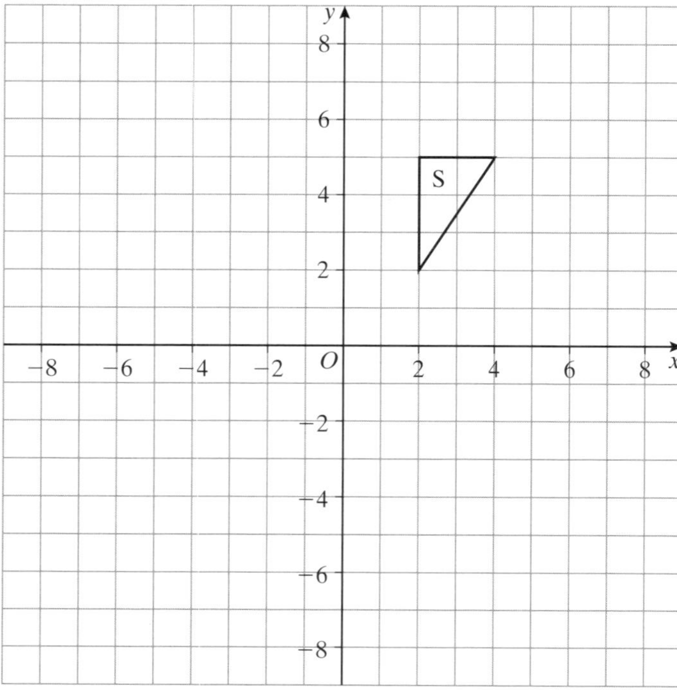

2

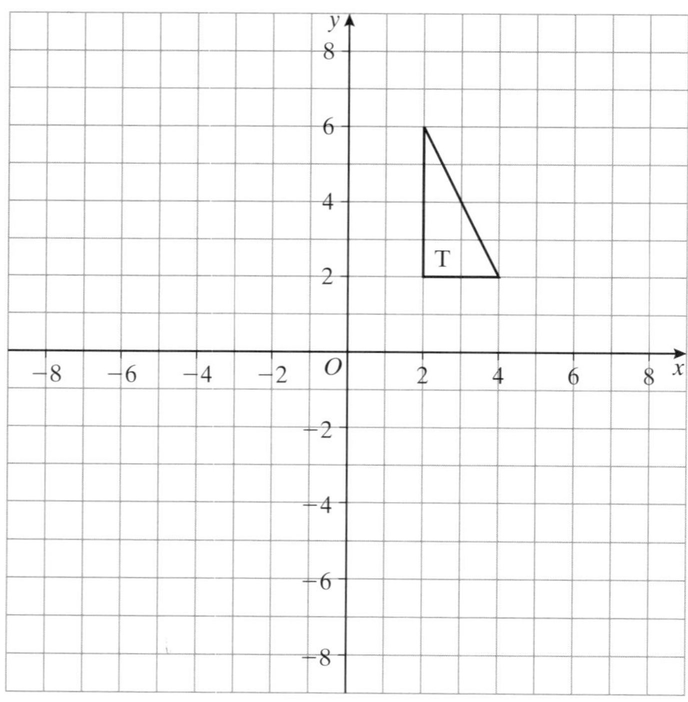

(continued)

 **Speed-up sheet**   GCSE Maths for Edexcel: Higher Teacher's Resource © Hodder Murray 2006

3

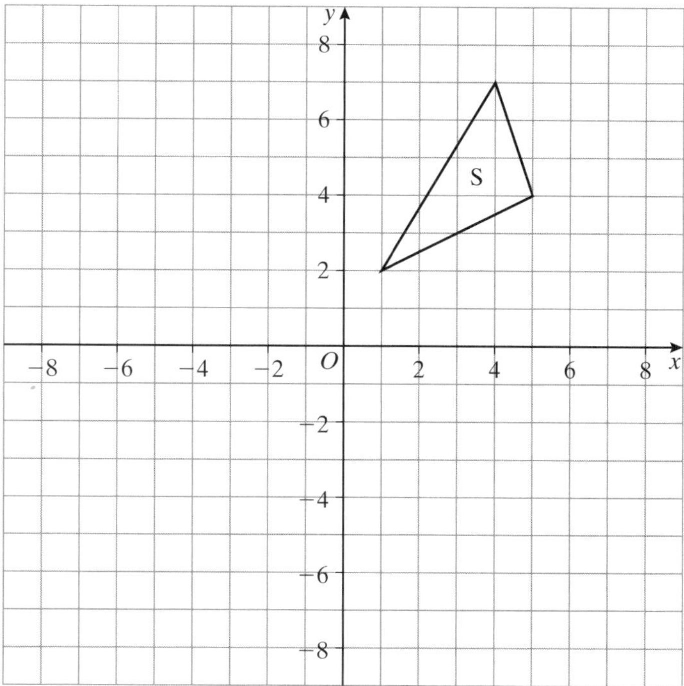

4

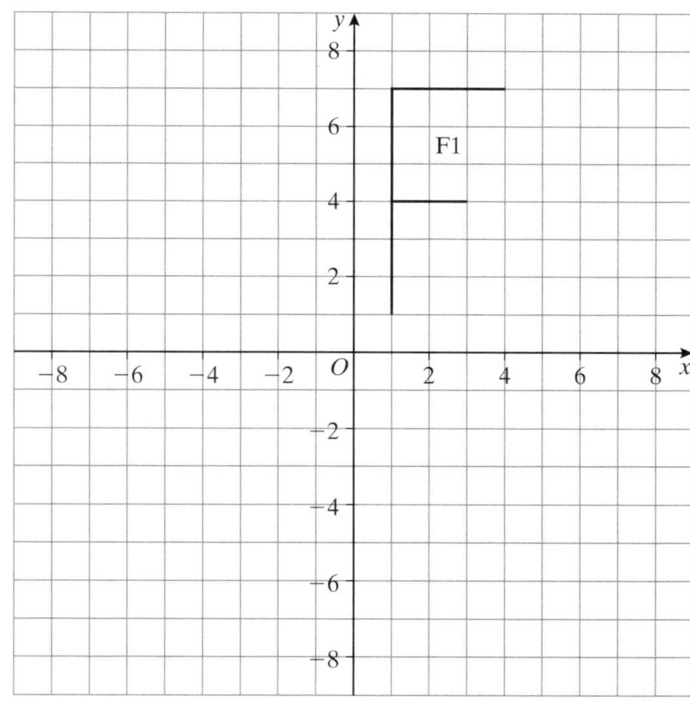

5

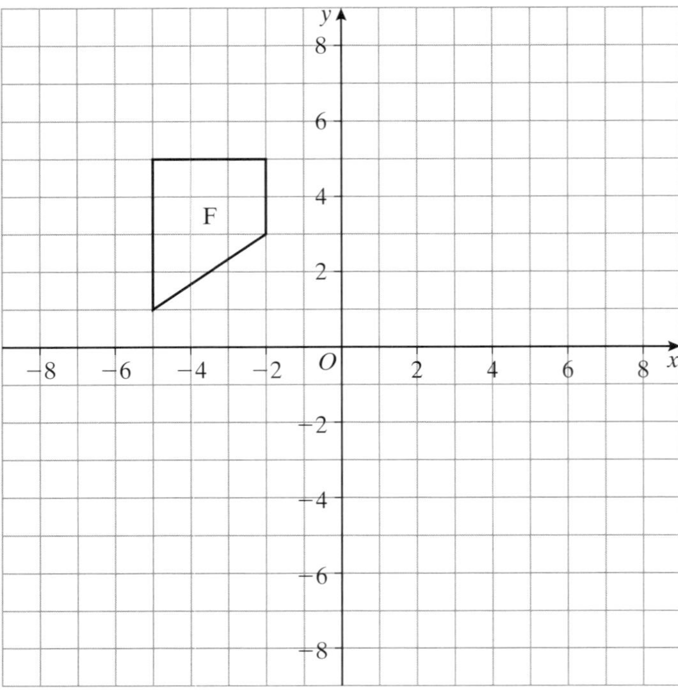

6

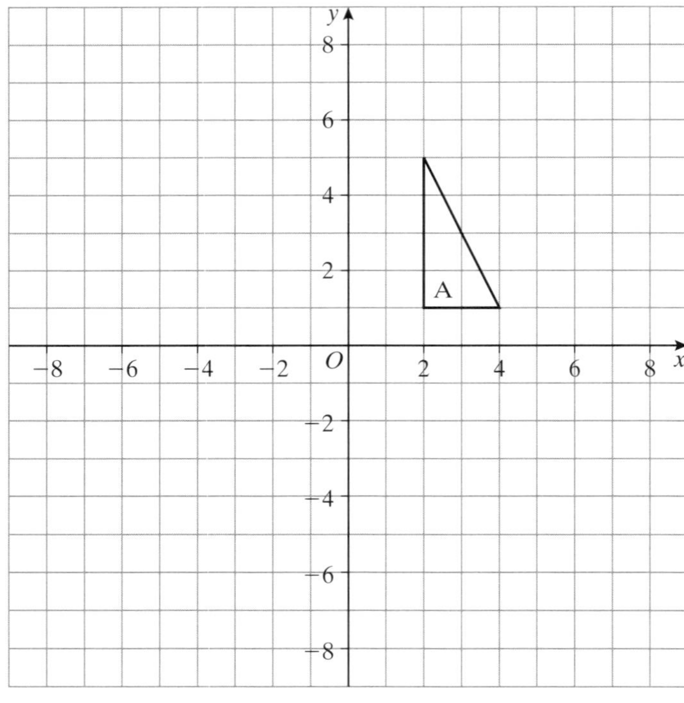

▶▶ EXERCISE 15.4

1

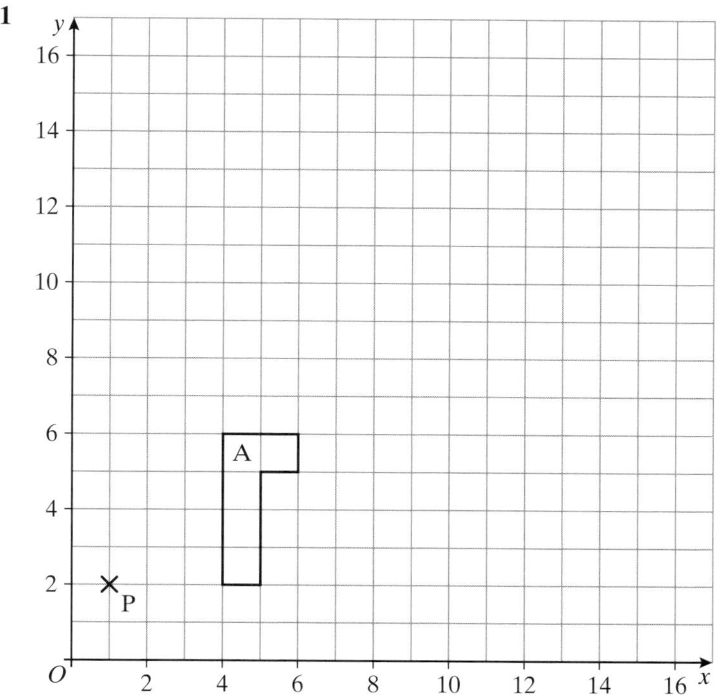

2

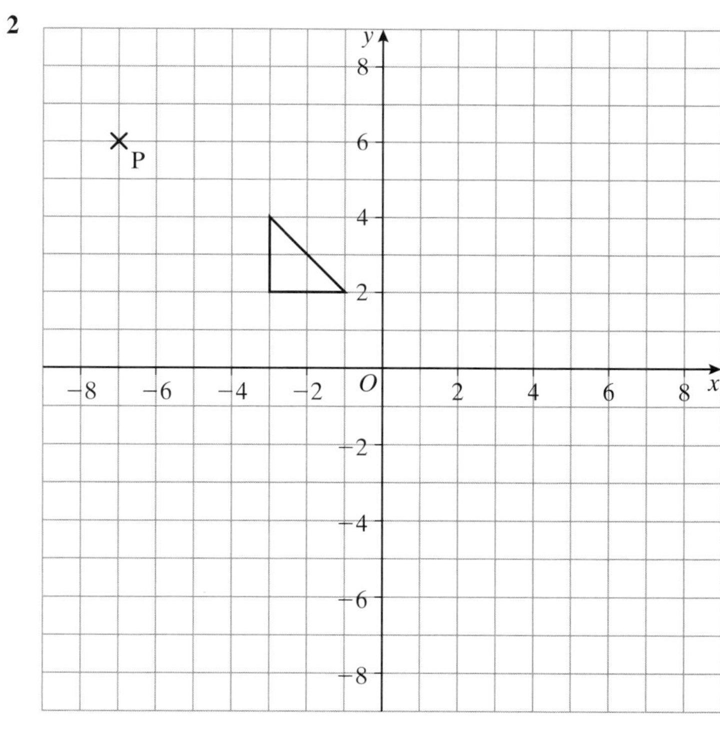

(continued)

▶▶ EXERCISE 15.4 (continued)

3

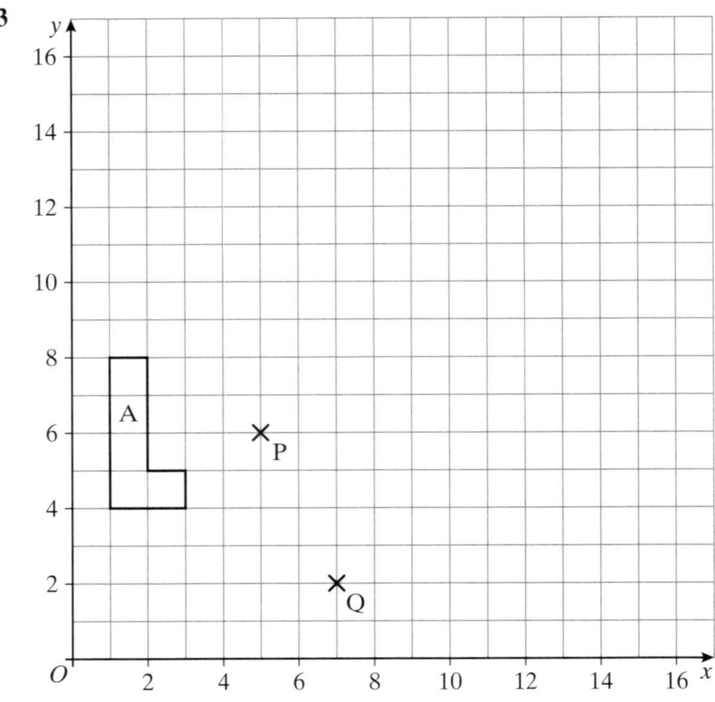

5

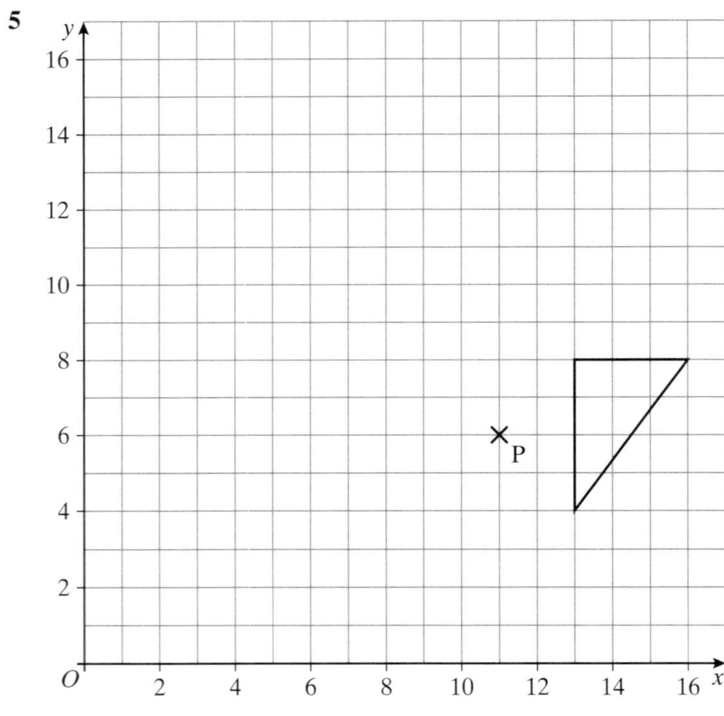

 **Speed-up sheet**

GCSE Maths for Edexcel: Higher Teacher's Resource © Hodder Murray 2006

▶▶ REVIEW EXERCISE 15

1
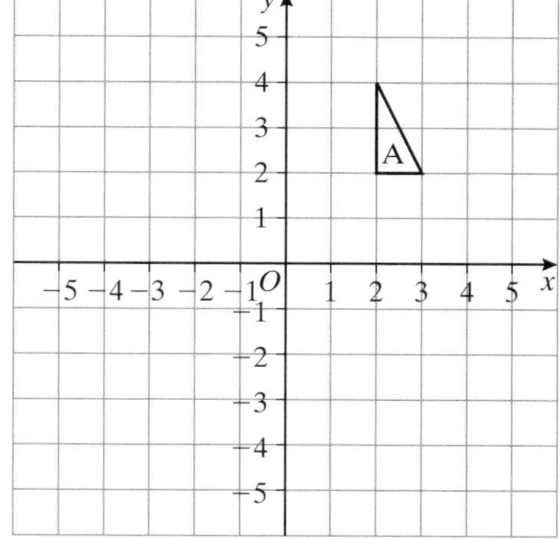

[Edexcel]

4
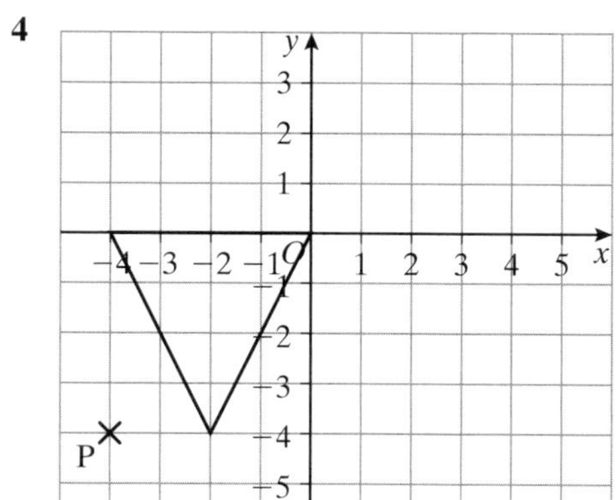

[Edexcel]

5
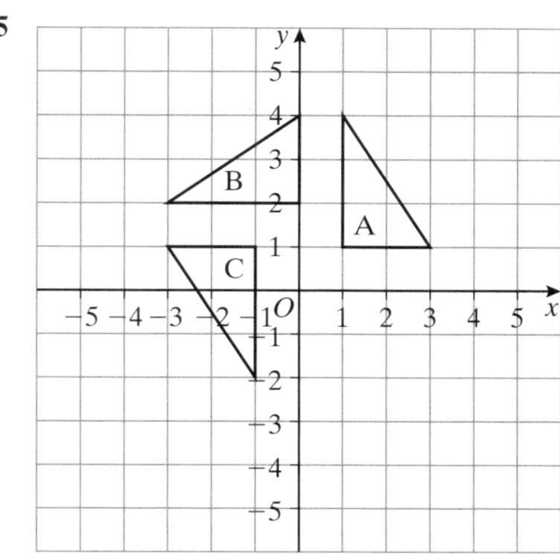

[Edexcel]

6
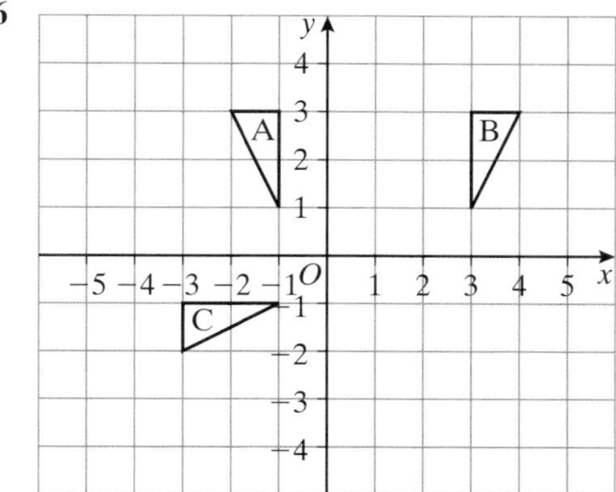

[Edexcel]

10
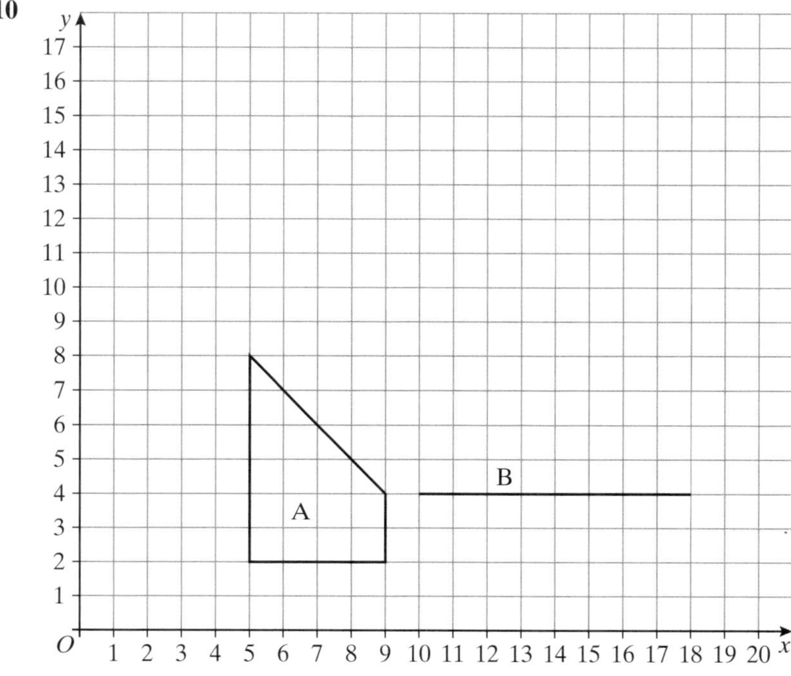

[Edexcel]

(continued)

12

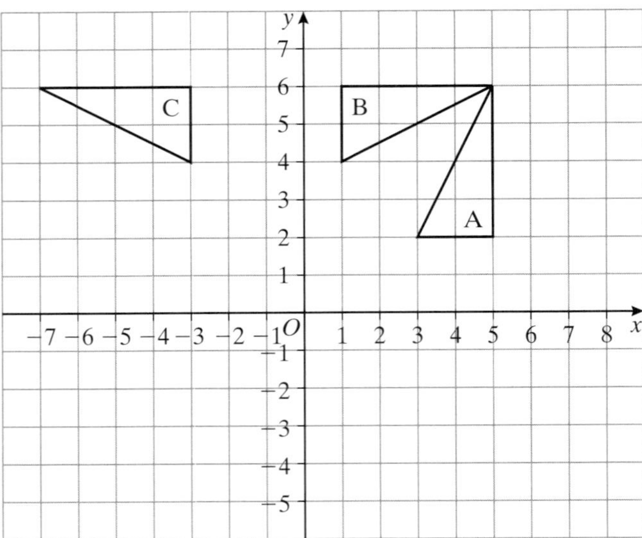

[Edexcel]

 **Speed-up sheet**

GCSE Maths for Edexcel: Higher Teacher's Resource © Hodder Murray 2006

▶▶ EXERCISE 18.2

3

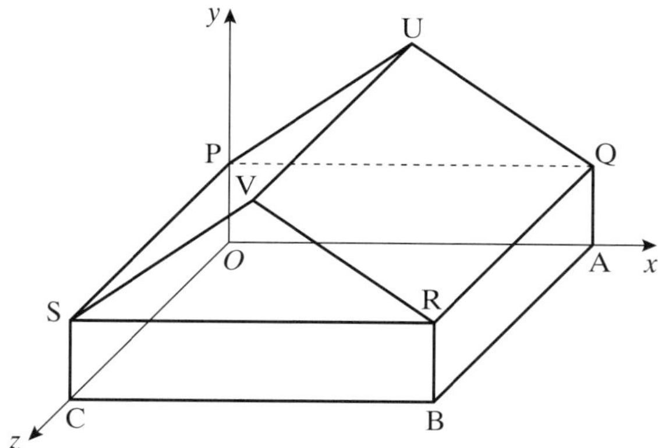

▶▶ EXERCISE 18.5

1

$2a + 3b$	$3ab$	$a + b + c$	$2a^2c$	$2a^2 + bc$	$ab(b + 2c)$

[Edexcel]

2

Expression	Length	Area	Volume	None of these
$\pi a^2 b$				
$\pi b^2 + 2h$				
$2ah$				

[Edexcel]

4

Expression	Length	Area	Volume	None of these
$3rl$				
$\dfrac{2(r + l)^2}{h}$				
$\dfrac{4\pi r^4}{3l}$				

[Edexcel]

5

$\dfrac{ab}{h}$	$2\pi b^2$	$(a + b)ch$	$2\pi a^3$	πab	$2(a^2 + b^2)$	$\pi a^2 b$

▶▶ REVIEW EXERCISE 18

2

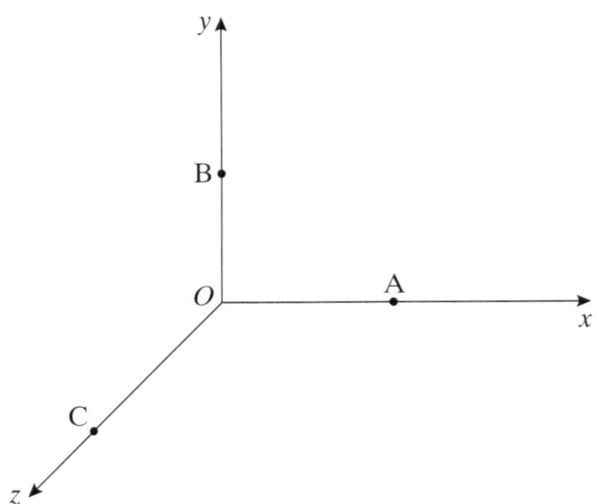

9

$\dfrac{\pi r^2}{x}$	$\pi(r + x)$	$\pi r + r$	$\dfrac{\pi r^3}{x}$	$\pi r^2 + rx$	$\dfrac{r^2}{\pi x}$

πr^3	$\pi r^4 + \pi x$	$\dfrac{\pi r^4}{x}$	$\pi r^2 + \pi rx$

[Edexcel]

Speed-up sheet

▶▶ STARTER 19

1 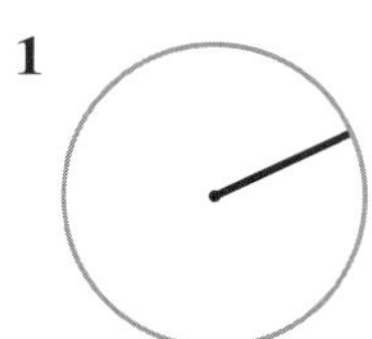
....

2 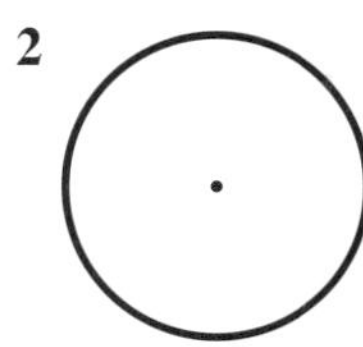
....

3 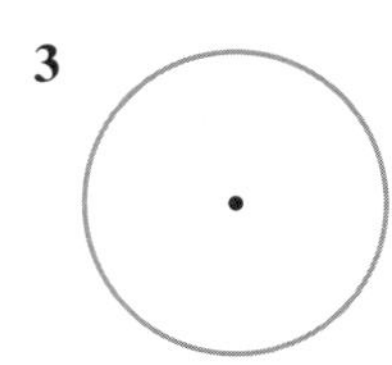
....

4 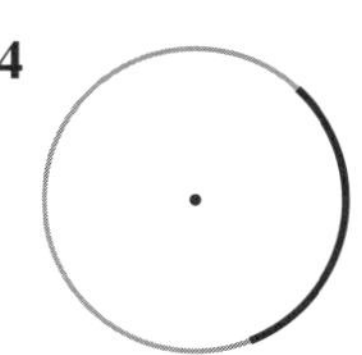
....

5
....

6 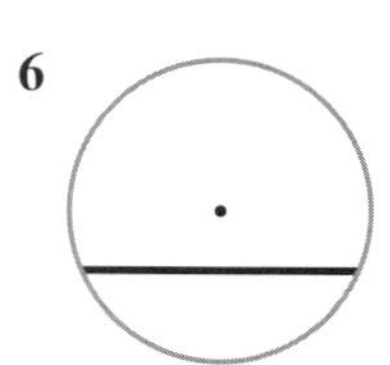
....

7 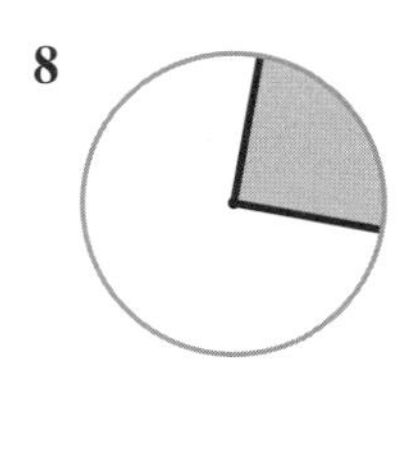
....

8 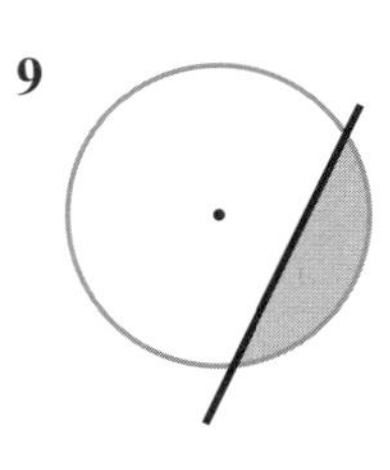
....

9
....

▶▶ STARTER 20

Enrico's sheet	
0 to 50	
51 to 100	
101 to 150	

Kwame's sheet

0 _______________
1 _______________
2 _______________
3 _______________
4 _______________
5 _______________
6 _______________
7 _______________
8 _______________

etc.

Sunita's sheet

71 to 80

81 to 90

91 to 100

101 to 110

111 to 120

etc.

▶▶ EXERCISE 20.2

5

	German	Spanish	Italian	Total
Male	25			40
Female			13	
Total		30	23	100

[Edexcel]

6

	Liked coffee	Did not like coffee	Total
Boys			
Girls			
Total			

[Edexcel]

▶▶ REVIEW EXERCISE 20

7

	School dinners	Sandwiches	Other	Total
Boys	12	3		16
Girls	8		2	
Total				30

[Edexcel]

  **Speed-up sheet**   GCSE Maths for Edexcel: Higher Teacher's Resource © Hodder Murray 2006

1

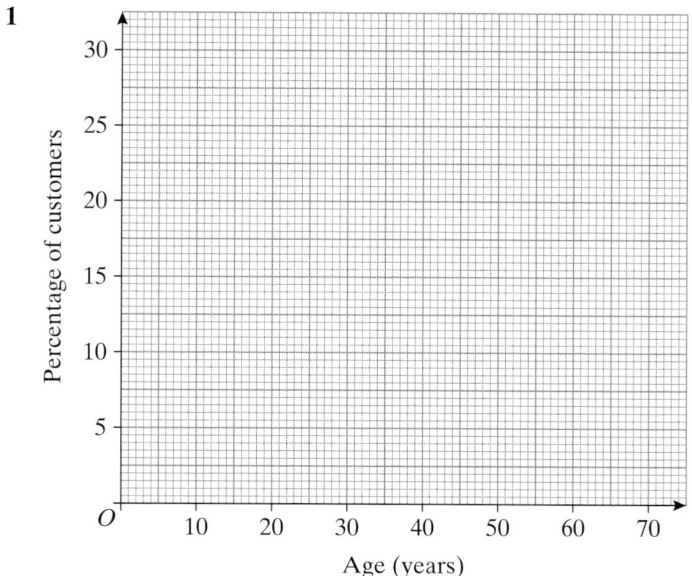

[Edexcel]

2

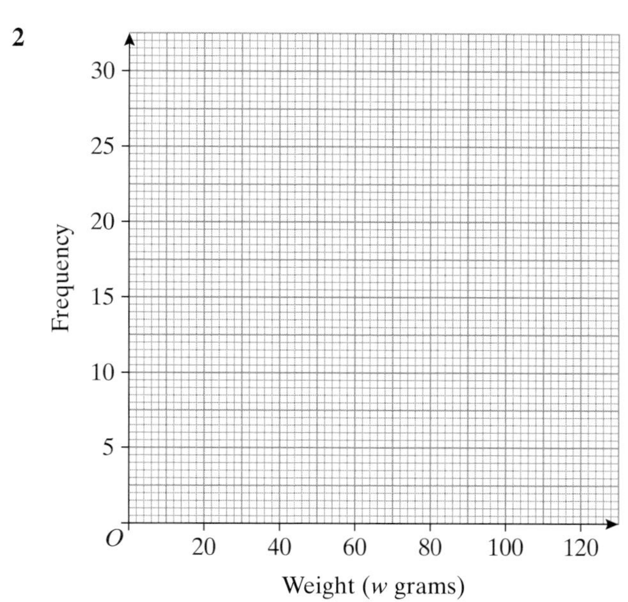

[Edexcel]

3

Height (h cm)	Frequency
$140 \leqslant h < 150$	15
$150 \leqslant h < 160$	
$160 \leqslant h < 165$	20
$165 \leqslant h < 170$	
$170 \leqslant h < 180$	
$180 \leqslant h < 190$	12
$190 \leqslant h < 210$	

[Edexcel]

(continued)

▶▶ EXERCISE 21.3 (continued)

4 a)

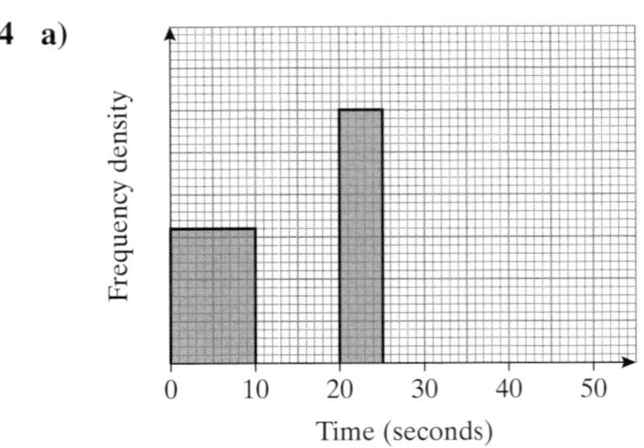

b)

Time (t seconds)	Frequency
$0 < t \leqslant 10$	10
$10 < t \leqslant 20$	
$20 < t \leqslant 25$	
$25 < t \leqslant 30$	
$30 < t \leqslant 50$	

[Edexcel]

5

Time (t hours)	Frequency
$0 < t \leqslant 1$	20
$1 < t \leqslant 2$	28
$2 < t \leqslant 4$	34
$4 < t \leqslant 8$	52
$8 < t \leqslant 16$	

[Edexcel]

▶▶ EXERCISE 21.4

2

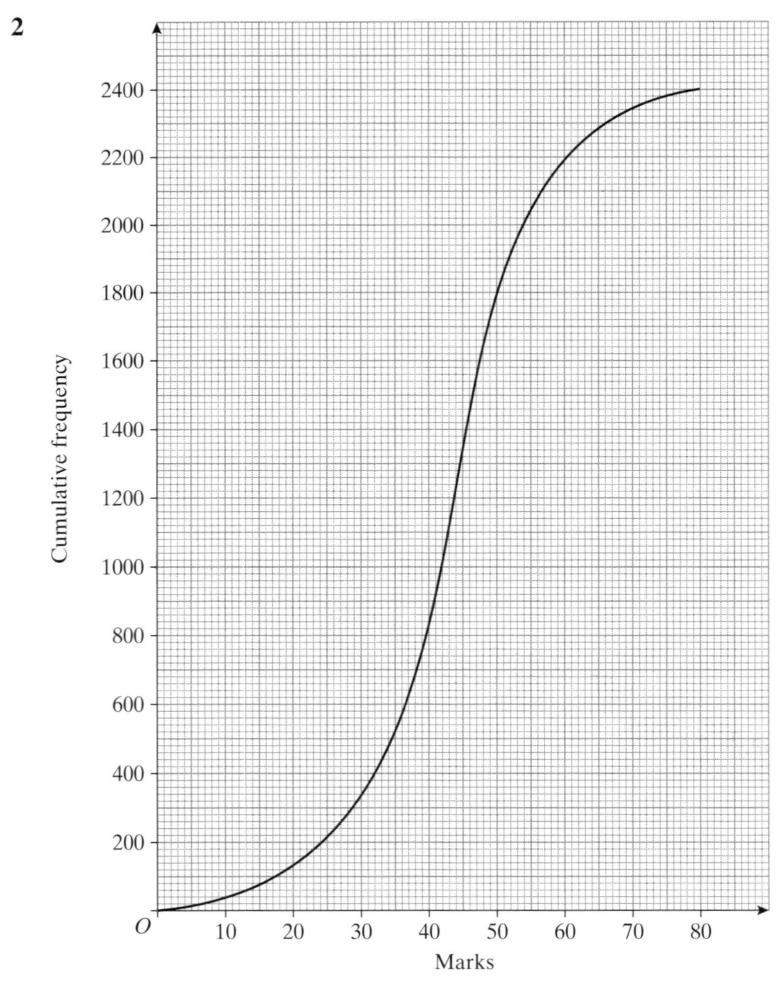

[Edexcel]   (continued)

3

Waiting time (t seconds)	Cumulative frequency
$0 \leqslant t < 50$	
$0 \leqslant t < 100$	
$0 \leqslant t < 150$	
$0 \leqslant t < 200$	
$0 \leqslant t < 250$	
$0 \leqslant t < 300$	

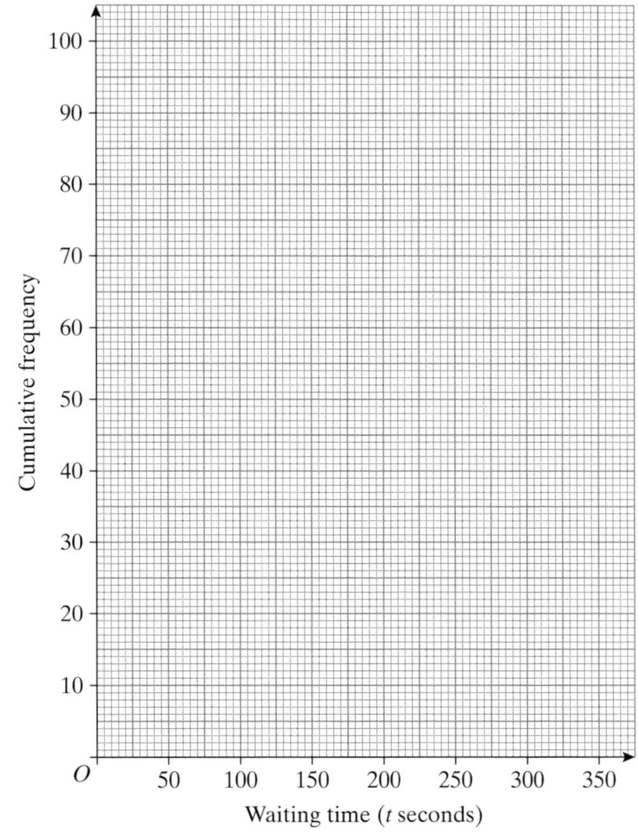

[Edexcel]

4

Hourly rate of pay (£x)	Cumulative frequency
$3.00 < x \leqslant 3.50$	1
$3.00 < x \leqslant 4.00$	2
$3.00 < x \leqslant 4.50$	4
$3.00 < x \leqslant 5.00$	7
$3.00 < x \leqslant 5.50$	19
$3.00 < x \leqslant 6.00$	2

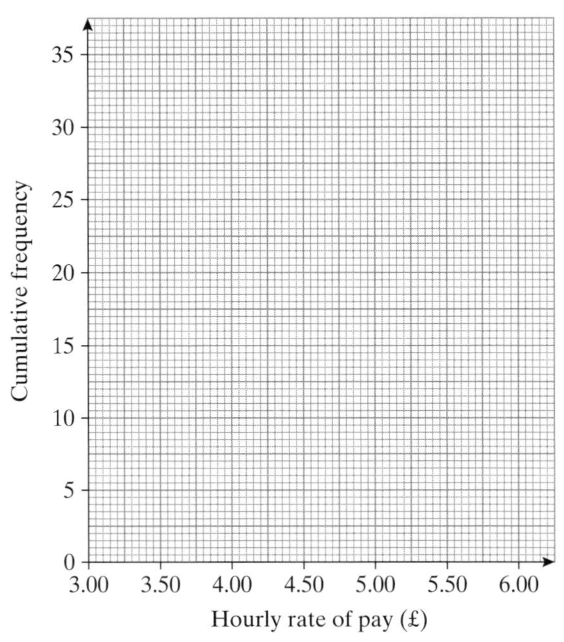

[Edexcel]

▶▶ EXERCISE 21.5

1

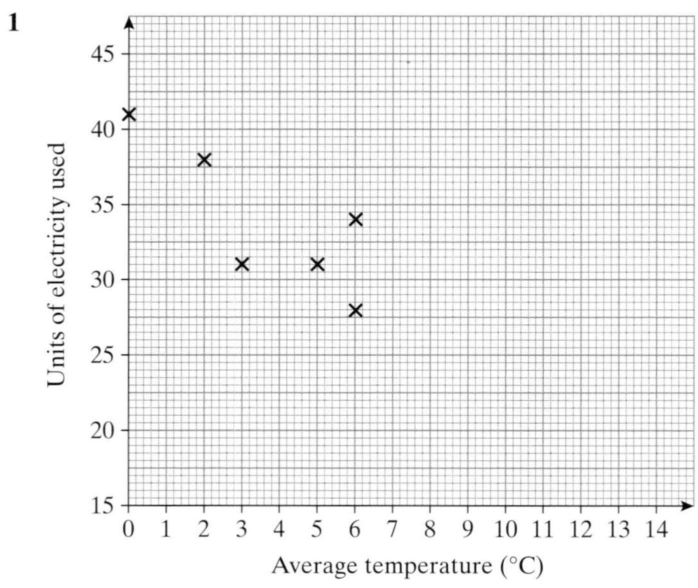

[Edexcel]

2

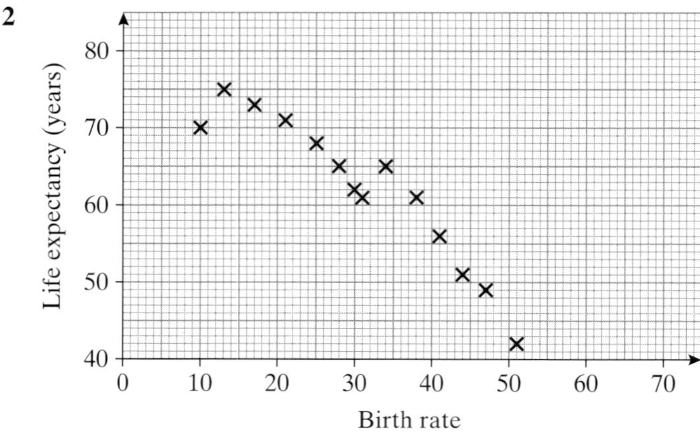

[Edexcel]

4

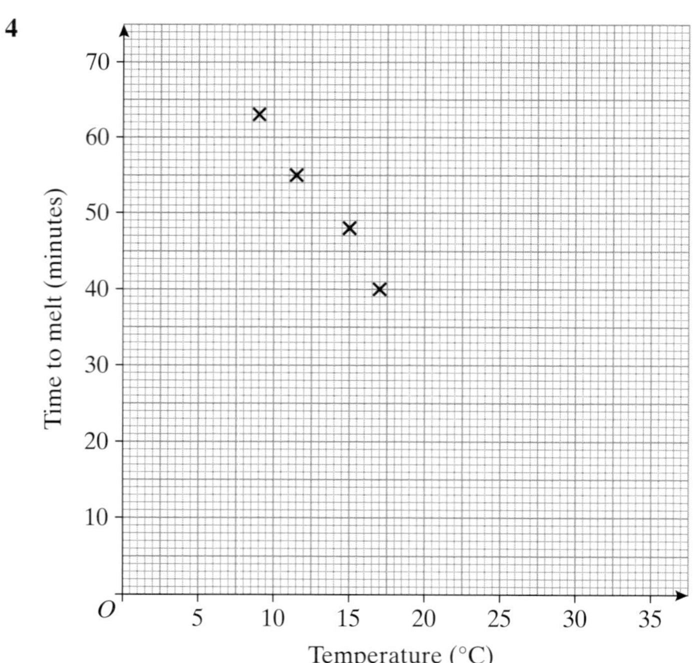

[Edexcel]

(continued)

5

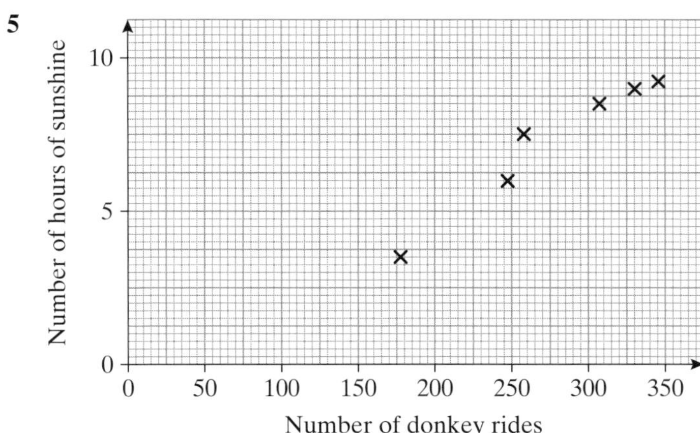

[Edexcel]

▶▶ REVIEW EXERCISE 21

6

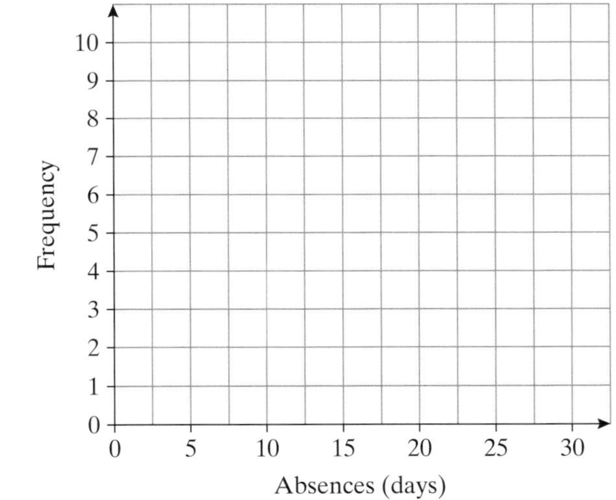

[Edexcel]

7

Time (t minutes)	Frequency
$0 < t \leqslant 10$	
$10 < t \leqslant 15$	
$15 < t \leqslant 30$	
$30 < t \leqslant 50$	
Total	135

[Edexcel]

(continued)

8

Revision time (t minutes)	Frequency
$20 \leq t < 25$	20
$25 \leq t < 40$	
$40 \leq t < 60$	
$60 \leq t < 85$	
$85 \leq t < 95$	32

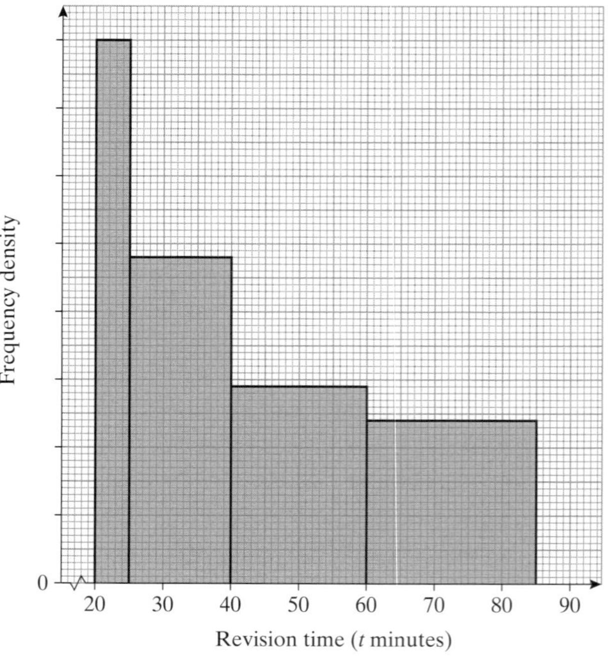

[Edexcel]

9

Age (x) in years	Frequency
$0 < x \leq 10$	160
$10 < x \leq 25$	
$25 < x \leq 30$	
$30 < x \leq 40$	100
$40 < x \leq 70$	120

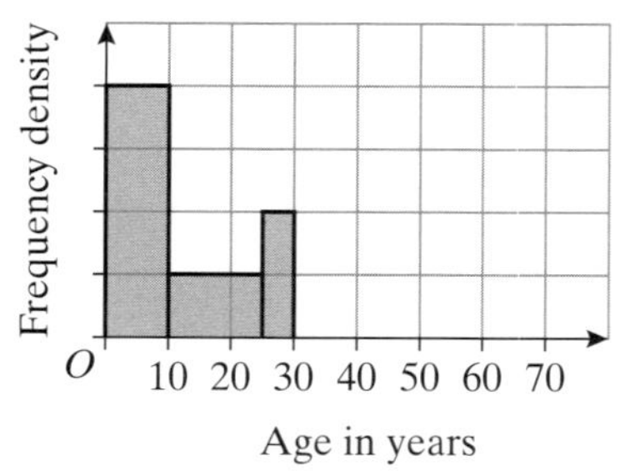

[Edexcel]

10

Number of hours worked (t)	Cumulative frequency
$0 < t \leq 30$	0
$0 < t \leq 40$	
$0 < t \leq 50$	
$0 < t \leq 60$	
$0 < t \leq 70$	
$0 < t \leq 80$	

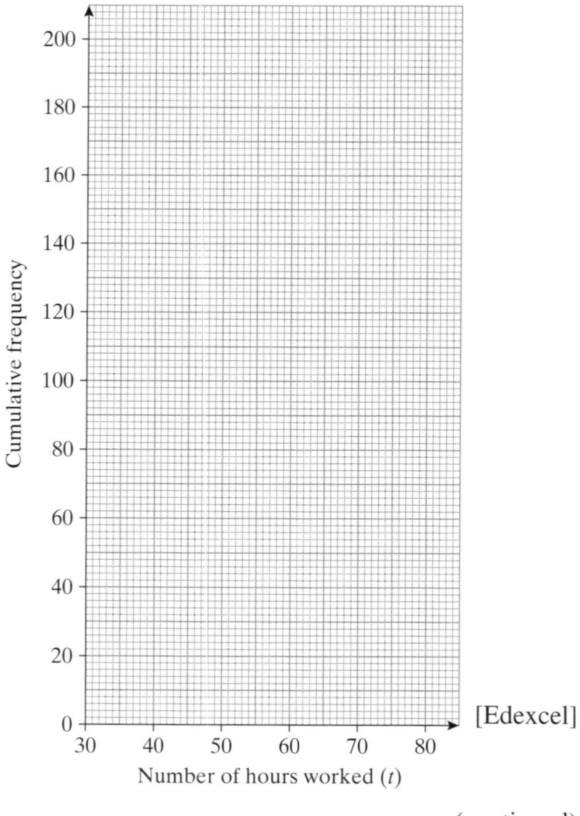

[Edexcel]

(continued)

Speed-up sheet

GCSE Maths for Edexcel: Higher Teacher's Resource © Hodder Murray 2006

11

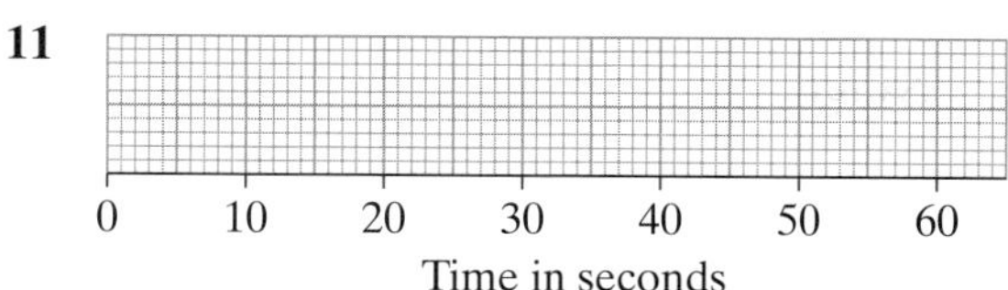

[Edexcel]

12

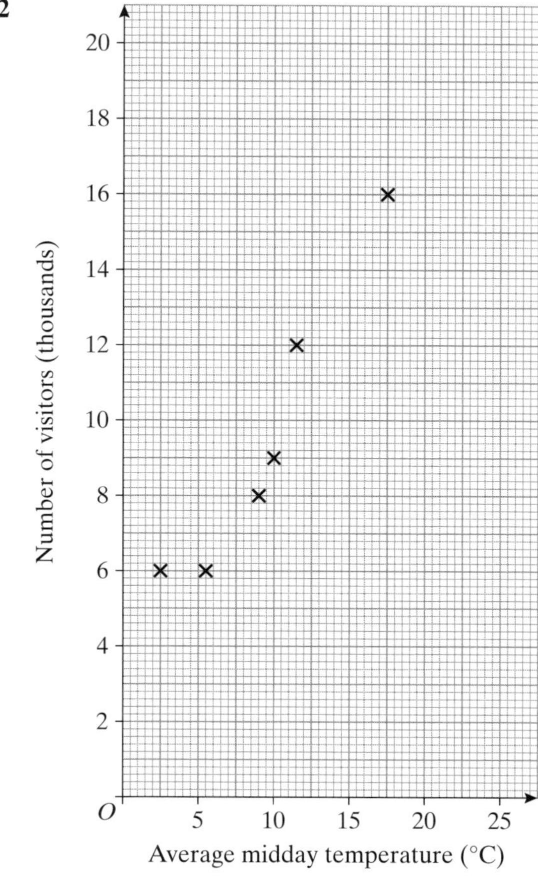

[Edexcel]

13

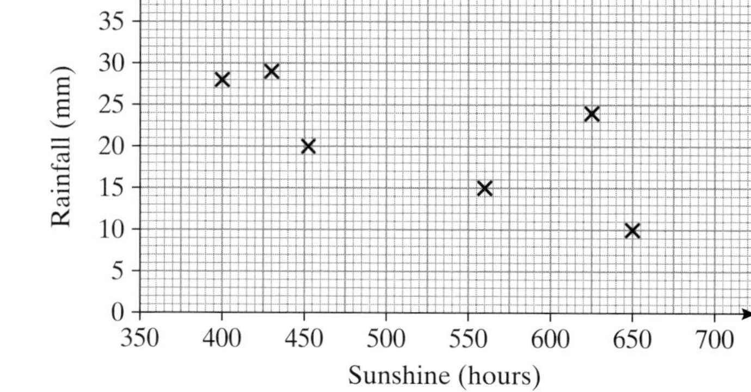

[Edexcel]

▶▶ EXERCISE 22.4

1

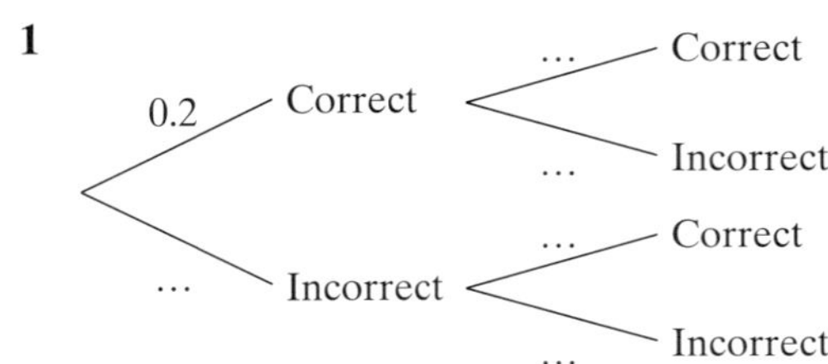

3

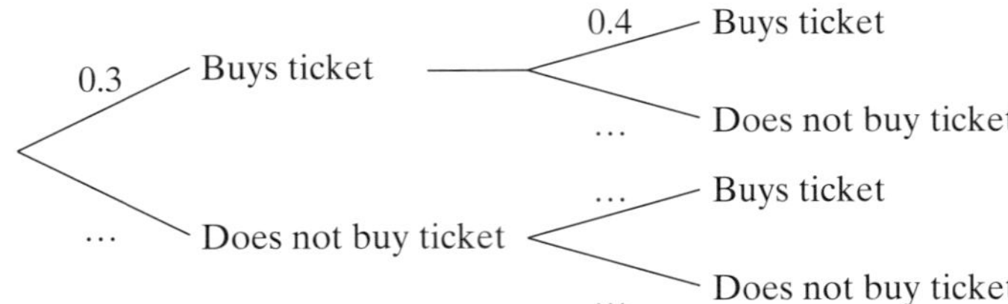

▶▶ REVIEW EXERCISE 22

5

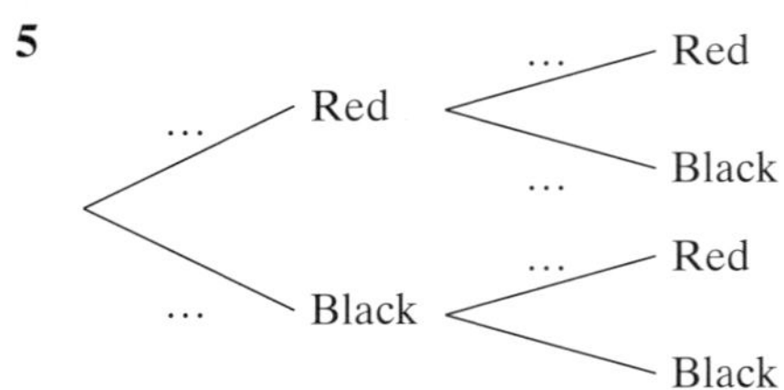

[Edexcel]

7

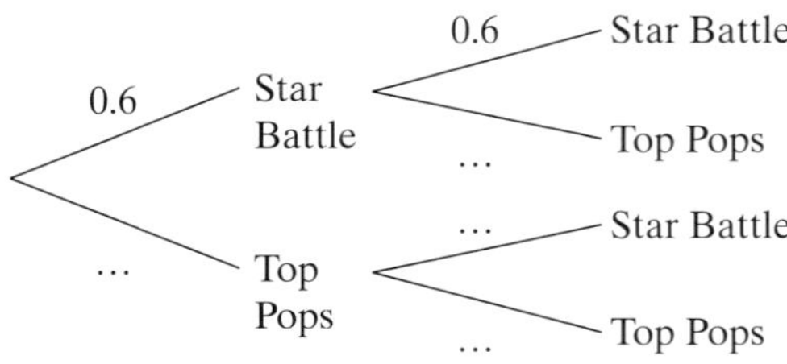

[Edexcel]

11

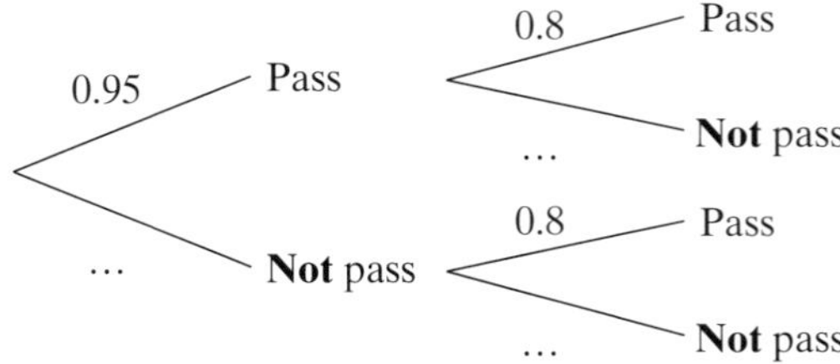

[Edexcel]

12

Spinner B

		1	2	3	4
Spinner A	×				
	1				
	2				8
	3				

[Edexcel]

 **Speed-up sheet**

▶▶ INTERNET CHALLENGE 23

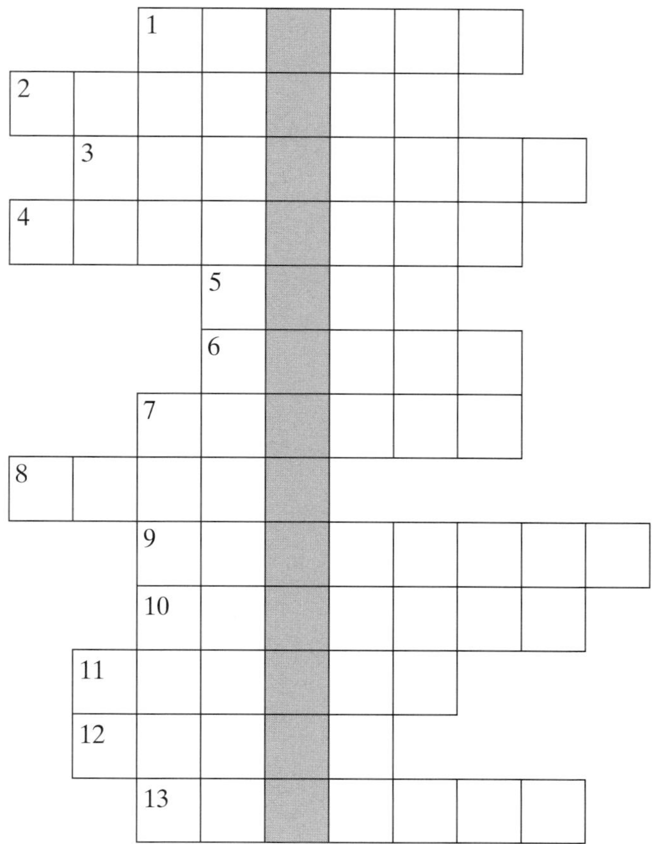

Clues

1 This British mathematician died 1954, and is considered the father of modern computer science.

2 6, 28 and 496 are examples of this type of number.

3 These indicate the part of a calculation that should be worked out first.

4 This quantity tells you how many times one number divides into another.

5 This graphics file format is often used as a way of compressing images for website display.

6 This computer chooses winning Premium Bond numbers.

7 Very large number, 100^{10}

8 A computing language devised in 1963 by Kemeny and Kurtz.

9, 12 ______________ _________ form is a convenient way of writing very large (or small) numbers.

10 After oxygen, this is the most abundant element in the Earth's crust.

11 Person who breaks into computer security systems.

12 See 9

13 Liquid _________ Display, or LCD, is used on most modern calculators.

1 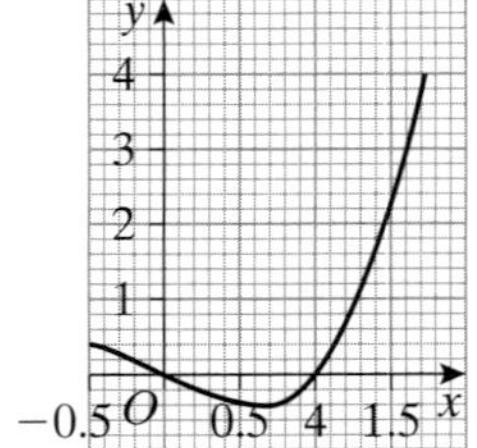

2 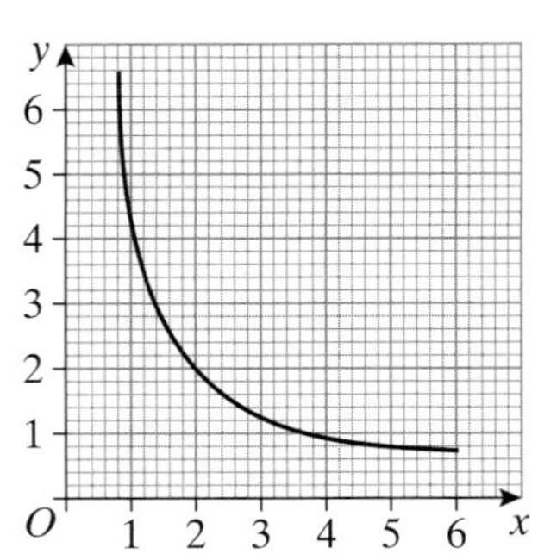

3 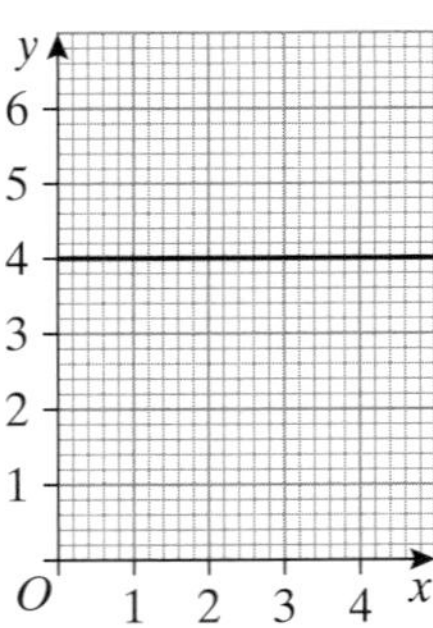

4 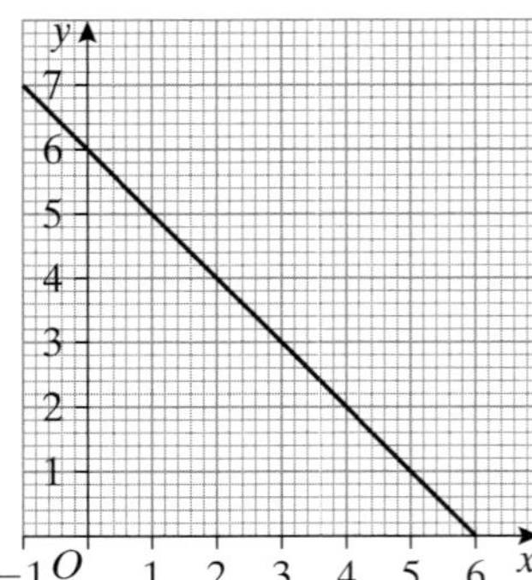

5 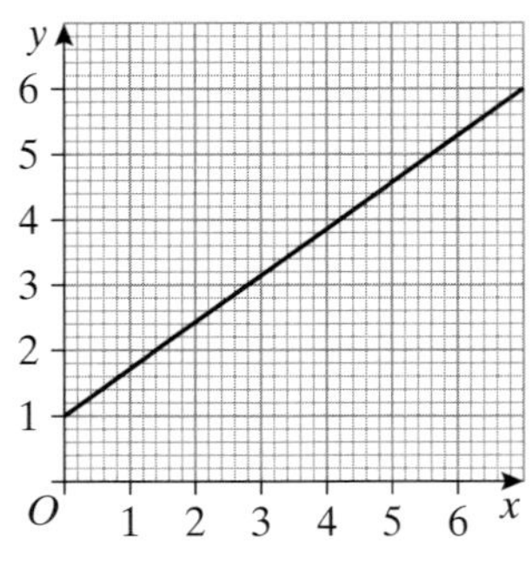

6 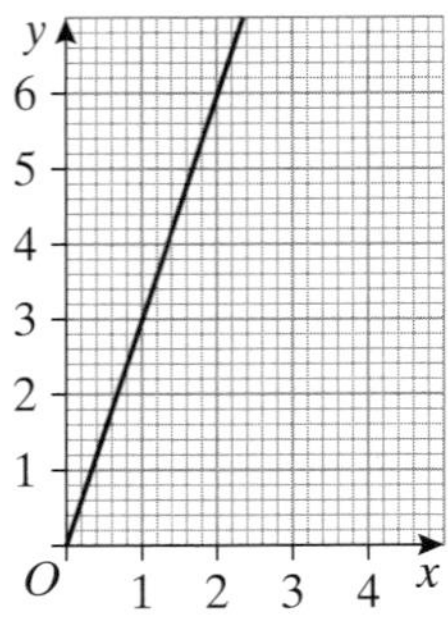

▶▶ **INTERNET CHALLENGE 24**

Planet	Mean distance, d, from the Sun (Earth–Sun = 1 unit)	Orbital period, T (years)
Mercury	0.387	
Venus	0.723	
Earth	1	1
Mars	1.524	
Jupiter	5.203	
Saturn	9.529	
Uranus	19.19	
Neptune	30.06	
Pluto	39.53	

▶▶ **INTERNET CHALLENGE 25**

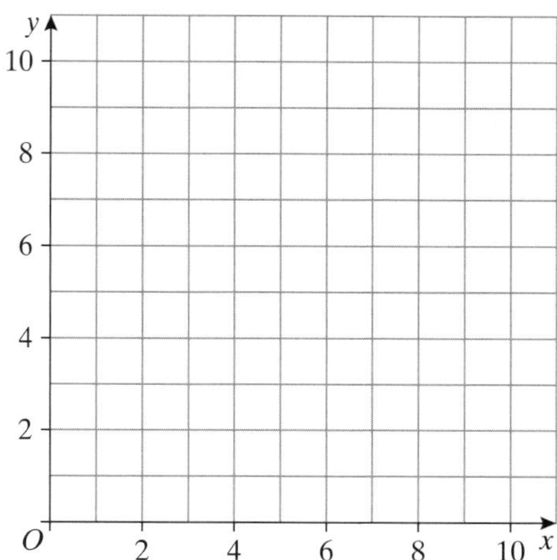

▶▶ STARTER 28

x	0	30	45	60	90	120
$\sin x$	0	0.5			1	

x	135	150	180	210	225	240
$\sin x$		0.5				

x	270	300	315	330	360	390
$\sin x$					0	

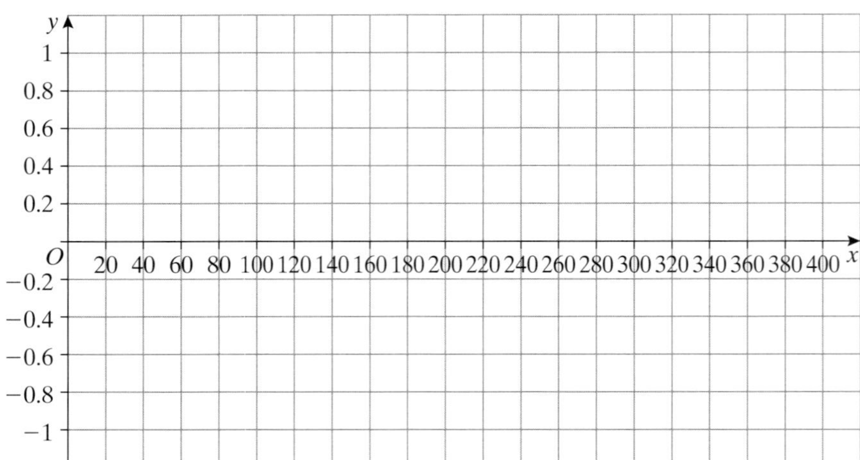

▶▶ EXERCISE 28.1

1

x	−3	−2	−1	0	1	2	3
y		−1					

2

x	−2	−1	0	1	2	3
y		−2			10	

3

x	−2	−1	0	1	2	3
y		5			11	

4

x	−2	−1	0	1	2	3
y						

5

x	−1	0	1	2	3	4	5
y							

6

x	−2	−1	0	1	2	3
y	0.25				4	

7

x	−2	−1	−0.5	0	0.5	1	2
y				not defined			

8

x	−1	0	1	2	3
y		1			

Speed-up sheet: Chapter 28

Speed-up sheet

1

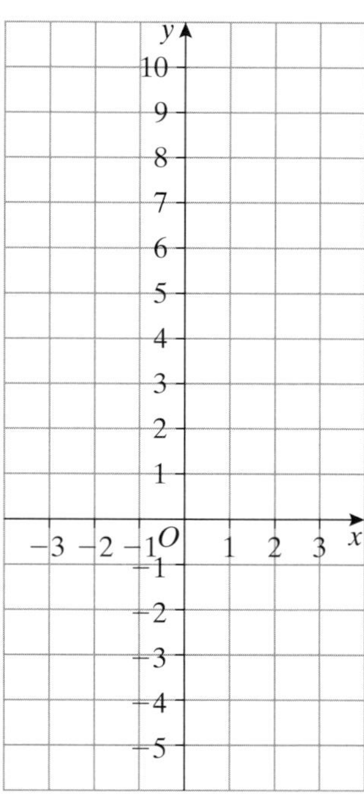

x	−3	−2	−1	0	1	2	3
y	6	1	−2	−3	−2	1	6

3

x	−3	−2	−1	0	1	2	3
y		5			−1		

2

x	−3	−2	−1	0	1	2	3
y			−3			12	

(continued)

4

x	-3	-2	-1	0	1	2	3
y	-15				-3		

5

x	-1	0	1	2	3	4
y		1				16

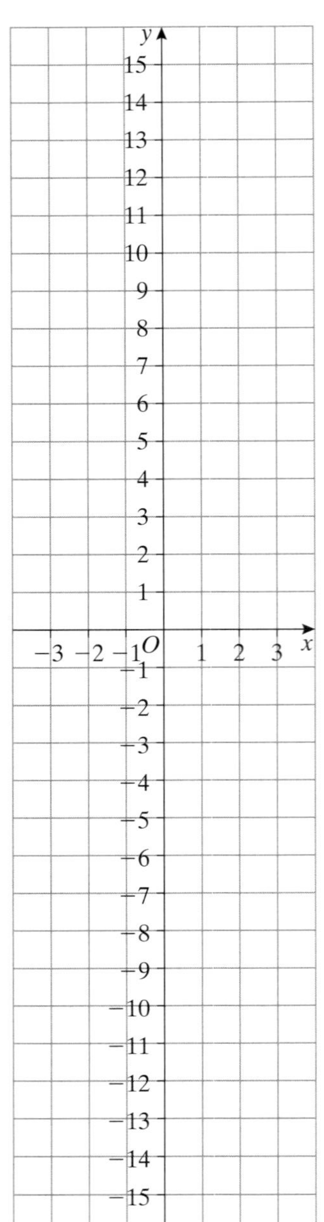

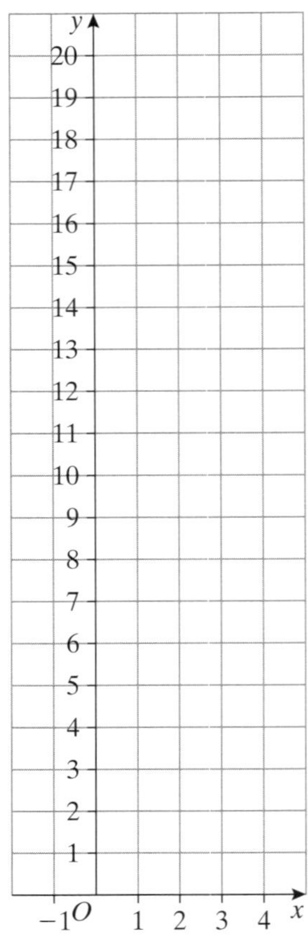

6

x	-3	-2	-1	0	1	2	3
y		4		8			-1

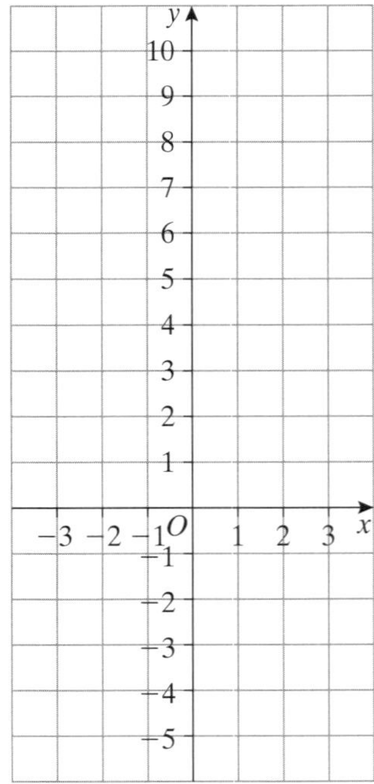

(continued)

7

x	-3	-2	-1	0	1	2	3	4
y	-4			not defined		6		

8

x	-3	-2	-1	0	1	2	3	4
y	6		-6				0	

1

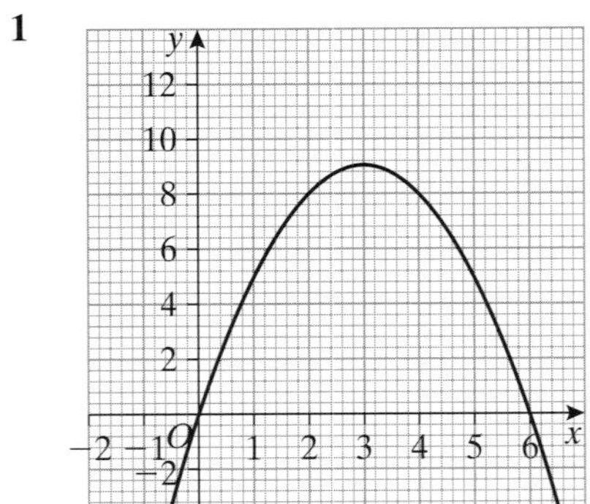

2

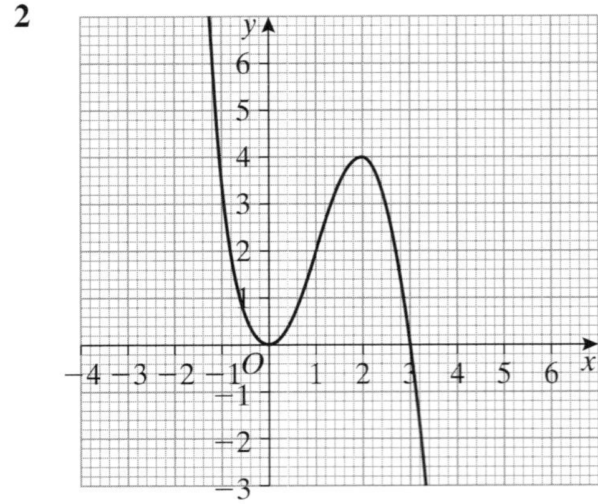

3

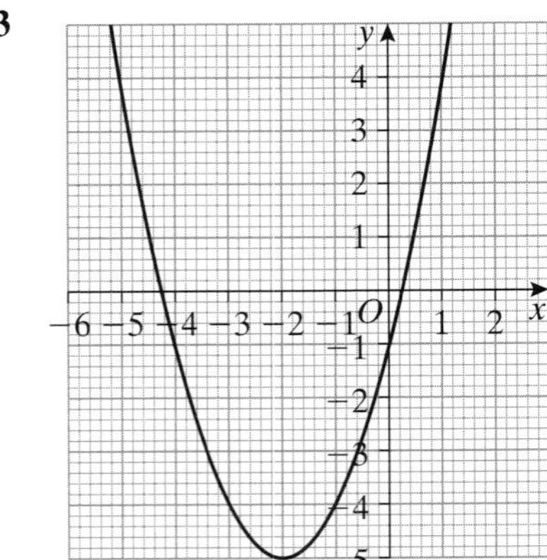

4

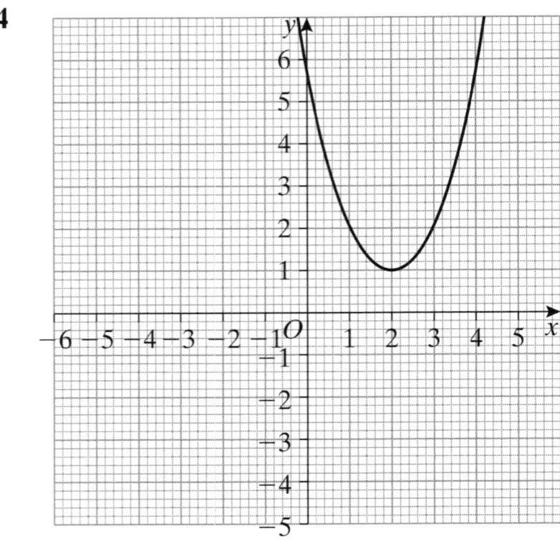

1

x	-3	-2	-1	0	1	2	3
y	18				2	8	

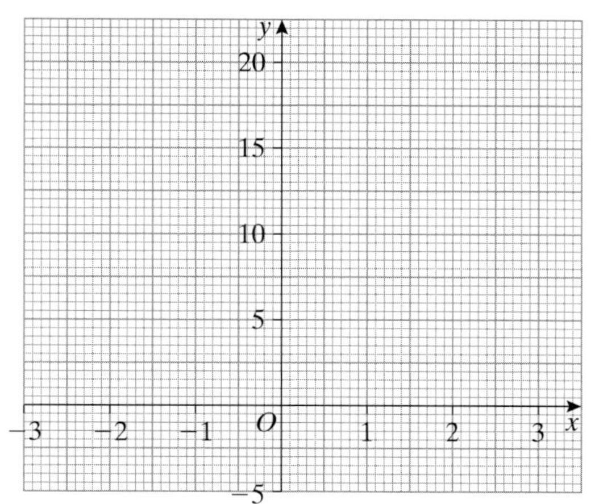

[Edexcel]

2

x	-3	-2	-1	0	1	2
$y = x^3 + 2$	-25					10

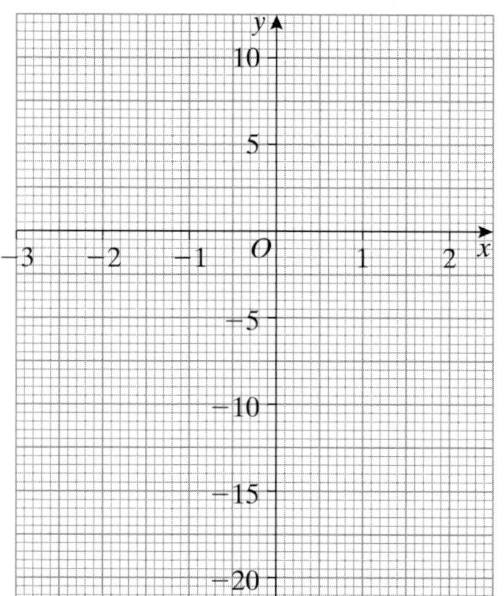

[Edexcel]

4

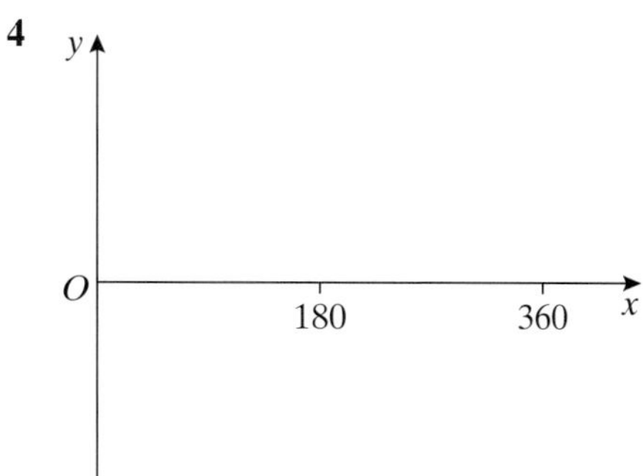

8

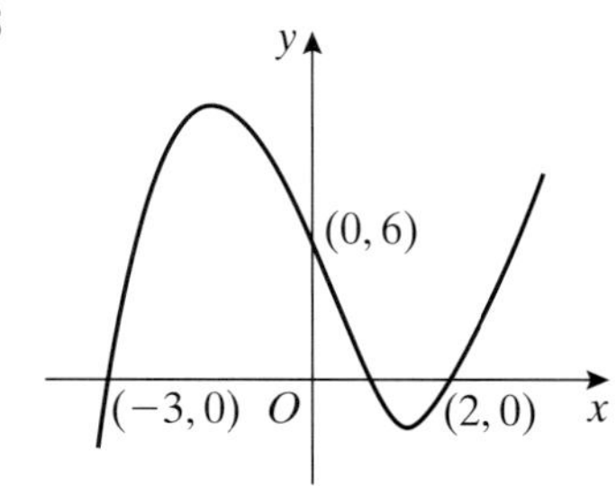

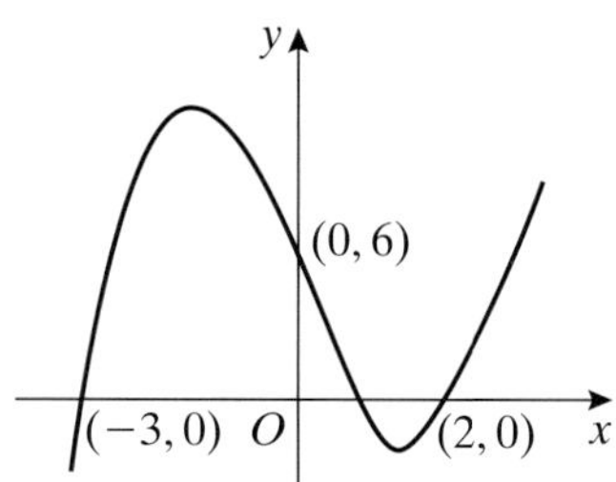

[Edexcel]

▶▶ EXERCISE 32.2

1

1974	1975												1976
Dec	Jan	Feb	Mar	Apr	May	Jun	Jul	Aug	Sep	Oct	Nov	Dec	Jan
17	13	14	6	4	1	0	0	0	3	2	5	8	11
	14.7	11	8										

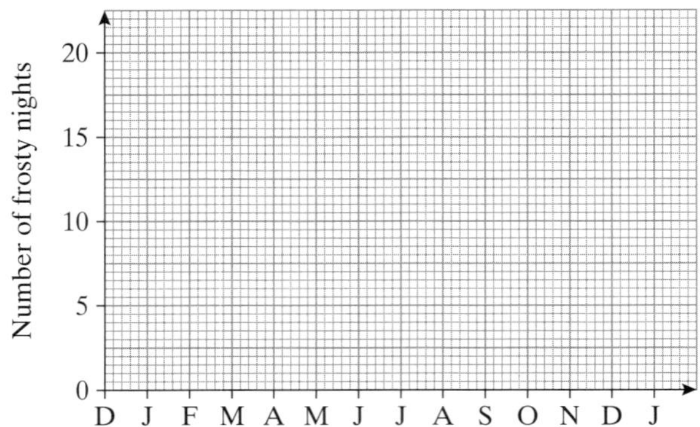

2

Week 1					Week 2					Week 3				
Mon	Tue	Wed	Thu	Fri	Mon	Tue	Wed	Thu	Fri	Mon	Tue	Wed	Thu	Fri
6	2	3	2	12	4	5	3	2	10	4	3	2	4	11
		5	4.6	5.2	5.2	5.2								

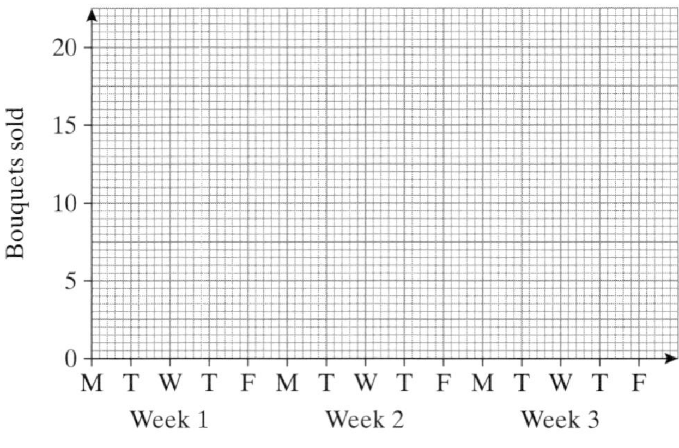

3

Year	1996				1997				1998			
Quarter	1st	2nd	3rd	4th	1st	2nd	3rd	4th	1st	2nd	3rd	4th
	9	23	28	7	10	28	29	6	13	25	31	9
		16.75	17									

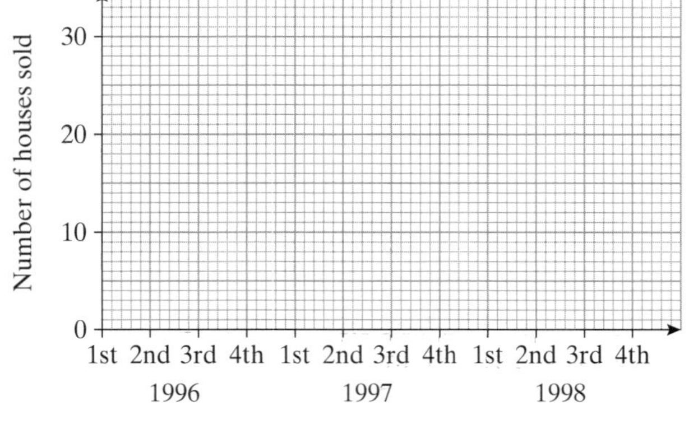

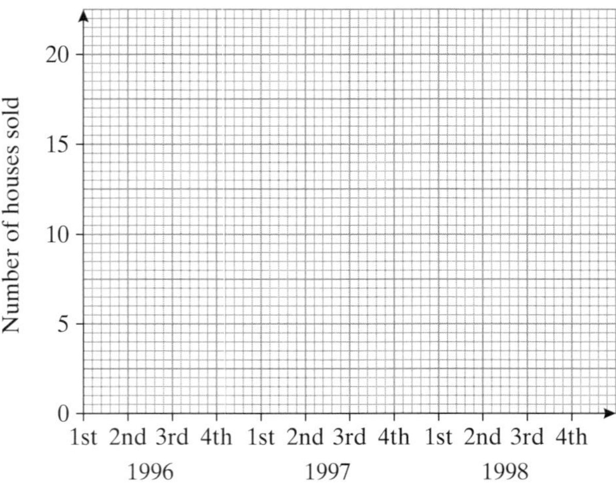

  **Speed-up sheet**   GCSE Maths for Edexcel: Higher Teacher's Resource © Hodder Murray 2006

▶▶ EXERCISE 32.3

1

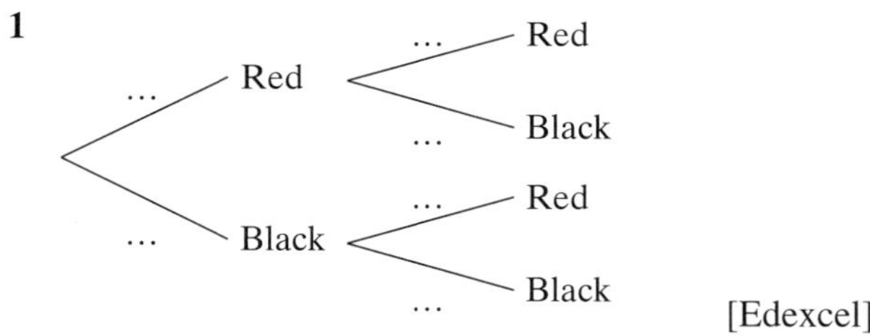

[Edexcel]

Answers

Chapter 1: Working with whole numbers

Starter 1

Here are possible solutions for the numbers 1 to 25; others are also possible.

1 $\dfrac{44}{44}$

2 $\dfrac{4 \times 4}{4 + 4}$

3 $\dfrac{4 + 4 + 4}{4}$

4 $4 + 4 \times (4 - 4)$

5 $\dfrac{4 \times 4 + 4}{4}$

6 $4 + \dfrac{4 + 4}{4}$

7 $4 + 4 - \dfrac{4}{4}$

8 $4 + 4 + 4 - 4$

9 $4 + 4 + \dfrac{4}{4}$

10 $\dfrac{44 - 4}{4}$

11 $\dfrac{4}{4} + \dfrac{4}{4}$

12 $4! - (4 + 4 + 4)$

13 $\dfrac{4! \times \sqrt{4} + 4}{4}$

14 $4 \times 4 - 4 + \sqrt{4}$

15 $4 \times 4 - \dfrac{4}{4}$

16 $4 + 4 + 4 + 4$

17 $4 \times 4 + \dfrac{4}{4}$

18 $4 \times 4 + 4 - \sqrt{4}$

19 $4! - 4 - \dfrac{4}{4}$

20 $4 \times 4 + \sqrt{4} + \sqrt{4}$

21 $4! - 4 + \dfrac{4}{4}$

22 $4 \times 4 + 4 + \sqrt{4}$

23 $4! - \dfrac{\sqrt{4} + \sqrt{4}}{4}$

24 $4 \times 4 + 4 + 4$

25 $4! + \dfrac{\sqrt{4} + \sqrt{4}}{4}$

It is possible to find solutions all the way up to 100, but some unusual combinations of symbols might be needed, including

$$4! = 24, \quad \frac{4}{0.4} = 10, \quad \frac{4}{0.\dot{4}} = 9, \quad \sqrt{\frac{4}{0.4}} = 3$$

Exercise 1.1

1 130	**2** 557	**3** 191	**4** 587
5 125	**6** 200	**7** 207	**8** 45
9 398	**10** 177	**11** 363	**12** 1138
13 263	**14** 472	**15** 3934	**16** 1408
17 3884	**18** 939	**19** 166	**20** 1875

Exercise 1.2

1 432	**2** 1016	**3** 708	**4** 1816
5 2282	**6** 2100	**7** 2012	**8** 3987
9 5112	**10** 6885	**11** 6468	**12** 20 580
13 68 414	**14** 26 670	**15** 17 248	**16** 40 495
17 1265	**18** 2388	**19** £6.42	**20** 76 800

Exercise 1.3

1 47	**2** 195 r 2	**3** 298 r 4
4 9615	**5** 255	**6** 291
7 272 r 2	**8** 57 r 1	**9** 452
10 399	**11** 221	**12** 236
13 199	**14** 144	**15** 343 r 1
16 135 r 10	**17** 125 grams	**18** £250
19 14 or 15	**20** 11 or 12	

Worksheet 1.3

1 £90

2 a) 7 coaches   b) £1225

3 11 880 feet

4 £10.404

5 a) 16p   b) (i) 89p   (ii) 11p

6 a) £1020 per week is better

    b) (i) £5200   (ii) £1200

Exercise 1.4

1 -2	**2** 3
3 -1	**4** 3
5 2	**6** -9
7 5	**8** 12
9 10	**10** 0
11 8	**12** -7
13 -4	**14** -9
15 -18	**16** 20
17 -16	**18** -2
19 -6	**20** 12
21 $-5, -1, 0, 3, 8$	**22** $12, 9, 5, -4, -13$
23 4	**24** -6

Worksheet 1.4

1 a) $15°$ b) $7°$ c) $7°$

    d) $10°$ e) $5°$ f) $9°$

2 a) $-9°, -7°, -4°, -1°, 3°, 4°$

    b) $-8°, -2°, -1°, 0°, 2°, 7°$

3 a) $6, 2, 1, 0, -, -3$ b) $3, 1, 0, -2, -4, -8$

4

Day	Maximum temperature	Minimum temperature	Change in temperature
Monday	5	-2	7
Tuesday	8	-1	9
Wednesday	3	-3	6
Thursday	0	-4	4
Friday	11	-2	13

5 a) $-£150$ b) $-£50$

Exercise 1.5

1 2, 3, 5, 7, 11, 13, 17, 19, 23, 29, 31, 37

2 a) 4   b) 37

    c) 13 and 31   d) 7×13

3 a) $2^4 \times 5$ b) $2 \times 3^2 \times 5$ c) $2 \times 3^2 \times 5^2$

4   a) $2^2 \times 3^2$ b) 3^4 c) $2^4 \times 3^2$
    All use an even number of each factor
5 $a = 3, b = 7$

Exercise 1.6

1   a) 6   b) 15   c) 11
    d) 9   e) 1   f) 26
2   a) $2^2 \times 5, 2^5, 4$
    b) $2^2 \times 3^2, 2^2 \times 3 \times 5, 12$
    c) $2^4 \times 5, 2^2 \times 3^2 \times 5, 20$
    d) $2^3 \times 3^2, 2^2 \times 3^3, 36$
    e) $2^3 \times 3 \times 5, 3 \times 5 \times 13, 15$
    f) $2^4 \times 3^2, 2^3 \times 3^2 \times 5, 72$
3   a) 6   b) 12   c) 24
    d) 10   e) 6   f) 2

Exercise 1.7

1 60   **2** 208   **3** 90   **4** 200
5 144   **6** 60   **7** 154   **8** 150
9 180   **10** 220   **11** 144   **12** 84
13 180   **14** 210   **15** 108   **16** 165
17   a) $2^2 \times 3 \times 5, 2^2 \times 3 \times 7$
    b) 420
18   a) $2 \times 3 \times 11, 3^2 \times 11$ b) 198   c) 33
19   a) $2 \times 5, 2^2 \times 3^2, 2^3 \times 7$ b) 2   c) 2520
20   a) $2^3 \times 5, 2^4 \times 3, 2^3 \times 3 \times 5^2$ b) 8   c) 1200
21   Every 20 days
22   In 120 days' time

Review Exercise 1

1 513   **2** 368   **3** 188
4 979   **5** 8   **6** 473
7 1797   **8** 99   **9** 38
10 530   **11** 1238   **12** 678
13 868   **14** 3312   **15** 3388
16 24 087   **17** 276   **18** 524
19 374   **20** 62 r 8   **21** −21
22 9   **23** 3   **24** 3
25 19   **26** 40   **27** −9
28 −32   **29** 28   **30** −27
31   a) $2 \times 5 \times 7$ b) $2^2 \times 31$
    c) $2^5 \times 3$ d) $2^4 \times 3 \times 5$
32   a) 8   b) 168
33   a) 2   b) 220
34   a) $2^3 \times 3^2 \times 5$ b) 720
35   Lilian. The numbers have no factors in common other
    than 1, e.g. 15 and 4
36   480
37 $c = 4, d = 23$
38   a) $2^3 \times 3^2, 2^5 \times 3$ b) 24

Internet Challenge 1

1   There are 25 primes between 1 and 100.
    2, 3, 5, 7, 11, 13, 17, 19, 23, 29, 31, 37, 41, 43, 47,
    53, 59, 61, 67, 71, 73, 79, 83, 89, 97
2   There are 168 primes between 1 and 1000.
    2, 3, 5, 7, 11, 13, 17, 19, 23, 29, 31, 37, 41, 43, 47,
    53, 59, 61, 67, 71, 73, 79, 83, 89, 97, 101, 103, 107,
    109, 113, 127, 131, 137, 139, 149, 151, 157, 163,
    167, 173, 179, 181, 191, 193, 197, 199, 211, 223,
    227, 229, 233, 239, 241, 251, 257, 263, 269, 271,
277, 281, 283, 293, 307, 311, 313, 317, 331, 337,
347, 349, 353, 359, 367, 373, 379, 383, 389, 397,
401, 409, 419, 421, 431, 433, 439, 443, 449, 457,
461, 463, 467, 479, 487, 491, 499, 503, 509, 521,
523, 541, 547, 557, 563, 569, 571, 577, 587, 593,
599, 601, 607, 613, 617, 619, 631, 641, 643, 647,
653, 659, 661, 673, 677, 683, 691, 701, 709, 719,
727, 733, 739, 743, 751, 757, 761, 769, 773, 787,
797, 809, 811, 821, 823, 827, 829, 839, 853, 857,
859, 863, 877, 881, 883, 887, 907, 911, 919, 929,
937, 941, 947, 953, 967, 971, 977, 983, 991, 997

3   Primes have exactly two distinct factors (for
    example, 3 is 3×1. Thus 1 is not normally
    considered prime.
    Also the 'Fundamental Theorem of Arithmetic' states
    that any whole number may be factorised into a
    unique product of primes (for example, $10 = 2 \times 5$.
    If 1 were a prime, then 10 could also be written as
    $1 \times 2 \times 5$, and $1 \times 1 \times 2 \times 5$, and so on. So 1 does
    not behave in the same way as genuine prime
    numbers. Thus 1 is not a prime, – but it is not a good
    example of a non-prime either!
4   Tony Blair is the 52nd Prime Minister to have held
    the office since Sir Robert Walpole in 1721, and 52
    is clearly not prime. As and when his successor is
    elected, the answer will change, since 53 is prime.
5   Yes, there are infinitely many primes.
6   The largest known prime in 2005 was $2^{30402457}$: it
    requires over 9 million digits to be written out in
    full. Bigger ones are being found all the time, so be
    sure to check the internet regularly for the latest
    news.
7   No.
8   It passes from the North Pole to the South Pole, via
    Greenwich.
9   Check students' work.
10   It is conjectured that there are infinitely many adjacent
    prime pairs. The result has not yet been proved.

Chapter 2: Fractions and decimals

Starter 2

1   100 cm (1 metre)
2 $\frac{1}{8}$
3

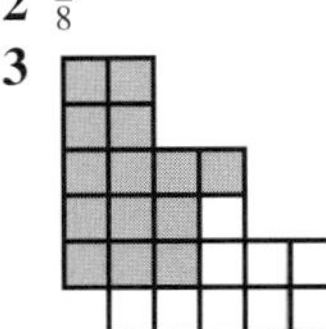

4   One person cuts and the other one chooses.

Exercise 2.1

1 $\frac{1}{4}$ **2** $\frac{3}{7}$ **3** $\frac{4}{41}$
4 $\frac{1}{4}$ **5** $\frac{1}{5}$ **6** $\frac{3}{4}$
7 $\frac{3}{5}$ **8** $\frac{3}{5}$ **9** $\frac{4}{7}$
10 $\frac{3}{5}$ **11** $\frac{7}{13}$ **12** $\frac{7}{9}$
13 $\frac{1}{2}, \frac{11}{20}, \frac{3}{5}, \frac{5}{8}$ **14** $\frac{2}{3}, \frac{3}{4}, \frac{5}{6}, \frac{7}{8}$ **15** $\frac{2}{3}, \frac{4}{5}, \frac{5}{6}, \frac{13}{15}$
16 $\frac{7}{9}$ **17** $\frac{9}{16}$ **18** $\frac{5}{6}$
19 $\frac{13}{24}$ **20** $\frac{3}{8}$ **21** $\frac{13}{20}$
22 $\frac{2}{9}$ **23** $\frac{9}{28}$ **24** $6\frac{1}{6}$

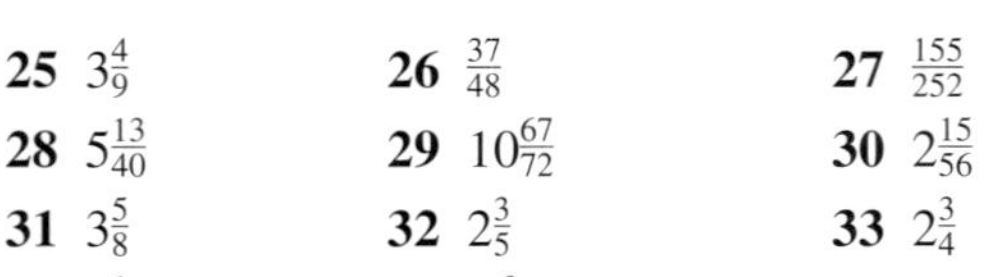

25 $3\frac{4}{9}$ **26** $\frac{37}{48}$ **27** $\frac{155}{252}$

28 $5\frac{13}{40}$ **29** $10\frac{67}{72}$ **30** $2\frac{15}{56}$

31 $3\frac{5}{8}$ **32** $2\frac{3}{5}$ **33** $2\frac{3}{4}$

34 $8\frac{1}{2}$ **35** $3\frac{3}{20}$

Worksheet 2.1

1 $4\frac{5}{12}$ km **2** $4\frac{9}{40}$ m **3** $7\frac{11}{12}$ km

4 $3\frac{3}{16}$ **5** $1\frac{27}{35}$ **6** $\frac{2}{3}$ hours

Exercise 2.2

1 $\frac{2}{5}$ **2** $\frac{5}{9}$ **3** $\frac{11}{20}$

4 $\frac{4}{9}$ **5** 15 **6** $\frac{1}{15}$

7 $\frac{5}{28}$ **8** 120 **9** $\frac{8}{15}$

10 $\frac{1}{2}$ **11** $\frac{15}{16}$ **12** $\frac{1}{18}$

13 $\frac{7}{16}$ **14** $\frac{1}{18}$ **15** $\frac{3}{4}$

16 1 **17** $\frac{5}{16}$ **18** 2

19 $\frac{9}{16}$ **20** $1\frac{1}{2}$ **21** $3\frac{1}{2}$

22 $\frac{8}{21}$ **23** $4\frac{2}{3}$ **24** $\frac{13}{14}$

25 3 **26** $1\frac{1}{17}$ **27** $\frac{1}{2}$

28 $5\frac{1}{7}$ **29** 3 **30** $\frac{5}{6}$

31 $20\frac{5}{8}$ **32** 6 **33** $31\frac{1}{2}$

34 12 **35** 100 **36** $\frac{1}{2}$

Worksheet 2.2

1 $14\frac{1}{16}$ cm^2 **2** $1\frac{4}{5}$ hours **3** $1\frac{7}{12}$ m

4 8 **5** $187\frac{1}{2}$ cm **6** 2286

Exercise 2.3

1 42.6 **2** 20.16 **3** 1.918 **4** 1.8

5 38.08 **6** 1.12 **7** 31.2 **8** 0.06

9 7.64 **10** 0.219 75 **11** 16 **12** 0.3

13 15.6 **14** 0.76 **15** 15.4 **16** 445

17 a) 14.352 **b)** 1435.2 **c)** 1 435 200

18 a) 1.728 **b)** 172.8 **c)** 14.4

19 a) 0.876 **b)** 87.6 **c)** 0.24

20 a) 0.347 **b)** 34 700 **c)** 902 200

Exercise 2.4

1 $\frac{6}{25}$ **2** $\frac{18}{25}$ **3** $\frac{3}{10}$

4 $\frac{5}{8}$ **5** $\frac{91}{100}$ **6** $\frac{1}{40}$

7 $1\frac{47}{50}$ **8** $\frac{19}{50}$ **9** $2\frac{1}{8}$

10 $\frac{303}{1000}$ **11** $\frac{7}{9}$ **12** $\frac{29}{99}$

13 $1\frac{1}{3}$ **14** $\frac{1733}{3330}$ **15** $\frac{43}{99}$

16 $\frac{6}{11}$ **17** $\frac{107}{333}$ **18** $1\frac{113}{330}$

19 0.625 **20** $0.\dot{4}28\,57\dot{1}$ **21** $0.\dot{4}$

22 0.45 **23** 0.28

24 Wrong. $0.3\dot{5} \times 2 = 0.707\,070\ldots$ which is not $0.\dot{7}$

Exercise 2.5

1 3.142 **2** 3.1416 **3** 16.2

4 0.24 **5** 14.1 **6** 14.8

7 6.2240 **8** 1.90 **9** 15.4

10 20 **11** 14.3 **12** 359 300

13 370 **14** 10 **15** 0.0021

16 11.0 **17** 34.5 cm **18** 21 500 mm^2

19 Not possible, last week must be at least 24 500, the week before not more than 24 450, so no overlap is possible.

20 No. The age is only given to the nearest million years so Charlie's calculation is not valid.

Worksheet 2.5A

1 a) 38.22 **b)** 382.2 **c)** 3.822

2 a) 11 210 **b)** 11.21 **c)** 112.1

3 a) 205.9 **b)** 20.59 **c)** 205.9

4 a) 140.4 **b)** 14 040 **c)** 14.04

5 a) 159.9 **b)** 159.9 **c)** 15.99

Worksheet 2.5B

1 10 **2** 150 **3** 300

4 9 **5** 50 **6** 0.05

7 80 **8** 100 **9** 1800

10 7 **11** $30 \times 40 = 1200$

12 $80 \div 20 = 4$ **13** $500 \div 25 = 20$

14 $60 + 80 = 140$ **15** $4 \times 5 = 20$

16 $90 - 30 = 60$ **17** $520 + 600 = 1120$

18 $1000 \times 700 = 700\,000$ **19** $1 \times 0.5 = 0.5$

20 $6 \div 3 = 2$ **21** 10

22 10 **23** 4

24 2 **25** 200

Worksheet 2.5C

1 2.65 and 2.75 **2** 3.75 and 3.85

3 4.85 and 4.95 **4** 16.15 and 16.25

5 12.05 and 12.15 **6** 100.35 and 100.45

7 6.95 and 7.05 **8** 49.95 and 50.05

9 0.65 and 0.75 **10** 9.85 and 9.95

11 6.335 and 6.345 **12** 1.815 and 1.825

13 2.495 and 2.505 **14** 10.315 and 10.325

15 0.865 and 0.875 **16** 1.065 and 1.075

17 0.005 and 0.015 **18** 4.995 and 5.005

19 12.335 and 12.345 **20** 29.995 and 30.005

21 74.5 seconds and 75.5 seconds

22 a) 72 cm **b)** 70 cm

23 a) 2.35 cm **b)** 2.45 cm

24 155.0025 cm^2

25 24 500 people

26 12.5 kg

Review Exercise 2

1 $\frac{13}{20}$ **2** $1\frac{4}{15}$ **3** $3\frac{11}{24}$

4 $4\frac{11}{20}$ **5** 5 **6** $23\frac{1}{3}$

7 $3\frac{3}{4}$ **8** 14 **9** $\frac{7}{12}, \frac{5}{8}, \frac{3}{4}, \frac{5}{6}$

10 $\frac{3}{5}, \frac{2}{3}, \frac{7}{10}, \frac{11}{15}$ **11** 67.2 **12** 7.5

13 0.45 **14** 0.24 **15** 41

16 1.048 **17** 4.85 **18** 44

19 5.4 **20** 4.8 **21** $\frac{5}{8}, 0.65, 0.\dot{6}, \frac{3}{4}$

22 a) $\frac{7}{8}$ **b)** $\frac{5}{11}$

23 a) 0.067, 0.56, 0.6, 0.605, 0.65

b) $-10, -6, -4, 2, 5$

c) $\frac{2}{5}, \frac{1}{2}, \frac{2}{3}, \frac{3}{4}$

24 $\frac{5}{12}$ **24** $\frac{2}{5}, \frac{1}{2}, \frac{2}{3}, \frac{3}{4}$ **25** $\frac{1}{5}$

26 For example, $\frac{1}{3}$ and $\frac{3}{8}$

27 855.4 kg

28 a) $\frac{31}{40}$ **b)** $2\frac{11}{12}$

29 a) 69.3   **b)** 6.93   **c)** 0.0693
30 a) 100.5 mm   **b)** 101.5 mm
31 a) 119.31   **b)** 119 310   **c)** 1.23
32 $\frac{22}{99}$ **36 a)** $\frac{3}{11}$
33 a) $\frac{13}{99}$
    b) $\frac{254}{495}$
34 a) $\frac{4}{11}$
    b) $2\frac{15}{110}$
35 $2\frac{1}{15}$

36 b) Let $X = 0.0\dot{x}$
So $X = 0.0x0x0x...$
       $100\,X = x.0x0x0x...$
Subtracting gives:
       $99X = x$
So $X = \dfrac{x}{99}$

Internet Challenge 2

1 $\frac{1}{6}$ **2** $\frac{11}{20}$ **3** $\frac{1}{4} + \frac{1}{12}$ **4** $\frac{1}{2} + \frac{1}{3} + \frac{1}{12}$ **5** $\frac{1}{2} + \frac{1}{6} + \frac{1}{9}$

6 No, but the Erdos–Strauss conjecture claims that fractions of the form $\frac{4}{n}$ may be written as the sum of three Egyptian fractions. The conjecture has been checked for every individual case up to $n = 10^{14}$.

7 Emily will write $\frac{5}{8}$ as $\frac{1}{2} + \frac{1}{8}$.

For each of the first four sacks, divide the contents of the sack into two equal piles, giving eight piles of $\frac{1}{2}$ a sack each. For the fifth sack, divide its contents into eight equal piles, giving $\frac{1}{8}$ of a sack each. Give each chicken coop one of the $\frac{1}{2}$ sack piles and one of the $\frac{1}{8}$ sack piles. The idea is that it is easier to judge splitting a sack into equal piles ($\frac{1}{2}$ or $\frac{1}{8}$) than to divide it asymmetrically into $\frac{3}{8}$ and $\frac{5}{8}$.

Chapter 3: Ratios and percentages

Starter 3
288

Exercise 3.1
1 $2:5$ **2** $2:3$ **3** $7:11$
4 $3:13$ **5** $4:5:7$ **6** $3:4:6$
7 $3:4:7$ **8** $2:5$ **9** $5:7:8$
10 $1:4:5$ **11** £15, £20, £25   **12** 60, 100, 140
13 £50, £150, £250   **14** $24, $36, $84
15 24, 48, 60   **16** £50, £125, £175
17 £48, £60, £108   **18** $5:10:25 \rightarrow 1:2:5$
19 a) £720   **b)** £3360
20 a) £960   **b)** £1600
21 150 g flour, $\frac{3}{8}$ tsp salt, 75 g suet, $\frac{1}{2}$ tsp herbs

Worksheet 3.1
1 a) $1:1.5$ **b)** $1:1.25$ **c)** $1:1.3$
   d) $1:0.8$ **e)** $1:2.25$ **f)** $1:1.375$
   g) $1:0.6$ **h)** $1:1.75$ **i)** $1:0.125$
2 a) $3.5:1$ **b)** $2.75:1$ **c)** $1.6:1$
   d) $2.5:1$ **e)** $0.75:1$ **f)** $0.6:1$
   g) $2.2:1$ **h)** $0.8:1$ **i)** $2.4:1$
3 a) $10:7$ **b)** $1:0.7$

Exercise 3.2
1 $\frac{2}{5}$ **2** $\frac{9}{25}$ **3** $\frac{1}{20}$ **4** $\frac{3}{10}$
5 $\frac{33}{100}$ **6** $\frac{1}{3}$ **7** $\frac{1}{8}$ **8** $\frac{1}{200}$
9 75%   **10** 40%   **11** 30%   **12** 85%
13 84%   **14** 95%   **15** $87\frac{1}{2}\%$ **16** $7\frac{1}{2}\%$
17 66%, $\frac{2}{3}$, 0.67, 69%, 0.7, $\frac{5}{7}$

18 a) 84%   **b)** 155   **c)** 71%
19 70% and 71.1% so both scores are about the same
20 20%

Exercise 3.3
1 271.2   **2** 382.2   **3** 1380, 1490.4
4 £35.24   **5** 14%
6 a) £11 480   **b)** £5190
7 2 hours 21 minutes (or 141 minutes)
8 a) £39 520   **b)** £39 125   **c)** $1.04 \times 0.99 = 1.0296$
9 a) 35%   **b)** 9%
10 a) $2344   **b)** £1360

Worksheet 3.3
1 £45   **2** £60   **3** £77   **4** £108
5 £324   **6** £254.40   **7** £240   **8** £414
9 £78.75   **10** £161.25
11 £212.50   **12** 180 cm², 44 cm²
13 4.84 m   **14** £384
15 176 cm²   **16** 15.62(5) m

Exercise 3.4
1 a) £280   **b)** £85   **2 a)** £66   **b)** £650
3 £76   **4 a)** £20.40   **b)** £15.30
5 900 milliseconds
6 a) £9200   **b)** £3405
7 a) 160 pounds   **b)** 11 stone 6 pounds
8 2.86 million light years   **9** 15 400
10 a) 1200 by 775   **b)** Area reduced by 36%

Exercise 3.5
1 £540   **2** £56.48   **3** £13.50
4 £1604.06   **5** £170   **6** £750
7 31   **8** 806.69
9 a) Simple interest is better (£40 compared with £31.85)
   b) Compound interest is better (£218.76 compared with £200)
10 a) $4\% \times 25 = 100\%$ (which is the wrong calculation)
   b) $1.04^{18} = 2.026$ so 18 years is sufficient

Worksheet 3.5A
1 £60   **2** £30   **3** £14
4 £35   **5** £400   **6** £375
7 4 years   **8** 2 years   **9** 2.5%
10 4%   **11** £157.50
12 a) Tony £60, Gordon £28
   b) £32 difference

Worksheet 3.5B
1 a) 1.04   **b)** 1.09
   c) 1.12   **d)** 1.15
   e) 1.035   **f)** 1.0475
   g) 1.028   **h)** 1.0565
2 a) 1.06
   b) (i) £424
      (ii) £449.44
      (iii) £535.29 (to nearest penny)
3 £62.75   **4** £38.20   **5** £70.93
6 a) LFC £78.81, Magnet £79.64
   b) 83p
7 a) $3000 \times 1.09^4 = £4234.74$ (to nearest penny)
   b) £265.26

Review Exercise 3

1 $2:3$ **2** $54°$ **3** $8:10:15$
4 95% **5** £376 **6** 40%
7 a) £504 **b)** 80% **c)** £680
8 a) 12 grams **b)** 280 grams
9 a) 360 **b)** 22%
10 £9720
11 a) $3180\,\text{kg}$ **b)** £300
12 a) $16.8\,\text{cm}$ **b)** $25\,\text{cm}$
13 $68.77
14 200 g flour, 150 g almonds, 225 g sugar, 150 g butter, 10 pears
15 £40
16 a) £360 **b)** £288.26
17 a) £923.55 **b)** £1100
18 a) £493.50 **b)** 62.5%
19 a) £650 **b)** £465.66
20 £624.32
21 a) £35.70 **b)** £18.40
22 a) £6.37 **b)** £69.52
23 a) 4% **b)** £980
24 a) £5062.50 **b)** 0.4096
25 4 years

Internet Challenge 3

1 The RPI is an average measure of change in the prices of goods and services bought by the vast majority of households in the UK.
2 It is compiled and published monthly.
3 2.2% (December 2005), updates available at www.statistics.gov.uk/.
4 The Bank of England sets a rate at which it lends money to other financial institutions; this in turn affects the rates building societies and so on charge their customers.
5 It is reviewed monthly.
6 4.5% (January 2006), updates available at www.bankofengland.co.uk.
7 1920 to 1923
8 Old definition: one trillion $= 1\,000\,000\,000\,000\,000\,000$
New definition: one trillion $= 1\,000\,000\,000\,000$
9 The former Yugoslavia.
10 Hyperinflation

Chapter 4: Powers, roots and reciprocals

Starter 4

Task 1 **a)** 17 **b)** 14 **c)** 45
    d) 70 **e)** 92 **f)** 609
Task 2 **a)** XXI **b)** XXIV **c)** XXXIX
    d) CCXII **e)** CCCXIX **f)** XLVII
Task 3 Star Wars: 1977
    Lion King: 1994

Exercise 4.1

1 25 **2** 8 **3** 49
4 27 **5** 81 **6** 64
7 144 **8** 1000 **9** 12
10 15 **11** 4 **12** 6
13 14 **14** 5 **15** 9

16 10 **17** 361 **18** 3.24
19 213.16 **20** 729 **21** 4330.747
22 1.728 **23** 3.61 (3 s.f.) **24** 17.3 (3 s.f.)
25 2.88 (3 s.f.) **26** 3.68 (3 s.f.) **27** 1.58 (3 s.f.)
28 1.89 (3 s.f.)
29 8.49 or -8.49
30 3.362

Worksheet 4.1

1 41 **2** 20 **3** 52 **4** 9
5 8 **6** 16 **7** 24 **8** -15
9 5 **10** 52 **11** 29 **12** 27
13 24 **14** 12 **15** 48 **16** 5
17 153 **18** 5 **19** $\frac{91}{144}$ **20** 4

Exercise 4.2

1 81 **2** 1 000 000 **3** 6 **4** 10 000
5 729 **6** 32 **7** 2 **8** 1024
9 10 **10** 20 **11** 1728 **12** 2
13 249 000 **14** 9224 **15** 6.275 **16** 0.8145
17 1.817 **18** 1.86 **19** 1.445 **20** 3.46

Exercise 4.3

1 8 **2** 9 **3** 125 **4** 243
5 216 **6** 512 **7** 16 **8** 100 000
9 27 **10** 625 **11** 64 **12** 8

Exercise 4.4

1 $\frac{1}{9}$ **2** $\frac{1}{1000}$ **3** $\frac{1}{25}$ **4** $\frac{1}{4}$
5 $\frac{1}{81}$ **6** $\frac{1}{16}$ **7** $\frac{1}{32}$ **8** $\frac{1}{10}$
9 $\frac{1}{5}$ **10** $\frac{1}{400}$ **11** $\frac{5}{3}$ **12** $\frac{3}{4}$
13 $\frac{2}{5}$ **14** $\frac{25}{16}$ **15** $\frac{27}{8}$ **16** 2
17 $\frac{8}{3}$ **18** $\frac{9}{25}$ **19** $\frac{343}{1000}$ **20** $\frac{16}{9}$

Exercise 4.5

1 2^7 **2** 5^7 **3** 8^9
4 6^3 **5** 9 **6** 3
7 2^{12} **8** 3^6 **9** 3^2
10 2^4 **11** 3^2 **12** 6
13 32 **14** 3 **15** 1 000 000
16 16 **17** 64 **18** 243
19 16 **20** 49 **21** 27
22 1 **23** 1 000 000 **24** 1

Worksheet 4.5

1 a) 2^2 **b)** 2^4 **c)** 64
2 $2^5 \div 2^2 = 2^3$ **3** $2^3 \times 2^6 = 2^9$
4 $2^9 \div 2^4 = 2^5$ **5** $2^4 \div 2^6 = 2^{-2}$
6 $2^8 \times 2^2 = 2^{10}$ **7** $2^{10} \div 2^6 = 2^4$
8 $2^6 \times 2^3 = 2^9$ **9** $2^7 \div 2^7 = 2^0$
10 $2^8 \div 2^5 = 2^3$ **11** $(2^8 \times 2^4) \div 2^7 = 2^5$
12 $(2^9 \times 2^6) \div 2^{10} = 2^5$ **13** $\dfrac{2^7 \times 2^8}{2^6} = 2^9$
14 $\dfrac{2^3 \times 2^4 \times 2^6}{2^6 \times 2^7} = 2^0$ **15** $2^{10} \times 2^3 \div 2^9 = 2^4$

Exercise 4.6

1 3.5×10^5 **2** 4×10^4
3 3.52×10^8 **4** 1.93×10^7
5 7.65×10^2 **6** 4.5×10^{-3}
7 8×10^{-1} **8** 2.03×10^{-3}
9 8.27×10^{-10} **10** 3.3×10^{-4}

11 7 400 000   **12** 21 500 000
13 105 000   **14** 2 000 000 000
15 8400   **16** 0.005
17 0.000 002 5   **18** 0.000 000 100 4
19 0.000 000 000 083   **20** 0.000 505

Exercise 4.7

1 7.04×10^8 **2** 2.04×10^6
3 6.95×10^7 **4** 2.2×10^5
5 6×10^{17} **6** 4.2×10^{14}
7 5×10^2 **8** 7.5×10^8
9 150 000 000   **10** 2.98×10^{-5}
11 5.76×10^2 **12** 8×10^7
13 2.25×10^{14} **14** 1.65×10^{10}
15 1.87×10^{11}

Review Exercise 4

1 a) 125   **b)** 12   **c)** 225
2 a) $\frac{1}{10}$ **b)** 4   **c)** $\frac{1}{2}$
3 a) $\frac{1}{100}$ **b)** $\frac{1}{64}$ **c)** 4
4 a) 6^5 **b)** 3^3 **c)** 4^6
5 7.6×10^{-2}, 15 300, 3.2×10^8, 1.4×10^9
6 30
7 a) $\frac{1}{9}$ **b)** 6   **c)** 9   **d)** $\frac{27}{8}$
8 a) 1   **b)** $\frac{1}{16}$ **c)** 64
9 a) 64   **b)** 3   **c)** 12
10 a) 8.4×10^7 **b)** 2.1×10^{-5}
11 a) 125
   b) (i) 4.472 135 955
     (ii) 4.47
12 8.01×10^{10}
13 a) 4.2×10^5
   b) 2.4×10^{-6} grams
14 a) 1.44×10^6
   b) 1667
15 4.3×10^3 or -4.3×10^3

Internet Challenge 4

1 2×10^{30} **2** 5×10^9
3 4×10^6 **4** 6×10^3
5 2.25×10^8 **6** 2.998×10^8
7 5.5×10^{-7} **8** 10^{-9}
9 2.8×10^6 **10** 10^{11}

Chapter 5: Working with algebra

Starter Exercise

1 0   3 if misread as 3×1
2 2   0 if misread as $6 - (5 + 1)$
3 25   10 if misread as 5×2
4 14   20 if misread as $(2 + 3) \times 4$
5 9   7 if misread as $(4 + 10) \div 2$
6 9 -9 if misread as $-(3)^2$ or if an incorrect key
      sequence is used on a calculator

Exercise 5.1

1 8   **2** 200   **3** -14 **4** 36
5 20   **6** 14   **7** -22 **8** 100
9 36   **10** 10   **11** 7   **12** 3
13 a) 3625   **b)** 2500
14 a) 13   **b)** 3.61
15 a) 260   **b)** 18.5

Worksheet 5.1

1 a) 7   **b)** -1 **c)** -6 **d)** 7
   e) 8   **f)** 25   **g)** -13 **h)** -16
2 a) (i) $10°C$ **(ii)** $0°C$ **(iii)** $35°C$
   b) Tuesday by $2°C$
3 a) £200   **b)** £450
4 a) (i) 2   **(ii)** 30
   b) (i) 17   **(ii)** 108
   c) (i) 4   **(ii)** 18
   d) (i) 20   **(ii)** -15
   e) (i) 24   **(ii)** -4
   f) (i) -14 **(ii)** -63
   g) (i) 8   **(ii)** 99
5 a) 11   **b)** 5   **c)** 25

Exercise 5.2

1 k^3 **2** u^2 **3** x^5 **4** n^4
5 $2g^2$ **6** $5t^3$ **7** x^8 **8** y^7
9 z^9 **10** $10x^7$ **11** $24x^6$ **12** $2y^3$
13 $8y^6$ **14** $6z^5$ **15** $6x^9$ **16** $\frac{z^2}{2}$ or $\frac{1}{2}z^2$
17 x^8 **18** y^6 **19** $9z^6$ **20** $16x^{10}$
21 y^{30} **22** $16z^8$ **23** $4x^8$ **24** $125x^6$
25 $64x^3y^3$ **26** $36x^4y^2$ **27** $15x^5$ **28** $2y$
29 $27z^6$ **30** y **31** $40x^5$ **32** $3x^3$
33 $100x^{10}$ **34** x^6y^2 **35** x^9y^6 **36** $12x^5$

Worksheet 5.2

1 a^5 **2** y^6 **3** $2y^3$ **4** b^6
5 e^5 **6** $4f^2$ **7** h^6 **8** n^6
9 $9b^2$ **10** g^4 **11** $30p^2qr^3$ **12** $12p^3r^2$

Exercise 5.3

1 $7x + 20$ **2** $8x + 13$
3 $11x + 29$ **4** $8x + 25$
5 $22x + 16$ **6** $7x + 11$
7 $16x + 19$ **8** $7x + 4$
9 $17x + 1$ **10** $38x + 3$
11 $21x - 9$ **12** $2x + 10$
13 $3x + 3$ **14** $31x - 50$
15 $175x - 90$ **16** $6x + 12$
17 $6x + 33$ **18** $4x + 12$
19 $3x - 9$ **20** $5x - 3$

Worksheet 5.3

1 $10x + 15$ **2** $3k - 9$
3 $z^2 + 4z$ **4** $11n + 9$
5 $14a + 6$ **6** $5b^2 - 10b$
7 $5p + 4$ **8** $2q^2 + 11q - 3$

Exercise 5.4

1 $3x^2 + 13x + 12$ **2** $4x^2 + 13x + 10$
3 $2x^2 + 9x + 4$ **4** $2x^2 + 9x - 5$
5 $2x^2 - 4x - 6$ **6** $4x^2 + 24x + 11$
7 $3x^2 + 10x + 8$ **8** $6x^2 - 35x - 6$
9 $4x^2 + 4x - 15$ **10** $4x^2 + 51x - 13$
11 $6x^2 + 13x + 6$ **12** $8x^2 + 18x - 5$
13 $14x^2 - 15x - 9$ **14** $x^2 - 7x + 12$
15 $6x^2 + 5x - 6$ **16** $2x^2 - 11x + 15$
17 $x^2 - 49$ **18** $4x^2 - 9$
19 $x^2 + 6x + 9$ **20** $9x^2 - 24x + 16$

Answers

Worksheet 5.4

1 $ab + 2b + 4a + 8$
2 $c^2 + 5c + 6$
3 $y^2 + 3y - 4$
4 $x^2 + 2x - 3$
5 $q^2 + 2q - 8$
6 $x^2 + 10x + 25$
7 $m^2 - 16m + 64$
8 $2p^2 + 3p + 1$
9 $3n^2 + 7n + 2$

Exercise 5.5

1 $x(x + 6)$
2 $2x(x + 3)$
3 $2x(x + 3y)$
4 $y(y - 10)$
5 $2y(y - 5)$
6 $3x(2 + 3x)$
7 $4(3y^2 + 2)$
8 $4y(3y + 2)$
9 $g(f + 3g)$
10 $3y(3y + 4)$
11 $x^4(5x - 4)$
12 $6x^2(2 - x)$
13 $7a(2a + 3b)$
14 $5y(x + 2)$
15 $2(7 + 5y)$
16 $3xy(5 - 3x)$
17 $4y^2(2 - 5y)$
18 $4y(3y - 2)$
19 $6(1 + 3x^2)$
20 $3pq(4q^2 - 4q + 5)$

Worksheet 5.5

1 $4(y - 3)$
2 $4(4x + 1)$
3 $5(x + 5y)$
4 $3p^2(1 + 2p)$
5 $10s(s^2 - 6)$
6 $t(t - 1)$
7 $2ab(a + 2b^2)$
8 $7p^2q(2 - 3q)$
9 $3x(2x^2 - x + 3)$

Exercise 5.6

1 $(x + 1)(x + 7)$
2 $(x + 2)(x + 7)$
3 $(x + 3)(x + 2)$
4 $(x + 5)(x + 6)$
5 $(x + 8)(x + 2)$
6 $(x - 1)(x - 3)$
7 $(x - 2)(x - 5)$
8 $(x - 6)(x - 5)$
9 $(x - 2)(x - 1)$
10 $(x - 3)(x - 4)$
11 $(x + 4)(x - 1)$
12 $(x + 3)(x - 2)$
13 $(x + 2)(x + 3)$
14 $(x + 1)(x - 5)$
15 $(x + 3)(x - 4)$
16 $(x - 2)(x - 6)$
17 $(x + 8)(x + 4)$
18 $(x - 9)(x + 8)$
19 $(x + 3)(x + 4)$
20 $(x + 4)(x - 11)$

Worksheet 5.6

1 $(x + 4)(x - 4)$
2 $(y + 10)(y - 10)$
3 $(t + 3)(t - 3)$
4 $(x + 2)(x + 1)$
5 $(y + 2)(y + 3)$
6 $(a + 3)(a + 4)$
7 $(m + 2)(m + 5)$
8 $(p + 3)(p + 6)$
9 $(q + 4)(q - 2)$
10 $(u + 5)(u - 5)$

Exercise 5.7

1 $(2x + 1)(x + 1)$
2 $(2x + 3)(x + 1)$
3 $(2x + 1)(x + 2)$
4 $(3x - 2)(x - 1)$
5 $(3x + 1)(x - 1)$
6 $(5x - 1)(x + 1)$
7 $(2x + 1)(x - 1)$
8 $(5x + 1)(x - 2)$
9 $(3x - 2)(x - 2)$
10 $(2x - 1)(x + 6)$
11 $(2x - 3)(x - 3)$
12 $(2x + 1)(3x - 1)$
13 $(3x + 5)(2x - 5)$
14 $(6x + 1)(2x + 1)$
15 $(3x + 2)(5x + 3)$
16 $(2x - 1)(2x - 1)$
17 $(6x - 1)(x - 2)$
18 $(2x + 7)(x + 1)$
19 $(2x + 3)(2x + 3)$
20 $(2x + 3)(x - 3)$

Exercise 5.8A

1 $(x + 1)(x - 1)$
2 $(y + 11)(y - 11)$
3 $(x + 9)(x - 9)$
4 $(y + 20)(y - 20)$
5 $3(x + 5)(x - 5)$
6 $2(x + 3)(x - 3)$
7 $7(y + 3)(y - 3)$
8 $10(x + 2)(x - 2)$
9 $3(x + 3)(x - 3)$
10 $4(y + 5)(y - 5)$

Exercise 5.8B

1 $(x + 1)(x + 5)$
2 $x(x + 8)$
3 $(y + 4)(y + 11)$
4 $(x + 6)(x + 5)$
5 $x(x + 7)$
6 $(y - 2)(y + 5)$
7 $(4x - 1)(x - 2)$
8 $(y - 6)(y + 5)$
9 $(x - 1)(x - 2)$
10 $(x - 3)(x - 5)$
11 $(y + 4)(y - 4)$
12 $5y(x - 2y)$
13 $(2x - 1)(2x - 3)$
14 $7(y + 10)(y - 10)$
15 $(x + 6)(x - 4)$
16 $(2y + 5)(y - 2)$
17 $4z(z - 1)$
18 $(2x + 1)(x + 1)$
19 $3(x + 2)(x - 2)$
20 $(2x - 1)(x + 3)$

Exercise 5.9

1 $P = 3x$
2 $T = 30n$
3 $T = 60x + 5y$ or $T = 5(12x + y)$
4 $P = 500 + 10m$
5 $w = \dfrac{A}{l}$
6 a) $n + 11$ b) $T = 2n + 11$
7 a) $15x$ b) $T = 15x + 25y$
8 a) £15 b) $C = 5 + 2n$
9 a) 18 MB b) $S = 128 - 0.3n$ c) 426
10 a) $(a - 2)(b - 2)$ b) $V = (a - 2)(b - 2)$

Worksheet 5.9

1 a) $m + 10$ b) £28
2 a) $H + 9$ b) $H - 3$
  c) $2H - 6$ d) $H + 2$
  e) (i) 92 cm (ii) 104 cm (iii) 184 cm
3 a) $P = 2W + 2L$ b) $A = LW$ or $L \times W$
  c) 500 m d) 10 000 m^2
4 a) $C = 60 + 25H$ b) £122.50
5 $T = 60x + 100y$

Exercise 5.10

1 $r = \dfrac{A}{\pi l}$
2 $u = v - at$
3 $a = \dfrac{v - u}{t}$
4 $h = \dfrac{3V}{\pi r^2}$
5 $m = \dfrac{E}{c^2}$
6 $x = \dfrac{y - 3}{4}$
7 $x = 5(y - 3)$ or $x = 5y - 15$
8 $x = 5y - 3$
9 $h = \dfrac{2A}{b}$
10 $c = \sqrt{\dfrac{E}{m}}$
11 $y = \dfrac{A - x^2}{4x}$
12 $R = \dfrac{P}{I^2}$
13 $x = \dfrac{y}{m} + a$
14 $a = \dfrac{v^2 - u^2}{2s}$
15 $r = \sqrt{\dfrac{A}{4\pi}}$
16 $x = \sqrt{y + 9}$
17 $y = \sqrt{x^2 - z^2}$
18 $b = \dfrac{V}{ac}$
19 $r = \sqrt[3]{\dfrac{3V}{4\pi}}$
20 $u = \sqrt{v^2 - 2as}$

Review Exercise 5

1 a) 24 b) 32 c) 31 d) -40
2 a) 18 b) -8 c) 36 d) 10
3 a) -1 b) 13 c) 10
4 x^7
5 $3x^6$
6 $12x^5$
7 $5y^3$
8 $4z$
9 $3xy^3$

10 x^6
11 $25x^2y^4$
12 $9x^2$
13 $30x^6$
14 $10x^3$
15 $2y^2$
16 $7x + 16$
17 $5y + 13$
18 $8z - 7$
19 $13x + 5$
20 $10x - 14$
21 $2x + 8$
22 $30x - 2$
23 $2x - 16$
24 $10x$
25 12
26 $x^2 + 6x + 5$
27 $y^2 + 12y + 35$
28 $2z^2 + 9z + 4$
29 $x^2 - x - 20$
30 $2x^2 + 7x - 15$
31 $2x^2 - 3x + 1$
32 $3x^2 - 7x - 6$
33 $4x^2 - 9$
34 $x^2 - 16$
35 $5x^2 - 5$
36 $2x(12x + 5)$
37 $4y(4x - 5y)$
38 $(x + 7)(x + 3)$
39 $(y + 1)(y + 1)$ or $(y + 1)^2$
40 $(z + 8)(z - 8)$
41 $(2y - 1)(y + 5)$
42 $(2x - 1)(x - 4)$
43 $2x(6x - 5)$
44 $(2y + 3)(y + 2)$
45 $4(x + 3)(x - 3)$
46 $T = 26x + 19y$
47 **a)** $T = 5x + 3y$ **b)** $x = y$
48 **a)** $5n + 2(10 - n)$ **b)** $3n + 20$
49 $r = \dfrac{C}{2\pi}$
50 $a = \dfrac{2(s - ut)}{t^2}$
51 $r = \sqrt{\dfrac{A}{\pi}}$
52 $l = \dfrac{gT^2}{4\pi^2}$ or $l = g\left(\dfrac{T}{2\pi}\right)^2$
53 **a)** $12x$ **b)** $12x + 10y$
54 $w = ph + b$
55 $C = 20 - 4n$
56 **a)** $8p - 3q$ **b)** x^5 **c)** $2(2x + 3)$
    d) $x^2 + x - 6$ **e)** $2x^8$
57 **a)** y^7 **b)** $8x + 17$
    c) (i) $2(2a + 3)$ **(ii)** $3p(2p - 3q)$
58 **a)** Bryani, because $4 \times 3^2 = 4 \times 9 = 36$
    b) 64
59 **a)** $12a^5b^3$ **b)** $\dfrac{125p^9}{q^3}$ **c)** $\dfrac{4t^3}{u}$
60 **a)** $x^2 + 2x - 15$ **b)** $3a(2a - 3b)$
61 **a)** $x^2 + 2xy + y^2$ **b)** 25
62 **a)** p^9 **b)** $6q^6$
63 $x = \sqrt{5y - 4}$
64 **a)** k^3
    b) (i) $7x - 1$ **(ii)** $x^2 + 5xy + 6y^2$
    c) $(p + q)(p + q + 5)$
    d) m^8
    e) $6r^3t^6$

Internet Challenge 5

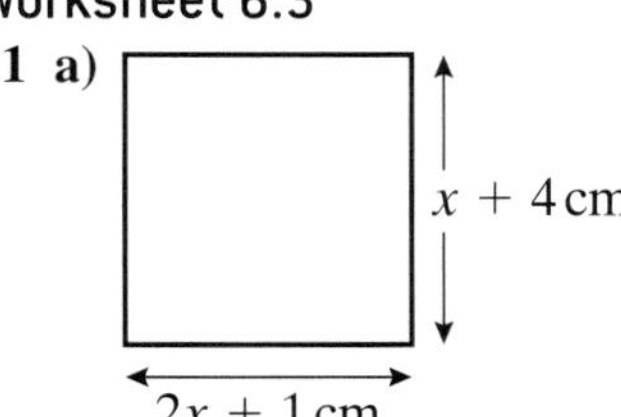

Chapter 6: Algebraic equations

Starter 6

Missing numbers clockwise from top:
2 14, 17 **3** 16, 7, 3 **4** 7, 10, 6

Exercise 6.1

1 Expressions: A, C
2 Equations: B, D, E, F, G, H, I, J (although D and H are actually identities)
3 Formulae: B, J
4 Identities: D, H

Exercise 6.2

1 4
2 7
3 9
4 30
5 12
6 2
7 4
8 -9
9 $2\frac{1}{2}$
10 $\frac{6}{5}$
11 16
12 -3
13 $\frac{4}{7}$
14 $\frac{8}{3}$
15 6
16 17
17 $\frac{1}{14}$
18 0
19 $\frac{5}{4}$ or $-\frac{5}{4}$
20 12 or -12

Exercise 6.3

1 3
2 1
3 4
4 -2
5 3
6 5
7 2
8 -2
9 0
10 11
11 $\frac{3}{4}$
12 $\frac{7}{5}$
13 $\frac{9}{2}$
14 $\frac{5}{8}$
15 $-\frac{1}{2}$
16 -2
17 $-\frac{1}{2}$
18 $\frac{5}{3}$
19 -5
20 $\frac{7}{3}$

Worksheet 6.3

1 **a)**

    b) The width and height of a square are the same.
    c) $x = 3$
    d) 7 cm by 7 cm
2 **a)** Check students' diagrams.
    b) 11 cm by 11 cm
3 **a)** Check students' diagrams.
    b) 14 mm by 14 mm
4 **a)** Check students' diagrams.
    b) 9.5 cm by 9.5 cm
5 **a)** Check students' diagrams.
    b) 4 mm by 4 mm

6 a)

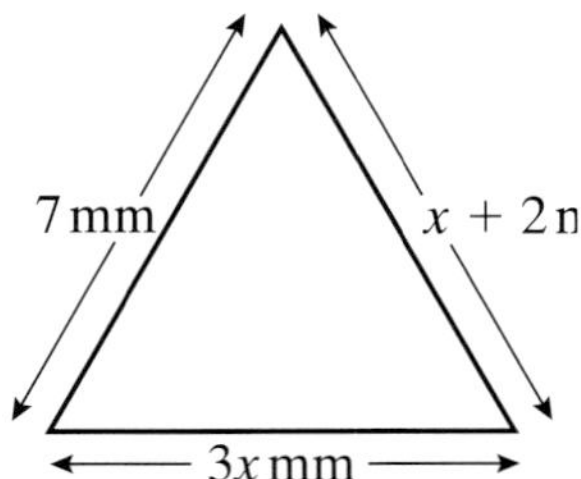

b) The perimeter is the distance all the way round an object.
c) $x = 2$
d) 4 mm, 6 mm, 7 mm
7 a) Check students' diagrams.
  b) 2 cm, 2 cm, 5 cm
8 a) Check students' diagrams.
  b) 3 m, 3 m, 4 m
9 a) Check students' diagrams.
  b) 5.5 cm, 3 cm, 5.5 cm
10 a) Check students' diagrams.
  b) 8 m, 12 m, 5 m
11 a) The angles in a triangle add up to $180°$.
  b) $x = 21$
  c) $42°, 84°, 54°$
12 $55°, 80°, 45°$
13 $50°, 60°, 70°$
14 $47.5°, 82.5°, 50°$
15 $48°, 66°, 66°$

Exercise 6.4

1 -3	**2** 7	**3** 2
4 6	**5** 10	**6** 4
7 4	**8** 3.5	**9** -4
10 4	**11** $\frac{1}{2}$	**12** 7

13 $2(x + 12) = 4x$, leading to $x = 12$
14 $n + 4 = 2(n - 5)$, leading to $n = 14$
15 a) $7x + 5 = 5(x + 7)$
  b) 15

Worksheet 6.4

1 1	**2** 2	**3** 8	**4** 4
5 2	**6** 3	**7** 1	**8** 2
9 5	**10** 4	**11** -1	**12** -2
13 -1	**14** -2	**15** -3	**16** -2
17 -2.8	**18** -2		
19 no unique solution		**20** -4	
21 $1\frac{1}{2}$	**22** $\frac{1}{2}$	**23** $-1\frac{1}{2}$	**24** $\frac{1}{2}$
25 $2\frac{1}{2}$	**26** $4\frac{1}{2}$	**27** $\frac{4}{3}$	**28** $-\frac{5}{8}$
29 $-\frac{1}{2}$	**30** $1\frac{2}{7}$	**31** 3	**32** 2
33 3	**34** -1	**35** $1\frac{4}{3}$	**36** -3
37 5	**38** $\frac{5}{8}$	**39** 0	**40** -9

Exercise 6.5

1 5	**2** $2\frac{1}{2}$	**3** 4	**4** 2
5 5	**6** 7	**7** -3	**8** 3
9 -1	**10** 5		

Worksheet 6.5

1 a) 6
  b) $6 \times \dfrac{x}{2} + 6 \times \dfrac{x}{3} = 6 \times 5$
    $3x + 2x = 30$
  c) $x = 6$

2 $x = 4$	**3** $x = 1$	**4** $x = 10$
5 $x = 20$	**6** $y = 30$	**7** $a = 12$
8 $b = 12$	**9** $y = 8$	**10** $y = 12$

11 a) 10
  b) $10 \times \dfrac{y}{5} + 10 \times \dfrac{y}{2} = 10 \times 7$
    $2y + 5y = 70$
  c) $y = 10$

12 $x = 16$	**13** $a = 4$	**14** $b = 4$
15 $x = 20$	**16** $a = 4$	**17** $x = 10$
18 $b = 48$	**19** $x = 8$	**20** $y = -20$

Exercise 6.6

1 4.3	**2** 1.4	**3** 1.8
4 4.5	**5** 3.4	**6** 1.6
7 4.1	**8** 1.3	**9** 3.1

10 a) If $x = 0$, $3x^2 - x - 1 = -1 \; (< 0)$
    If $x = 1$, $3x^2 - x - 1 = 1 \; (> 0)$
    So $3x^2 - x - 1 = 0$ somewhere between $x = 0$ and $x = 1$
  b) 0.77   **c)** -0.43

Worksheet 6.6

1 a)

x-value	Value of $x^2 + x$	Result
1	2	too low
2	6	too low
3	12	too low
4	20	too high

b) 3.5

2 a)

x-value	Value of $x^3 + 2x$	Result
0	0	too low
1	3	too low
2	12	too low
3	33	too low
4	72	too low
5	135	too high

b) 4.2

3 3.8	**4** 3.4	**5** 2.4
6 2.3	**7** 4.3	**8** 4.8

Review Exercise 6

1 a) formula   **b)** expression
  c) equation   **d)** identity

2 6	**3** -2	**4** 7
5 $\frac{1}{3}$	**6** $\frac{7}{4}$	**7** $\frac{8}{5}$ or $-\frac{8}{5}$
8 4	**9** 9 or -9	**10** 20
11 4	**12** 4	**13** 2
14 -2	**15** $\frac{7}{4}$	**16** -1
17 5	**18** -2	**19** 0
20 $\frac{5}{3}$	**21** 4	**22** -1
23 9	**24** 0	**25** 6
26 2	**27** 2	**28** 5
29 6	**30** $10\frac{2}{3}$	**31** 3.3

32 2.3

33 4.8

34 a) 1.6   **b)** -0.6

35 Glenn should have written -8 instead of $+8$ in the second line.
He would then get a final answer of 4.

36 Seyi is right.

37 a) $p = 3$ **b)** $r = -11$

38 $\frac{6}{5}$

39 a) 8   **b)** $6\frac{1}{2}$ **c)** $\frac{5}{8}$

40 a) 3   **b)** -2 **c)** $4\frac{1}{5}$

41 3.3

42 23

43 a) $3\frac{1}{2}$ **b)** 7

44 a) $x + 2$ **b)** $4x + 14$ **c)** $1\frac{1}{2}$

45 4.2

Internet Challenge 6

1 German

2 Braunschweig, 30 April 1777

3 77 years

4 Construction of the heptadecagon.

5 He added them in pairs: $1 + 100, 2 + 99, \ldots$ gives $101 \times 50 = 5050$.

6 Göttingen

7 Discovery of Ceres, the first known asteroid.

8 True (discovered by Gauss).

9 A polynomial of degree n will have exactly n solutions. For example, the equation $x^3 + 4x^2 + x - 6 = 0$ is of degree 3 (it has an x^3 term) and has 3 solutions. Note that some of the solutions may be duplicates, and some may only exist if you use complex numbers (which were also developed by Gauss).

10 Demagnetise it.

11 de Moivre; Normal distribution

12 Numbers containing two parts, a real part and a complex part based on the (imaginary) square root of minus 1.

13 When told his wife was dying.

14 The prince of mathematicians.

15 1855, Göttingen

16 Heptadecagon; no

17 A heptadecagon has 17 sides.

Chapter 7: Graphs of straight lines

Starter 7

1 Hint: The two squares that remain do not have to be the same size. Try removing two of the matches that meet in the centre of the original pattern.

2 Hint: Move the horizontal match half of its length to the right (or left).

3 Hint: Begin by removing the two upper left matches; place one of them to complete the fourth side of a square that now forms the face of the new fish.

4 Hint: Drag the right hand match slightly out, so a small square is formed where the matches meet.

Exercise 7.1

1 A $(1, 2)$, B $(-1, -2)$, C $(-2, 3)$, D $(3, -2)$, E $(-3, 0)$

2 a) M   **b)** $(0, 1)$ **c)** J   **d)** K
   e) $(6, -1)$ **f)** N   **g)** D

3

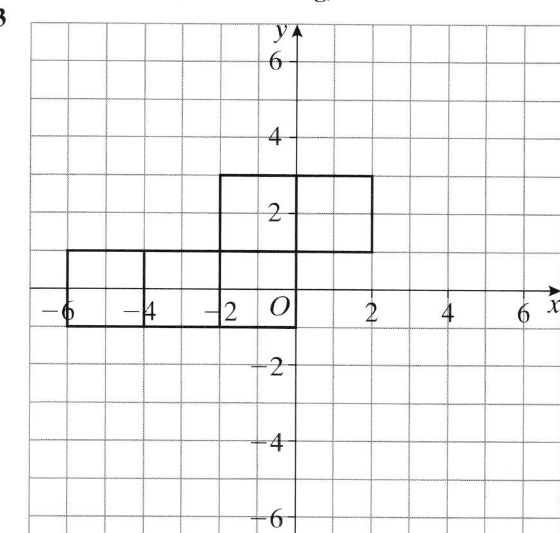

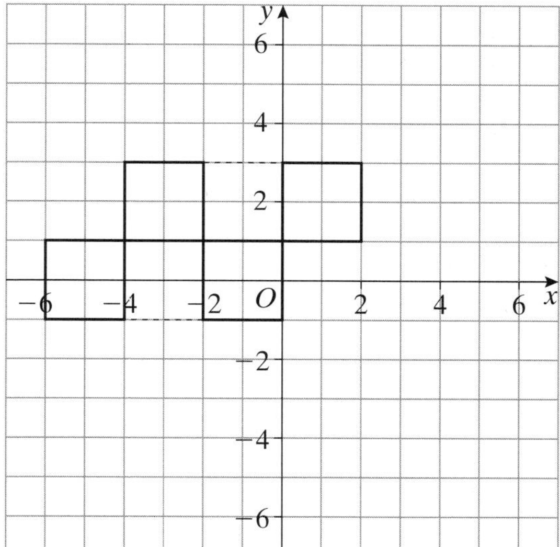

4

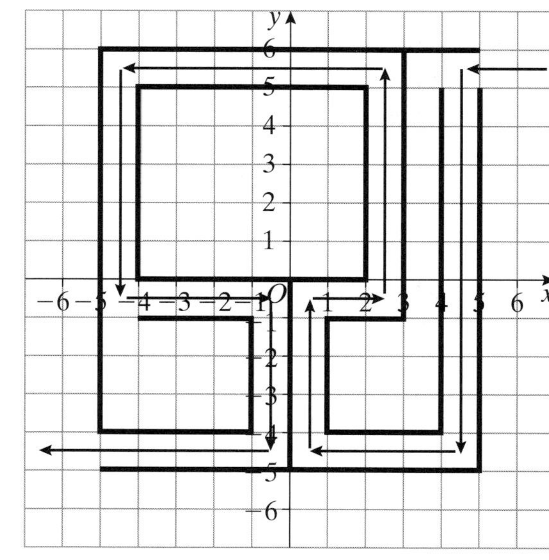

Exercise 7.2

1

x	-4	0	4
y	-7	1	9

$y = 2x + 1$

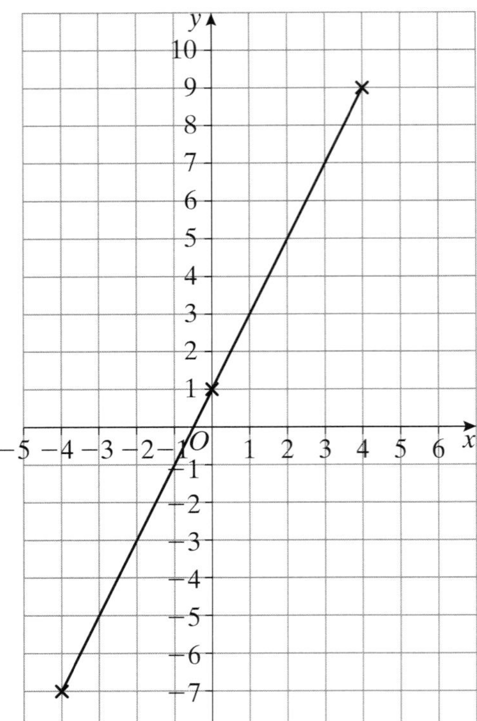

2

x	-5	0	5
y	-1	4	9

$y = x + 4$

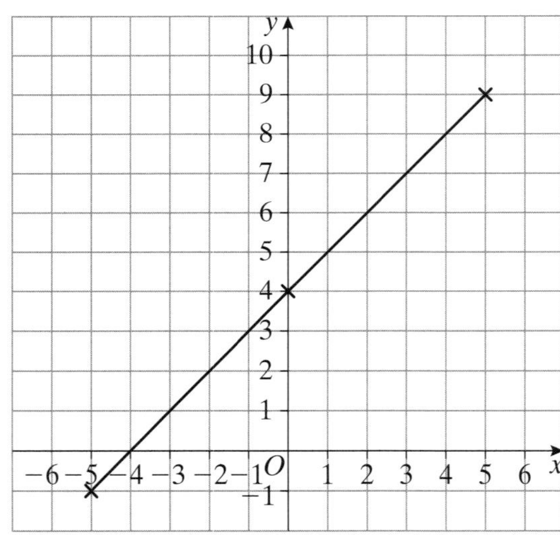

3

x	-4	0	5
y	-13	-1	14

$y = 3x - 1$

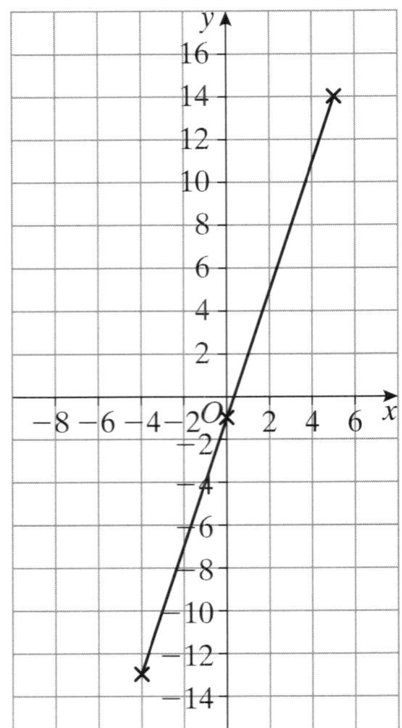

4

x	-2	0	1
y	-7	-3	5

$y = 2x - 3$

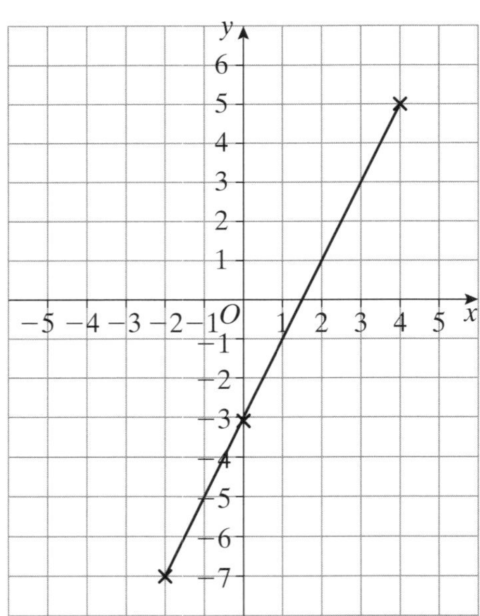

5

x	-6	0	4
y	1	4	6

$y = \frac{1}{2}x + 4$

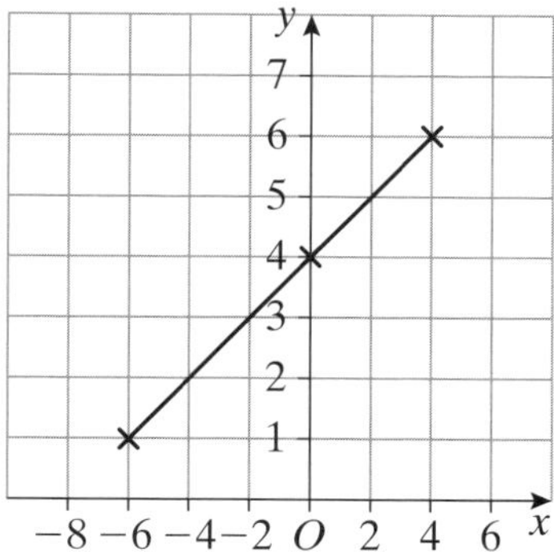

6

x	-5	0	5
y	-4	1	6

$y = x + 1$

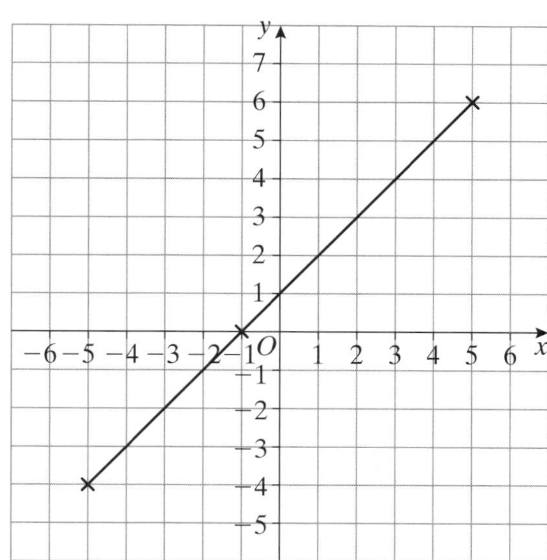

7

x	0	5	10
y	10	5	0

$x + y = 10$

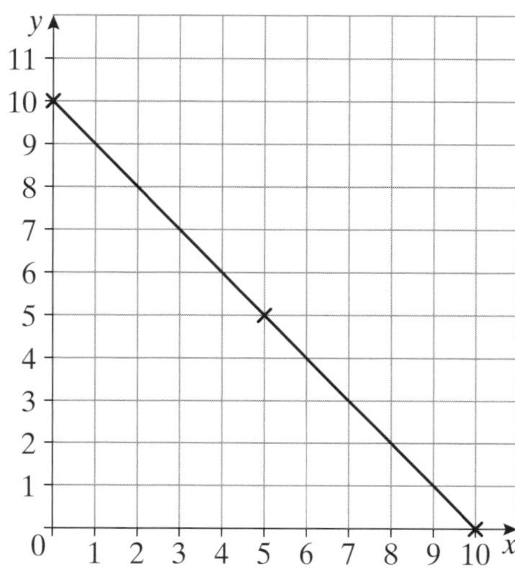

8

x	-2	0	4
y	9	5	-3

$2x + y = 5$

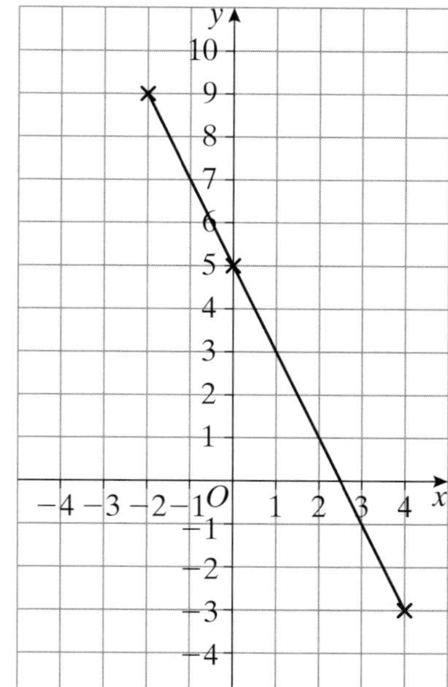

9

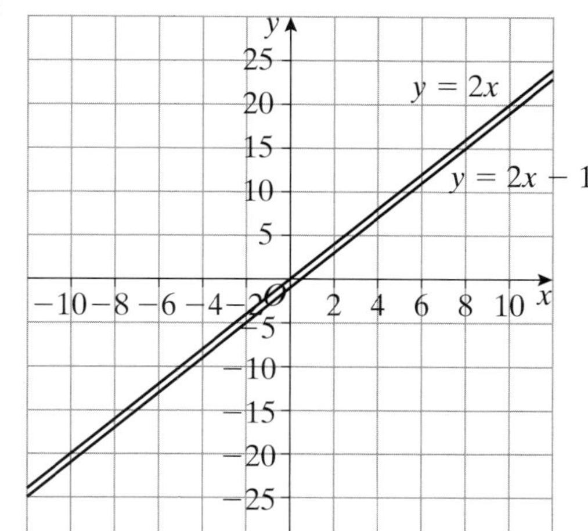

The graphs of the two lines are parallel.

10

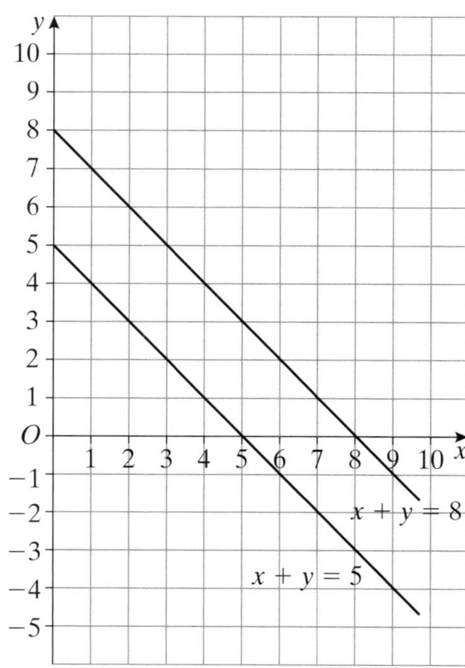

The graphs of the two lines are parallel.

Exercise 7.3

1 $m = 3, c = 1$ 2 $m = 1, c = 2$

3 $m = -2, c = 6$ 4 $m = \frac{1}{2}, c = 1$

5 $m = \frac{3}{4}, c = 2$ 6 $m = -\frac{2}{3}, c = 6$

7 $m = \frac{1}{2}, c = 1$ 8 $m = -\frac{1}{2}, c = 4$

Exercise 7.4

1 $y = 3x + 1$ 2 $y = x + 2$

3 $y = -2x + 6$ 4 $y = \frac{1}{2}x + 1$

5 $y = \frac{3}{4}x + 2$ 6 $y = -\frac{2}{3}x + 6$

7 $y = \frac{1}{2}x + 1$ 8 $y = -\frac{1}{2}x + 4$

9 a) P $(2, 2)$, Q $(8, 5)$

  b) $m = \frac{1}{2}, c = 1$

  c) $y = \frac{1}{2}x + 1$

10 a) $m = -\frac{1}{2}, c = 5$ b) $y = -\frac{1}{2}x + 5$

Exercise 7.5

1 a) $y = x - 6$ b) $y = -2x - 5$

  c) $y = \frac{1}{2}x + 1$ d) $y = \frac{1}{2}x + 2\frac{1}{2}$

  Lines c) and d) are parallel.

2 a) $y = 3x + 2$ is parallel to $y = 4 + 3x$.

    $y = 2x + 3$ is parallel to $y = 2x - 1$.

    $y = x + 2$ is parallel to $x - y = 5$.

    $2y = 8x - 3$ is parallel to $2y = 8x + 1$.

  b) The odd one out is $x + y = 2$; it is parallel to any line of the form $x + y = k$

3 a) $a = 5$

  b) $b = -5$, so the line has equation $y = 5x - 5$

4 a) $m = 4, c = 3$, so the line has equation $y = 4x + 3$

  b) $p = 15$

5 a) $y = 3x + 2$ b) $y = 3x - 5$

Review Exercise 7

1

x	-5	0	2
y	-1	4	6

$y = x + 4$

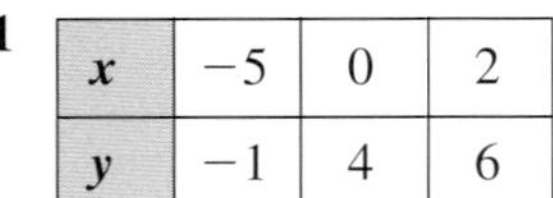

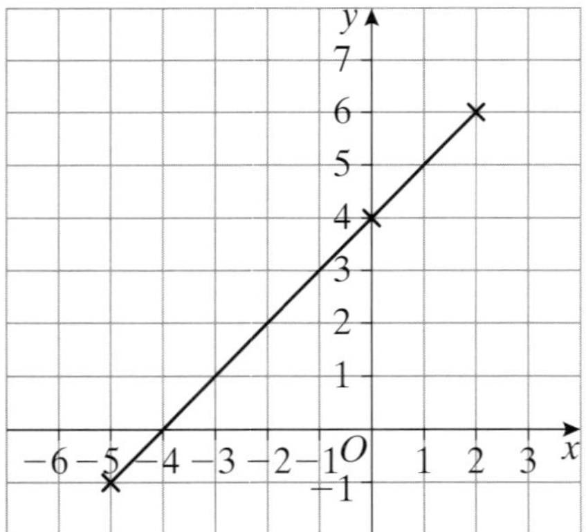

2

x	-2	0	6
y	0	1	4

$y = \frac{1}{2}x + 1$

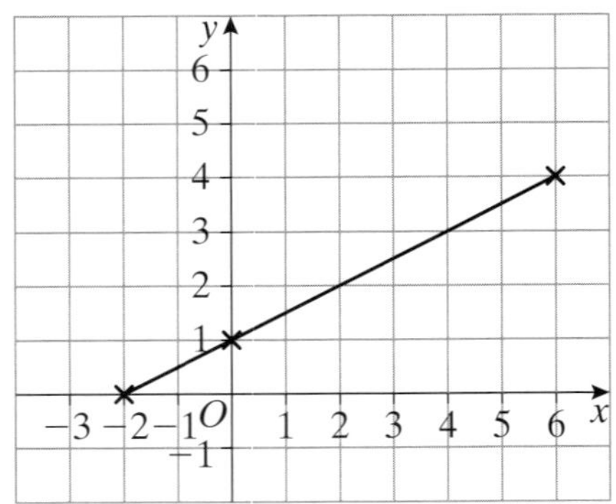

3

x	-6	0	6
y	-17	-5	7

$y = 2x - 5$

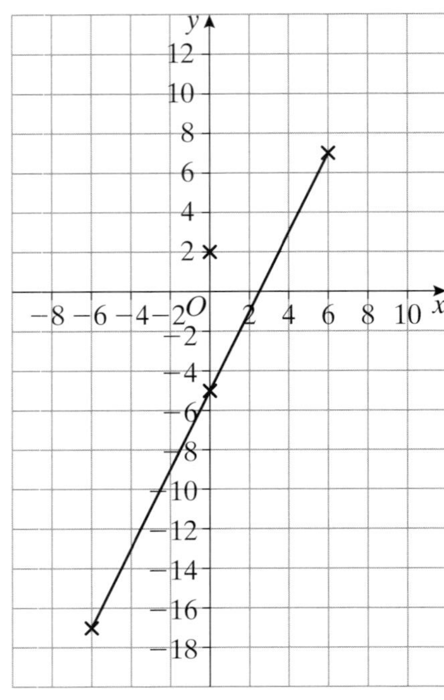

4

x	0	8	20
y	20	12	0

$x + y = 20$

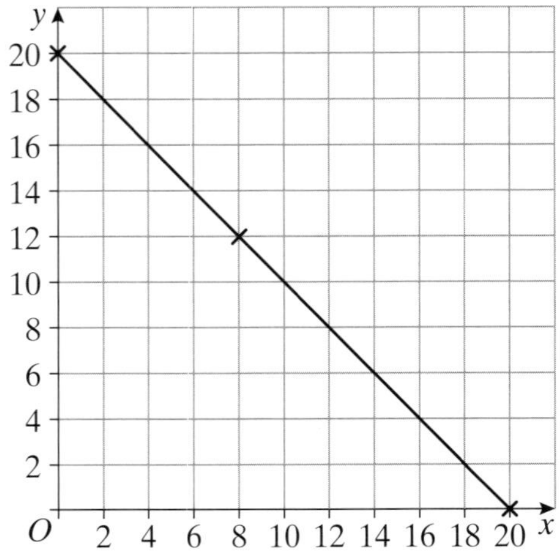

5 **a)** $m = \frac{1}{2}, c = 4 \quad y = \frac{1}{2}x + 4$
   b) $m = -1, c = 7 \quad y = -x + 7$
6 A: $y = 2x - 6$ B: $y = x + 2$
   C: $y = x$ D: $x = y + 6$
   E: $y = -\frac{1}{2}x + 5$ F: $x + y = 6$
7 $y = 4x + 7$
8 $y = 3x + 2$
9 $y = -2x + 5$
10 **a)** 8
    b) Any line of the form $y = \frac{1}{2}x + k$
    c) $x = 2y - 2$
11 **a)** $y = -\frac{1}{2}x + 3$
    b) $-\frac{1}{2}$
    c) $x + 2y = 14$ or $y = -\frac{1}{2}x + 7$
12 $y = 2x + 6$

Internet Challenge 7

1 Parallelogram
2 Trapezium
3 A prism whose faces are all parallelograms.
4 Blondie
5 Border between USA and Canada.
6 When any decision arises, all the possible outcomes occur, each in a separate 'parallel universe' hidden from the others.
7 'If a straight line crossing two straight lines makes the interior angles on the same side less than two right angles, the two straight lines, if extended indefinitely, meet on that side on which are the angles less than two right angles.'
   There are many other statements which are logically equivalent to the parallel postulate, including: 'Through a point not on a given line, exactly one line can be drawn in the plane parallel to the given line.'
8 On a (rather old) computer.
9 On a ski slope.
10 True
11 Yes (consider railway tracks going round a bend).
12 An electrician
13 **a)** All of them.
    b) Bristol, Dr Richard Gregory

Chapter 8: Simultaneous equations

Starter 8

Cherry = 5, lemon = 3, apple = 12, orange = 8, grapes = 7, banana = 4

Exercise 8.1

1 $x = 4, y = 1$ 2 $x = 3, y = 3$
3 $x = -1, y = 6$ 4 $x = -1, y = 5$
5 $x = -4, y = 1$ 6 $x = 2, y = 3$
7 $x = 10, y = 1$ 8 $x = 3, y = -1$
9 $x = 7, y = -2$ 10 $x = 4, y = 0$

Exercise 8.2

1 $x = 4, y = 3$ 2 $x = 1, y = -1$
3 $x = 2, y = 1$ 4 $x = 2, y = -1$
5 $x = 5, y = 2$ 6 $x = 1, y = 1$
7 $x = 5, y = -2$ 8 $x = 0, y = -2$
9 $x = 1, y = 4$ 10 $x = 2, y = 3$
11 $x = 1, y = -2$ 12 $x = 2, y = 1$

13 $x = 0, y = -1$ 14 $x = 10, y = -2$
15 $x = -1, y = -10$ 16 $x = 2, y = -2$
17 $x = 3, y = 1$ 18 $x = 4, y = -1$
19 $x = 2, y = 6$ 20 $x = -1, y = -3$
21 $x = 4, y = \frac{1}{2}$ 22 $x = 6, y = 1\frac{1}{2}$
23 $x = 3, y = -3$ 24 $x = 5, y = 0$
25 $x = 2, y = -7$ 26 $x = 3, y = 1\frac{1}{2}$
27 $x = \frac{1}{2}, y = -\frac{1}{2}$ 28 $x = 4, y = -1\frac{1}{4}$
29 $x = 7, y = -6$ 30 $x = 1\frac{1}{2}, y = -2\frac{1}{2}$
31 $x = 5, y = -3$ 32 $x = -2, y = -5$

Worksheet 8.2

1 $x = 2, y = 3$ 2 $x = 1, y = 3$
3 $x = \frac{1}{2}, y = 2$ 4 $x = 4, y = -2$
5 $x = 2\frac{1}{2}, y = 1$ 6 $x = -2, y = -3$
7 $x = -1\frac{1}{2}, y = 2$ 8 $x = -3, y = -3$
9 $x = 2, y = -2\frac{1}{2}$ 10 $x = -\frac{1}{2}, y = \frac{1}{2}$
11 $x = 4, y = 6$ 12 $x = 3, y = -4$
13 $x = -2, y = 3$ 14 $x = 2, y = 1\frac{1}{2}$
15 $x = -2\frac{1}{2}, y = 3$ 16 $x = 5, y = 4$
17 $x = 1, y = 2\frac{1}{2}$ 18 $x = -1, y = -1$
19 $x = 3, y = 7$ 20 $x = -2, y = -3$
21 $x = 1\frac{1}{2}, y = -2\frac{1}{2}$ 22 $x = 4, y = 4$
23 $x = \frac{1}{4}, y = 1$ 24 $x = 1\frac{1}{4}, y = -2$
25 $x = 5, y = 1$

Exercise 8.3

1
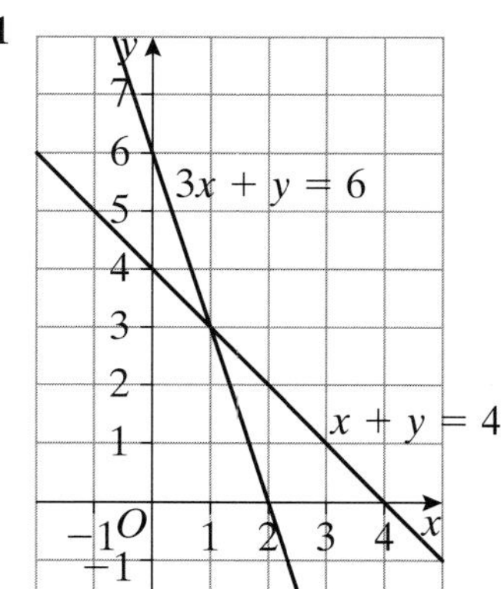

$x = 1, y = 3$

2
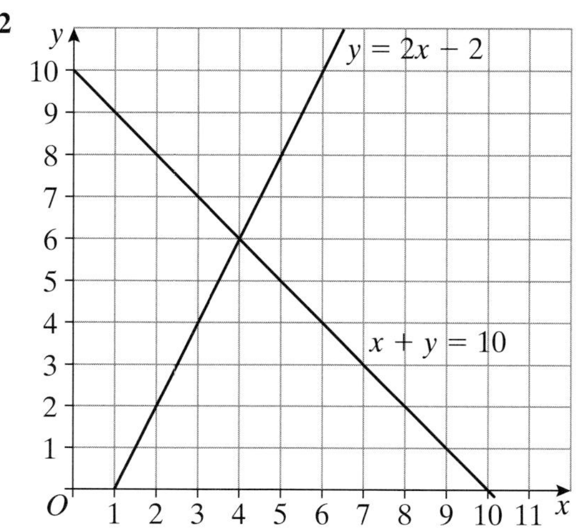

$x = 4, y = 6$

3

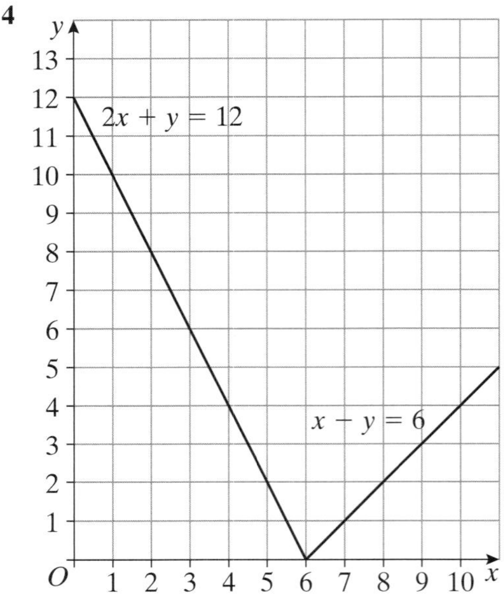

$x = 6, y = 2$

4

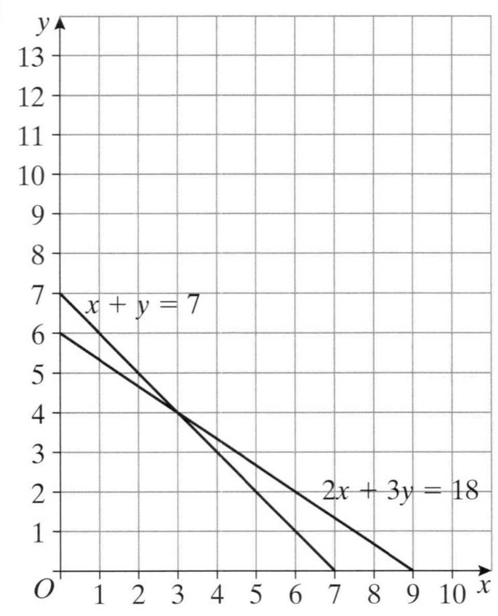

$x = 6, y = 0$

5

$x = 3, y = 4$

6

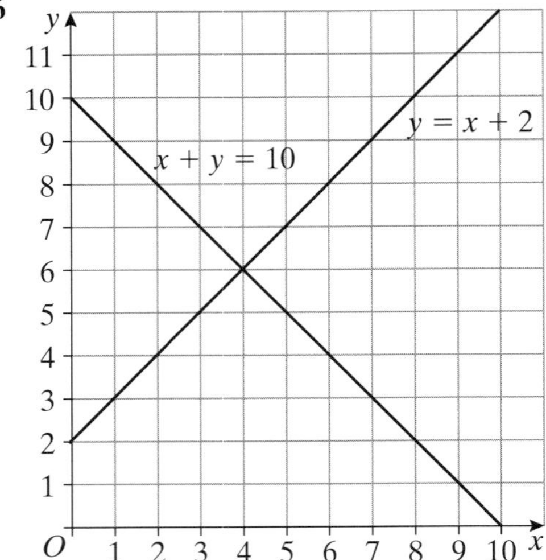

$x = 4, y = 6$

7

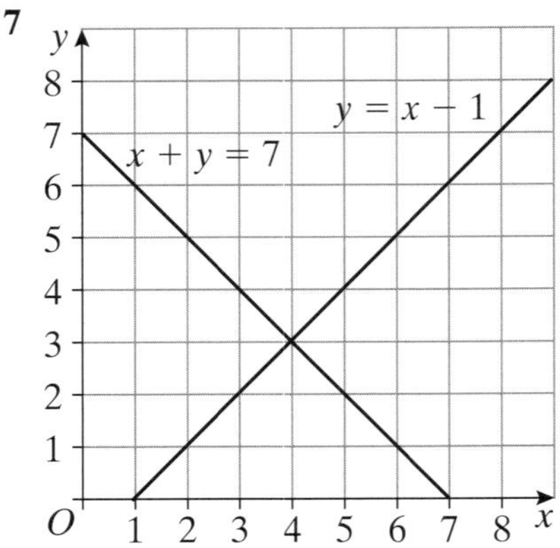

$x = 4, y = 3$

8

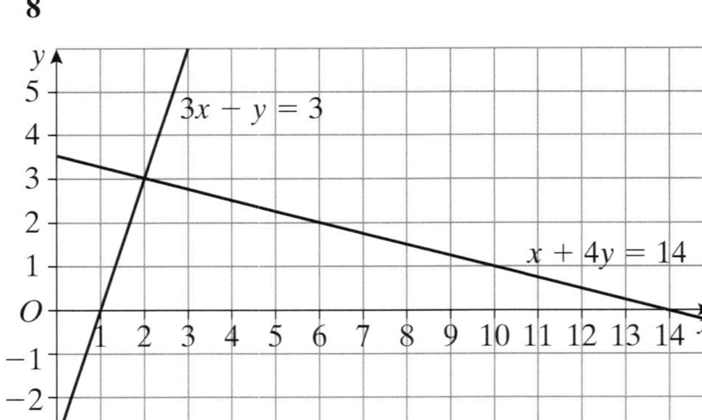

$x = 2, y = 3$

1

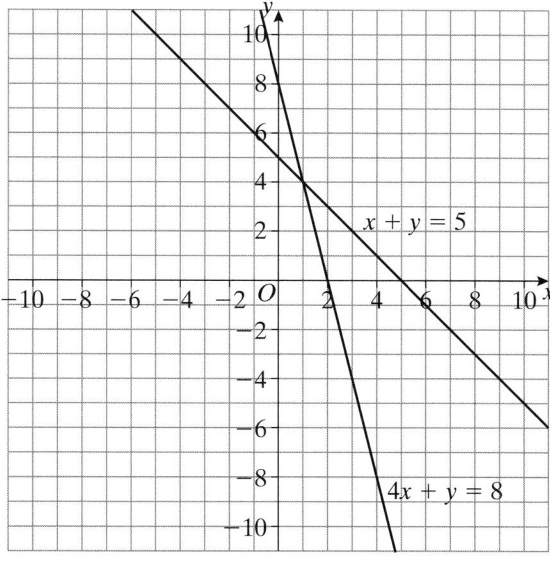

$x = 1, y = 4$

2

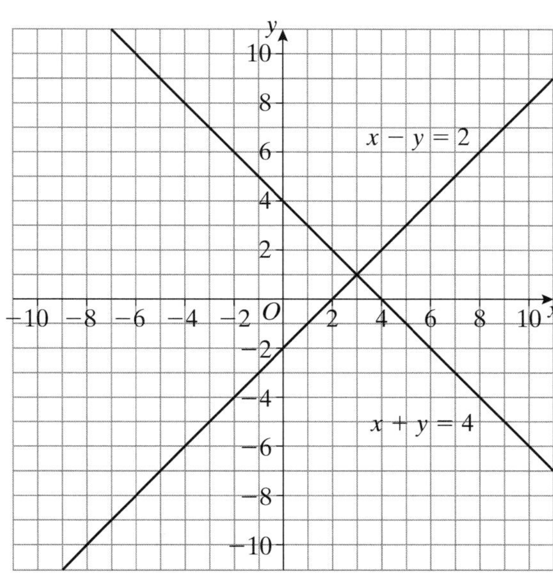

$x = 3, y = 1$

3

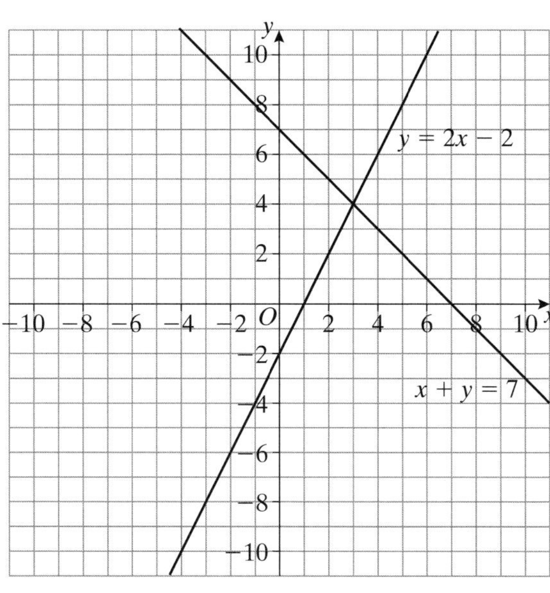

$x = 3, y = 4$

4

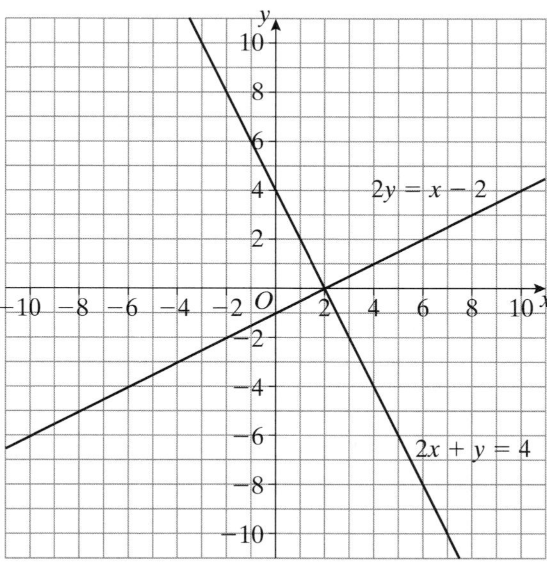

$x = 2, y = 0$

5

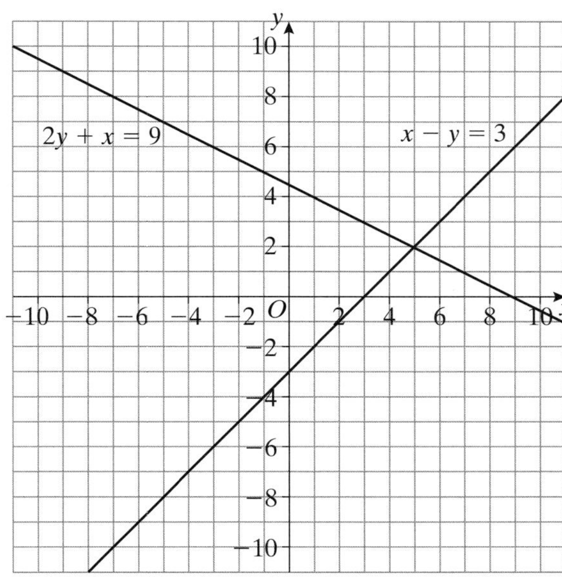

$x = 5, y = 2$

6

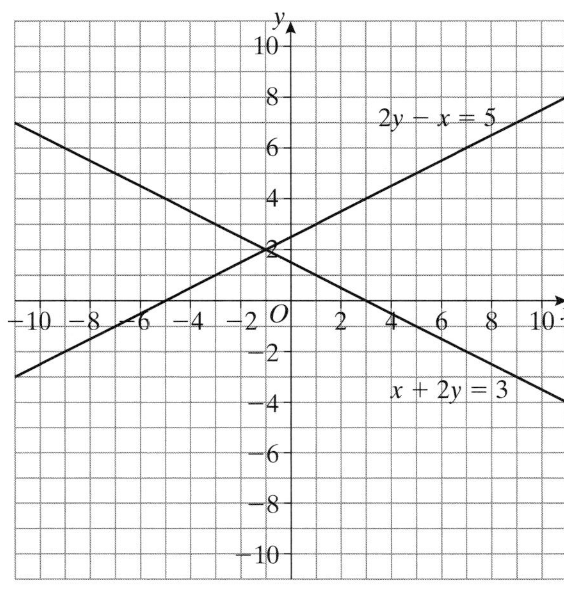

$x = -1, y = 2$

7

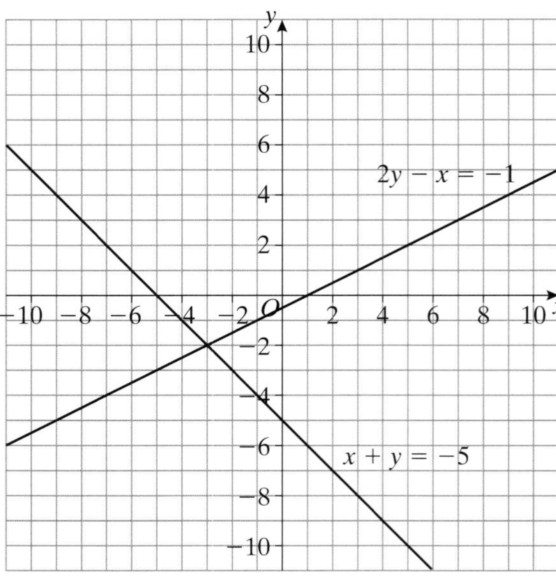

$x = -3, y = -2$

8

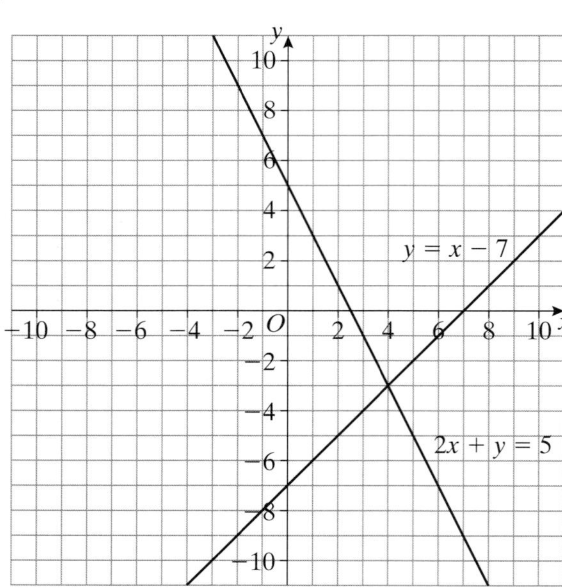

$x = 4, y = -3$

9

$x = 2.5, y = 3$

10

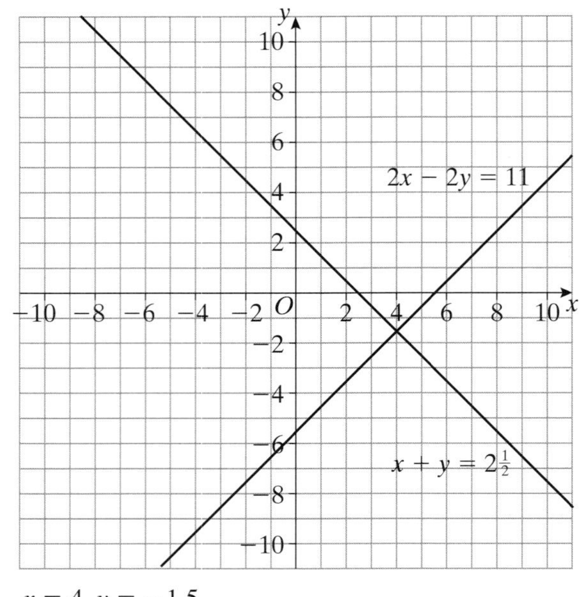

$x = 4, y = -1.5$

Exercise 8.4

1 a) $10x + 3y = 104$ $4x + y = 38$
 b) Shirt £5, jacket £18
2 a) $3x + 4y = 180$ $5x + 2y = 230$
 b) 40 passengers
3 a) If there are x A-level books and y GCSE books
 then $x + y = 160$ and $10x + 15y = 1800$
 b) 120 A-level books and 40 GCSE books
4 a) If there are x 2-litre cans and y 5-litre cans then
 $x + y = 500$ and $2x + 5y = 1420$
 b) 360 2-litre cans and 140 5-litre cans
5 Tomato plants 35 pence each, peppers 45 pence each

Worksheet 8.4

1 a) Tea 50p   **b)** Bacon roll £1.20
2 Seven 1p coins, nine 2p coins
3 Win 5 points
4 a) Adult £9   **b)** Child £5
5 Walk 3 mph, run 8 mph
6 a) Cup £3   **b)** Plate £5
7 a) CD £13   **b)** DVD £15
8 9 and 4
9 4 T-shirts
10 Dougal 4 and 8, friend 8 and 4
11 $x = 4, y = 2$
12 Pens 3, pencils 5
13 a) $5x + 2y = 190$ and $7x + 4y = 290$
 b) Train 30p, bus 20p
14 £1.20
15 a) $6x + 6y = 48$ **b)** $12x + 4y = 72$
 c) $x = 5, y = 3$ **d)** 13 cm by 11 cm

Review Exercise 8

1 $x = 4, y = 3$ **2** $x = 16, y = 0$
3 $x = -1, y = 5$ **4** $x = 1, y = -1$
5 $x = 4, y = 1$ **6** $x = 2, y = -1$
7 $x = 3, y = -2$ **8** $x = -1, y = -2$
9 $x = 5, y = \frac{1}{2}$ **10** $x = 2, y = -\frac{1}{2}$
11 $x = \frac{1}{2}, y = -4$ **12** $x = 2, y = -1\frac{1}{2}$

13

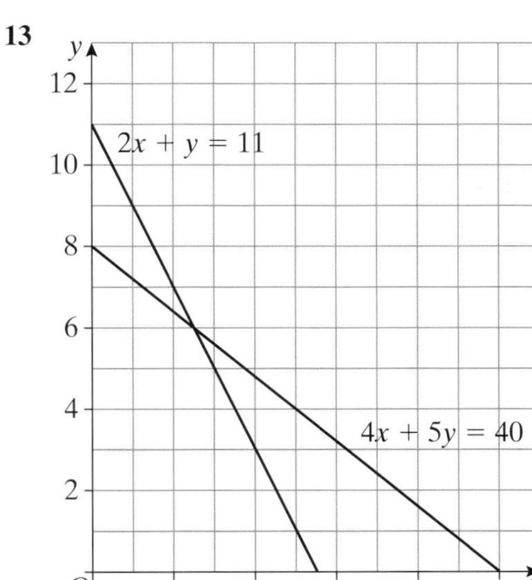

$x = 2\frac{1}{2}, y = 6$

14

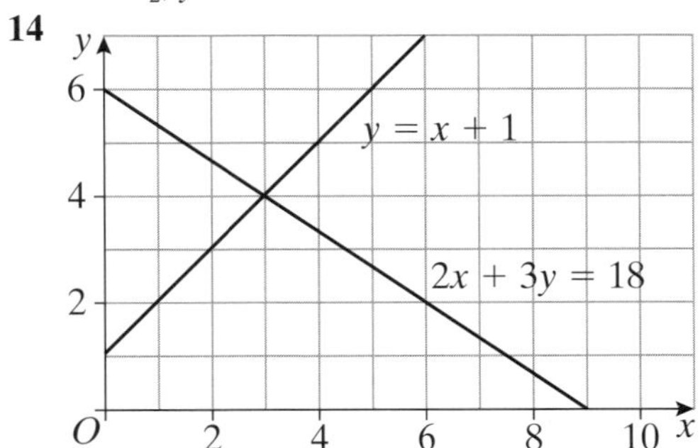

$x = 3, y = 4$

15

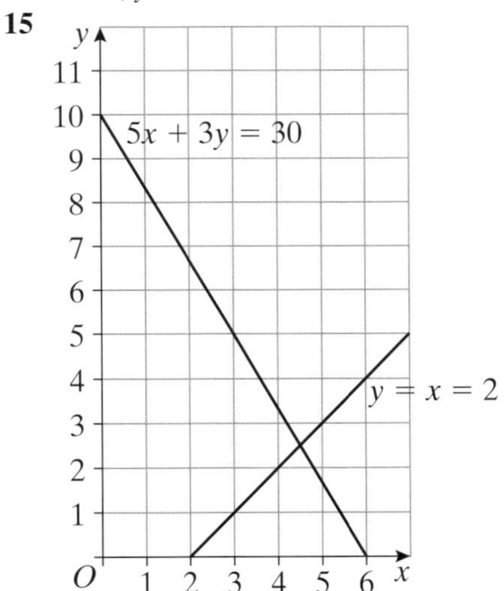

$x = 4.5, y = 2.5$

16 a) $5x + y = 207, 2x + 3y = 166$
 b) Cola 35p, orange 32p
17 a) $3c + 2s = 19, 4c + 5s = 30$
 b) $c = 5, s = 2$
 c) 42 minutes
18 a) $100x + 50y = 400, 150x + 100y = 650$
 b) $x = 3, y = 2$
 c) £10
19 $x = 2.5, y = -2$
20 $x = 4, y = -1$

Internet Challenge 8

1

1	15	14	4
12	6	7	9
8	10	11	5
13	3	2	16

2 A four by four magic square
3 Enter the number 1 in the middle of the bottom row.
 Work up through 2, 3, 4, …, moving right and down
 one cell each time; if this takes you outside the grid
 then move up or left (or both) by a number of squares
 equal to the dimension of the grid. If you reach a cell
 that is full then move up one cell.
 a) b) Check students' magic squares.
4 a)

52	61	4	13	20	29	36	45
14	3	62	51	46	35	30	19
53	60	5	12	21	28	37	44
11	6	59	54	43	38	27	22
55	58	7	10	23	26	39	42
9	8	57	56	41	40	25	24
50	63	2	15	18	31	34	47
16	1	64	49	48	33	32	17

5 a) The Loh-Shu
 b) It dates from 2800 BC, and so is nearly 4000 years
 old.

Chapter 9: Inequalities

Starter 9
The treasure is buried in the Swamp, at $(7, 9)$.

Exercise 9.1
1 4, 5, 6, 7, …   **2** 3, 4, 5, 6, …
3 …, $-1, 0, 1, 2, 3$ **4** …, $-1, 0, 1, 2$
5 1, 2, 3, 4, 5, 6   **6** 0, 1, 2, 3, 4, 5, 6
7 2   **8** 1
9 3, 4, 5   **10** 1, 2
11 1, 2, 3, 4, 5   **12** 1, 2
13 …, $-1, 0, 1, 2, 3, 4, 5$ **14** 4, 5, 6, 7, …
15 3, 4, 5   **16** 1, 2, 3, 4
17 1, 2, 3, 4   **18** 6, 7, 8, 9, 10
19 97, 98   **20** $-2, -1, 0, 1, 2$

Exercise 9.2
1 $x \geqslant 8$ **2** $x > 5$
3 $x < -3$ **4** $x < 2$
5 $x < 6$ **6** $x \leqslant 3$
7 $x \geqslant 5$ **8** $4 < x$
9 $2 \leqslant x$ **10** $x > -1$

11 $x \geqslant 4$ 12 $x < 4\frac{1}{2}$
13 $2 > x$ 14 $x \leqslant -4$
15 $x \leqslant 0$ 16 $-5 < x$
17 $x \geqslant 36$ 18 $x < 0$
19 $x \geqslant -1$ 20 $x < 3\frac{1}{2}$
21 $x \leqslant 14$ 22 $x > 5$
23 $x < 27$ 24 $x \geqslant 11$

Worksheet 9.2

1 $x < 4$ 2 $x > 8$
3 $x < 1\frac{1}{2}$ 4 $x < 2$
5 $x > 4$ 6 $x < 2$
7 $x > 1\frac{1}{2}$ 8 $x < 3$
9 $x \leqslant 2$ 10 $x > \frac{1}{2}$
11 $x < -2\frac{1}{2}$ 12 $x < 1\frac{1}{2}$
13 $x > -1$ 14 $x \geqslant 4$
15 $x > 1$ 16 $x > -1$
17 $x > 4$ 18 $x \geqslant 10$
19 $x \geqslant -0.75$ 20 $x < -1$
21 $x < 1$ 22 $x > 2$
23 $x \leqslant 4\frac{1}{2}$ 24 $x < -5$
25 $x < -3$ 26 $x \geqslant 10$
27 $x < 2$ 28 $x > \frac{1}{2}$
29 $x < -\frac{1}{2}$ 30 $x < 0$

Exercise 9.3

1 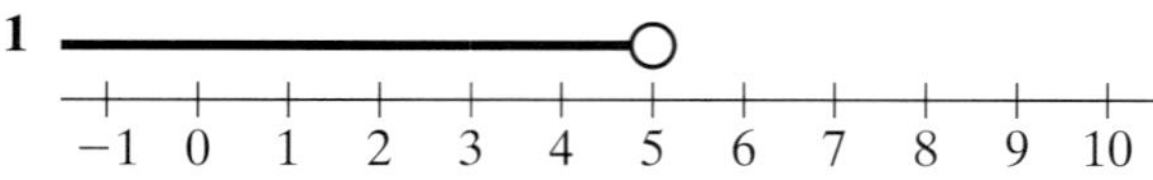

2 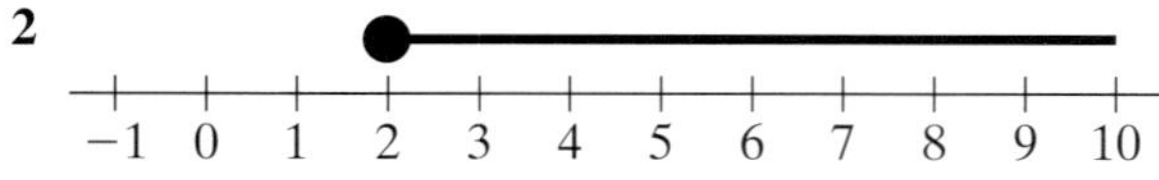

3 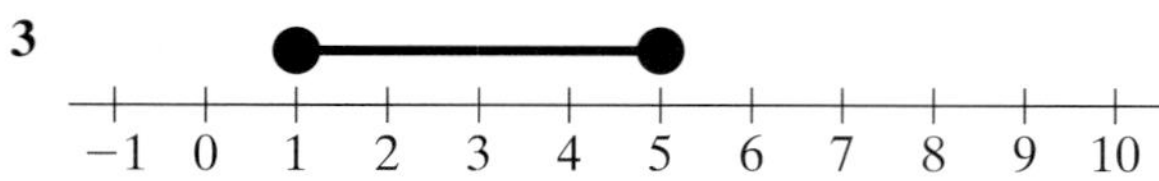

4 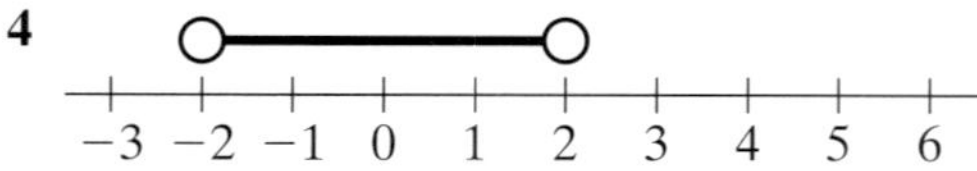

5 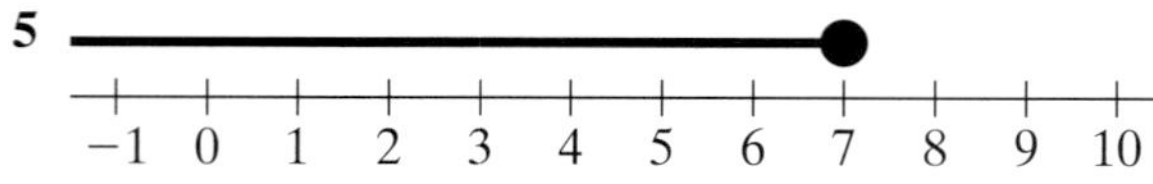

6 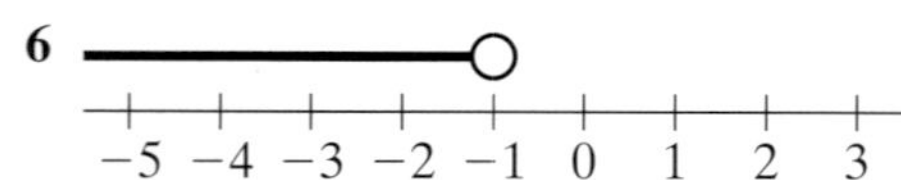

7 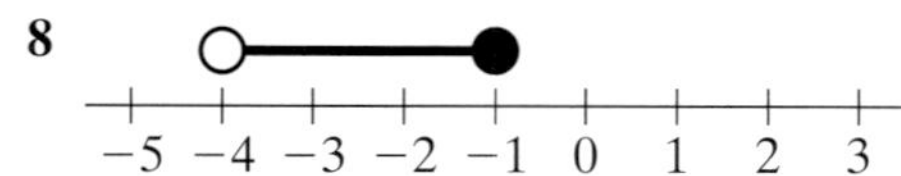

8 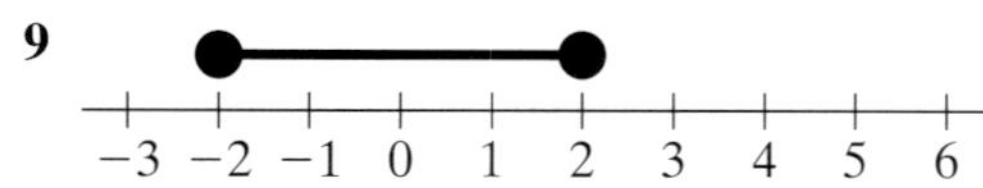

9 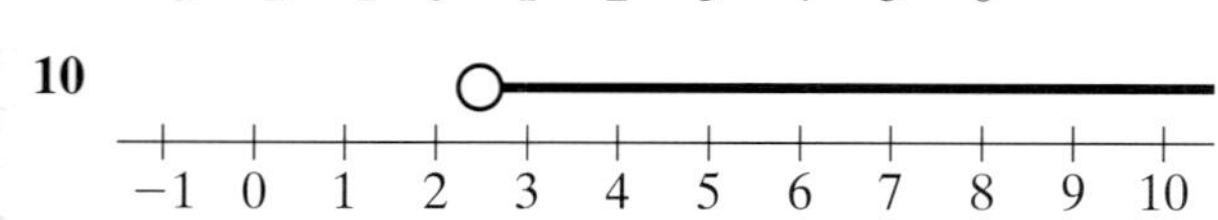

10 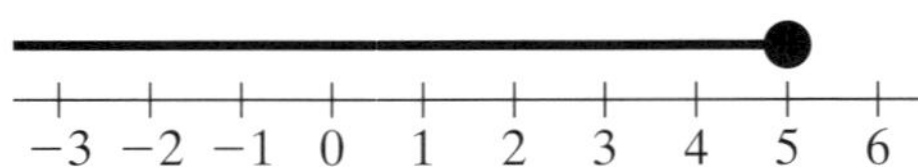

11 $x < 14$

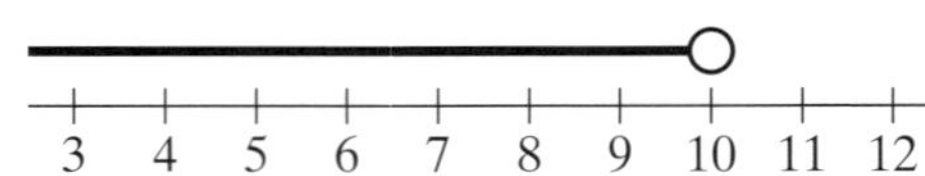

12 $x \leqslant 5$

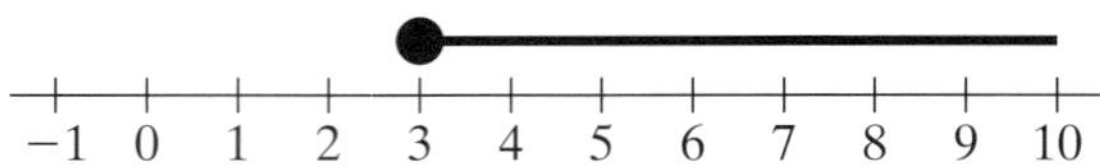

13 $x < 10$

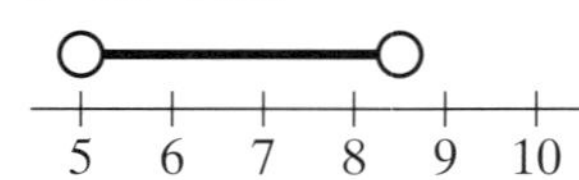

14 $x \geqslant 3$

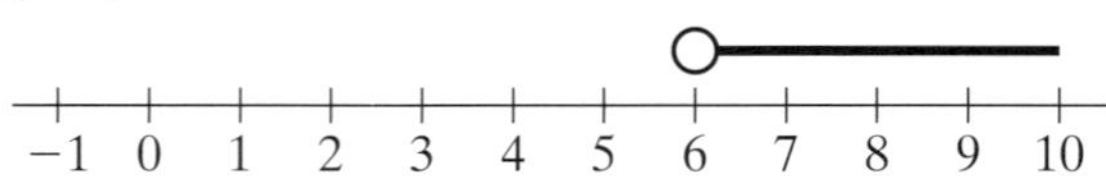

15 $5 < x < 8.5$

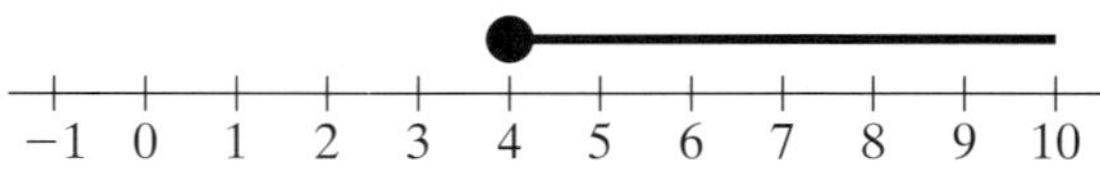

16 $x > 6$

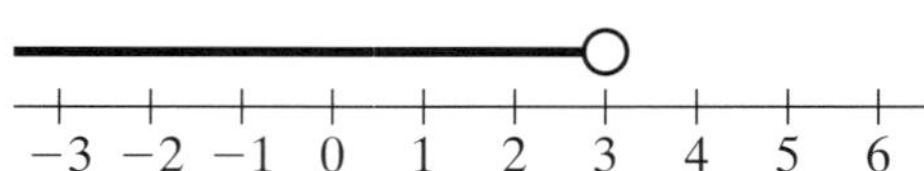

17 $x \geqslant 4$

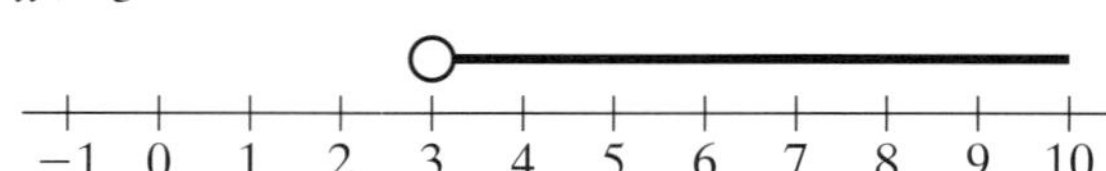

18 $x < 3$

19 $x > 3$

20 $x \leqslant -1$

Worksheet 9.3

1 $8, 9, 10, 11$
2 $5 < n < 9$; $6, 7, 8$
3 $7 \leqslant n < 14$; $7, 8, 9, 10, 11, 12, 13$
4 9
5 $-6, -5, -4, -3, -2, -1$
6 $2n + 1 \geqslant 7, n \geqslant 3$
7 $14 \leqslant x \leqslant 17, 15 \leqslant x \leqslant 19$; £15, £16, £17
8 $2(n - 3) > 8$; 8
9 $2x + 18 \geqslant 36$; 9
10 $-1, 0, 1, 2, 3$
11 $6, 12$
12 5
13 a) $-7 < T < -3$
   b) $-5 < T < -2$
   c) -4
14 $3, 4$
15 $-1\frac{1}{2} < x < 1\frac{2}{3}$

1

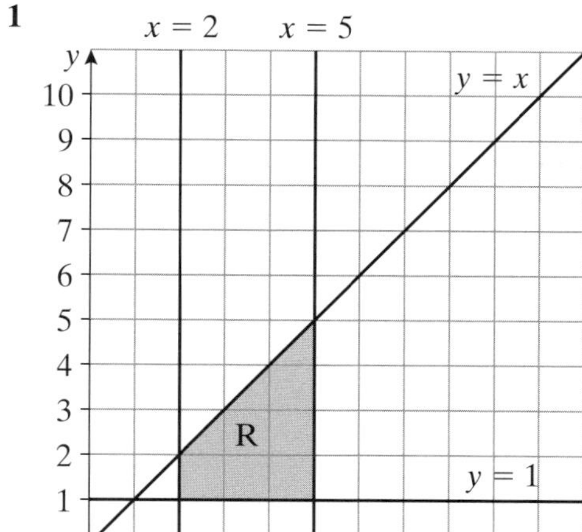

2

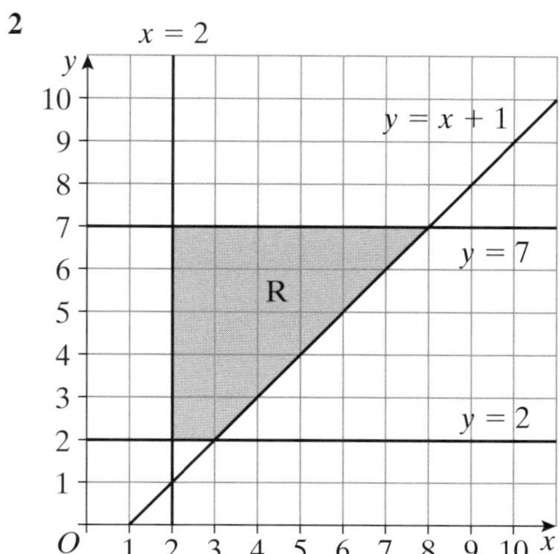

3

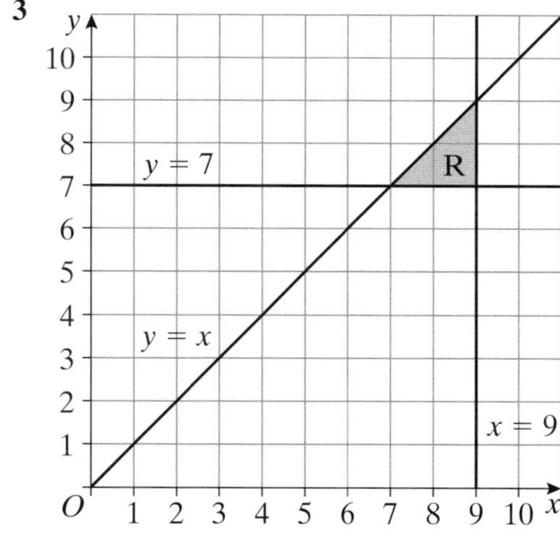

4

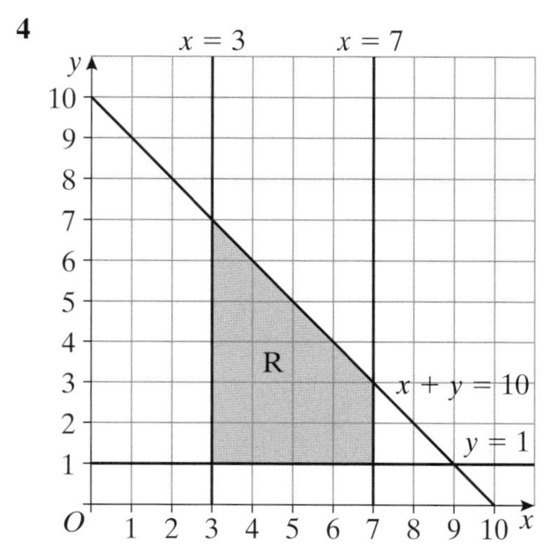

5 $x = 0$

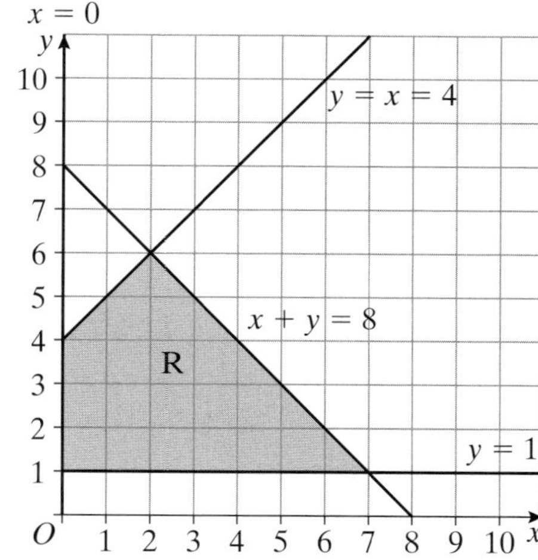

6 a) $L_1: y = 5$ **b)** $L_1: y > 5$
   $L_2: x = 7$ $L_2: x < 7$
   $L_3: y = x + 3$ $L_3: y < x + 3$

Worksheet 9.4A

1 a) b)

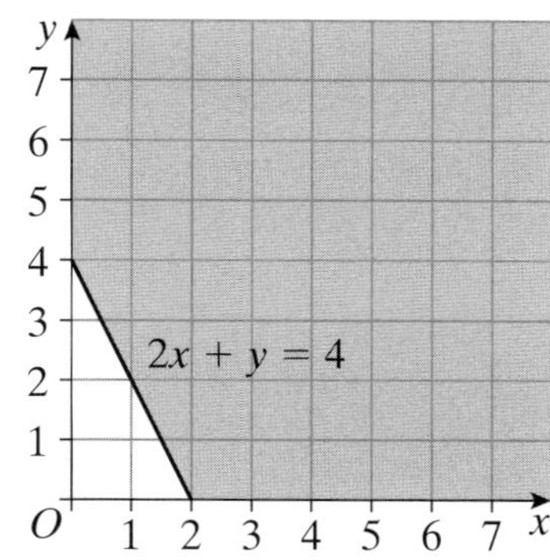

2

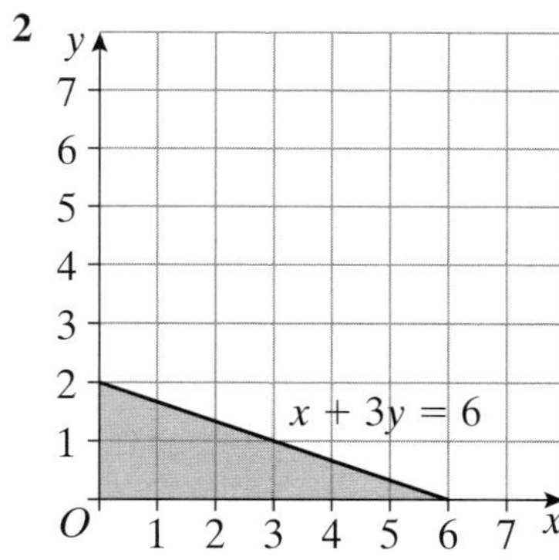

3

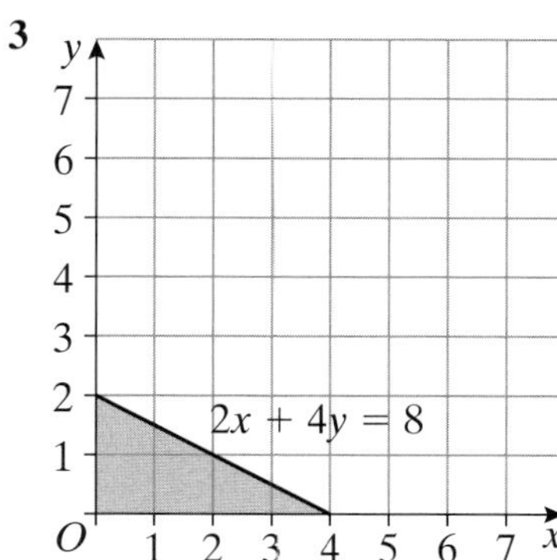

4

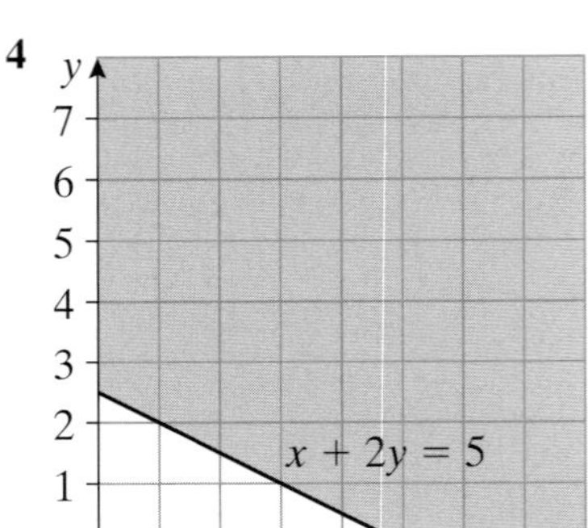

5

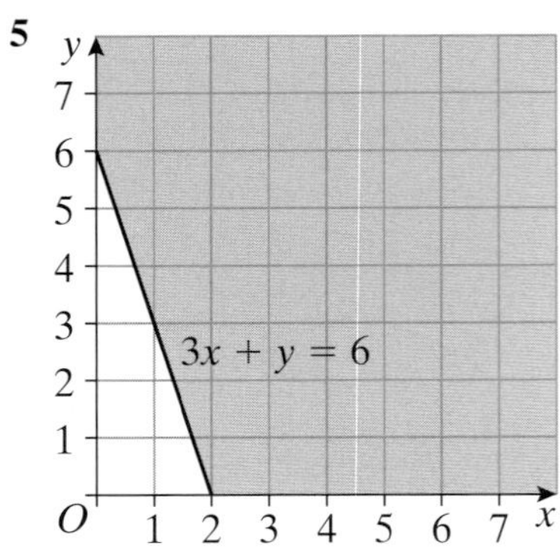

6

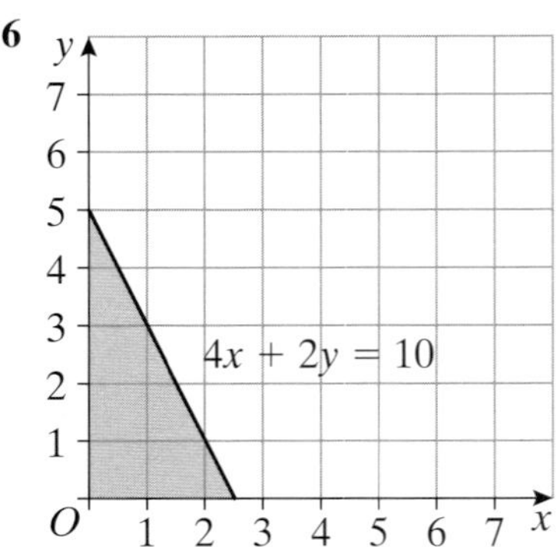

7

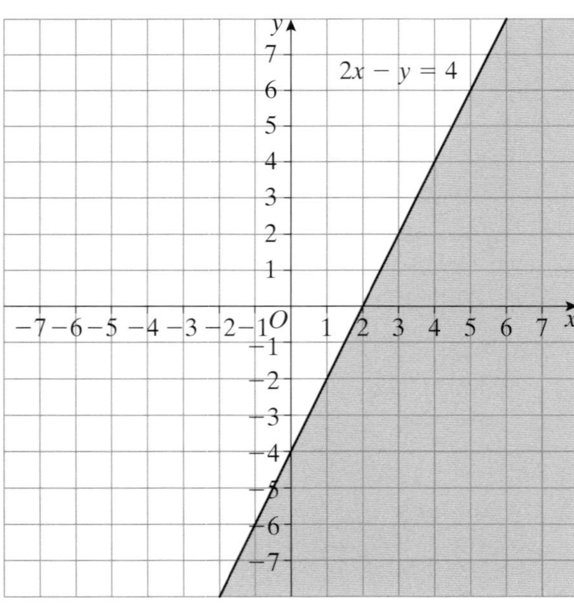

8

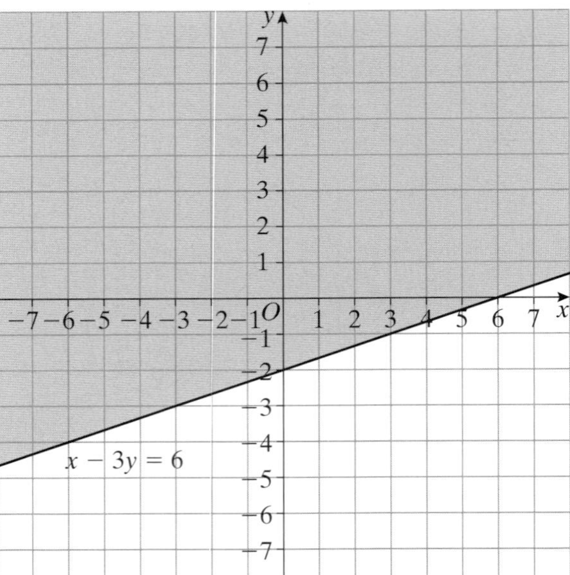

9

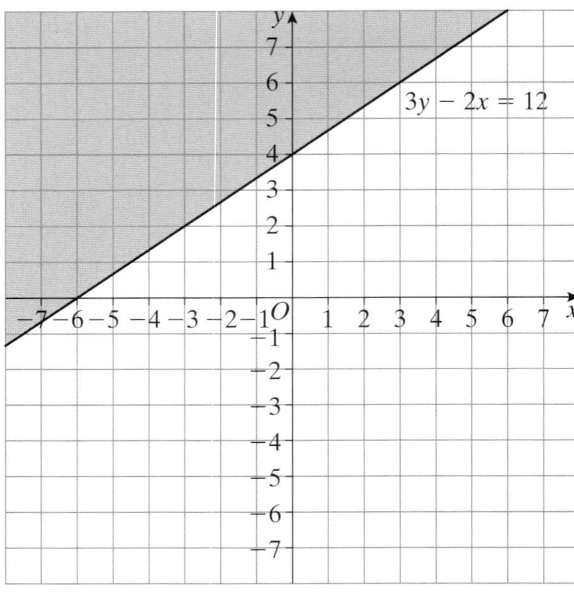

10

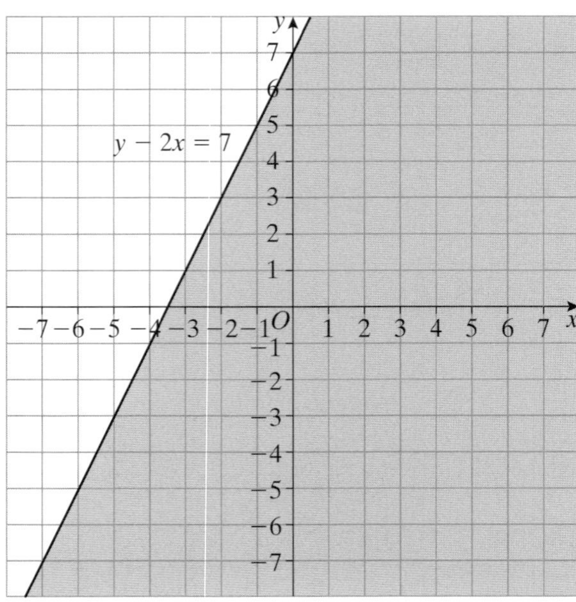

11

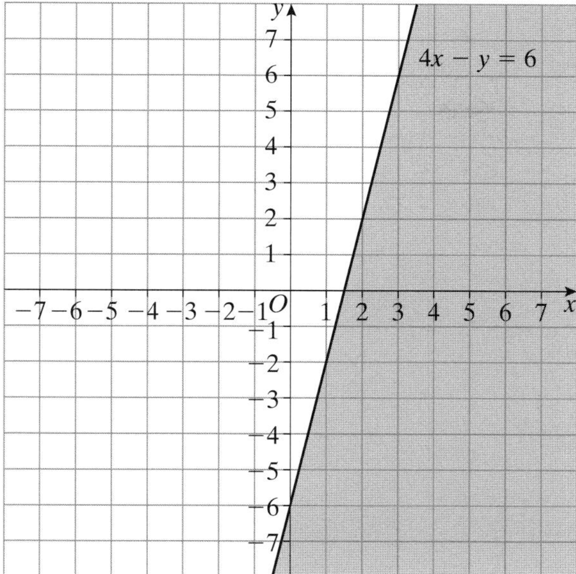

12

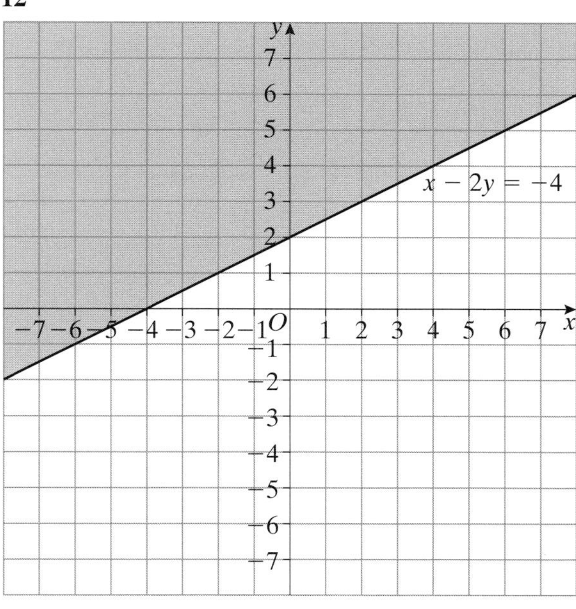

Worksheet 9.4B

1 a) b)

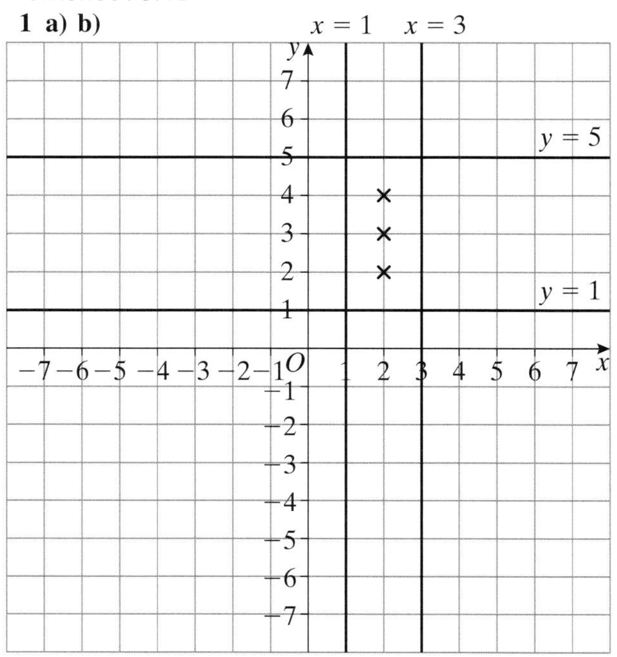

c) (2, 2), (2, 3), (2, 4)

2 a)

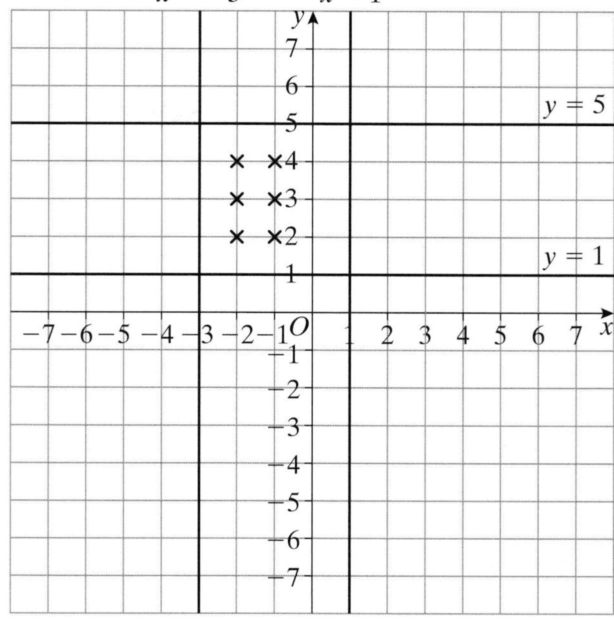

b) (−1, 2), (−1, 3), (−1, 4), (−2, 2), (−2, 3), (−2, 4)

3 a)

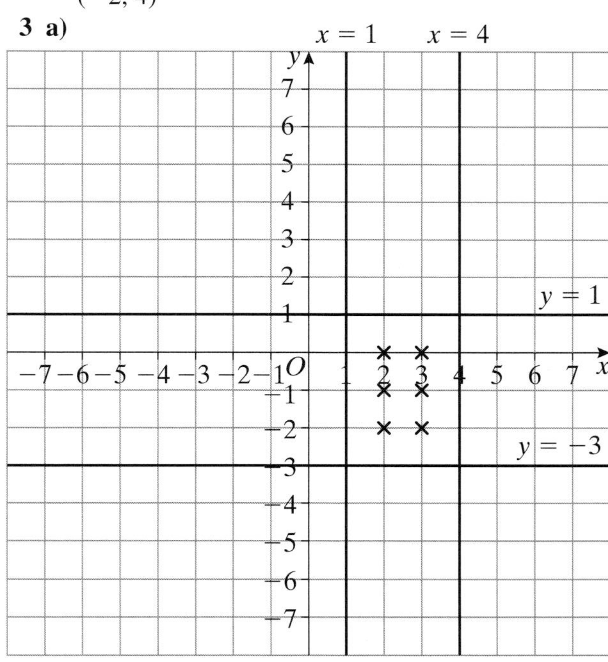

b) (2, −2), (2, −1), (2, 0), (3, −2), (3, −1), (3, 0)

4 a)

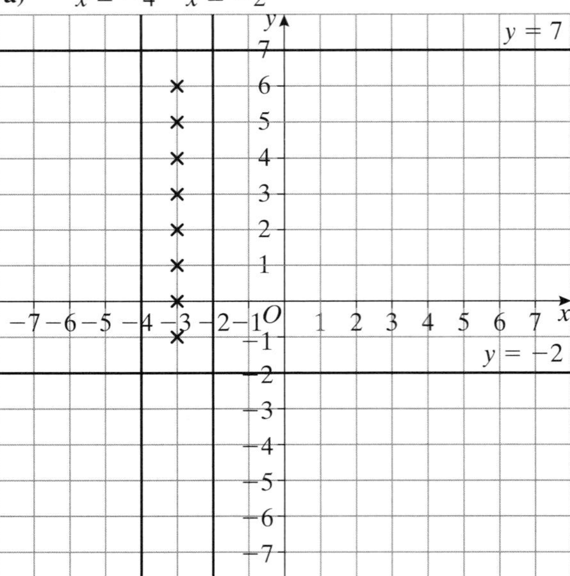

b) $(-3, -1)$, $(-3, 0)$, $(-3, 1)$, $(-3, 2)$, $(-3, 3)$, $(-3, 4)$, $(-3, 5)$, $(-3, 6)$

5 a)

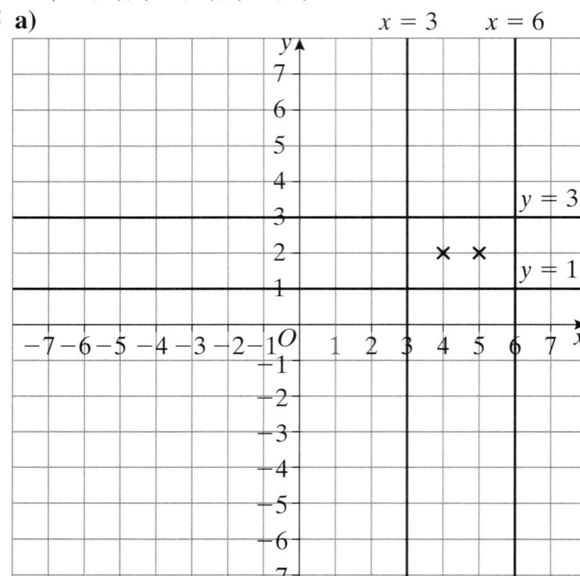

b) $(4, 2)$, $(5, 2)$

6 a)

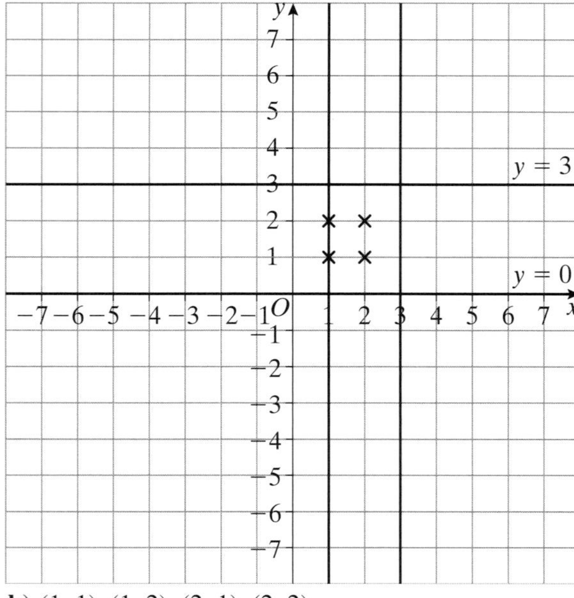

b) $(1, 1)$, $(1, 2)$, $(2, 1)$, $(2, 2)$

7 a)

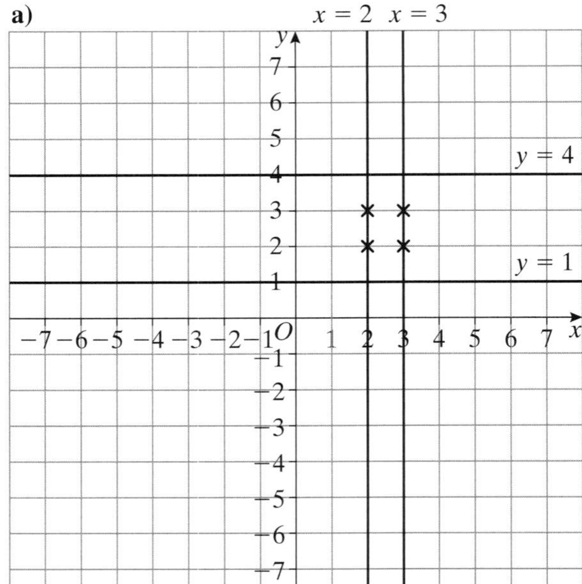

b) $(2, 2)$, $(2, 3)$, $(3, 2)$, $(3, 3)$

8 a)

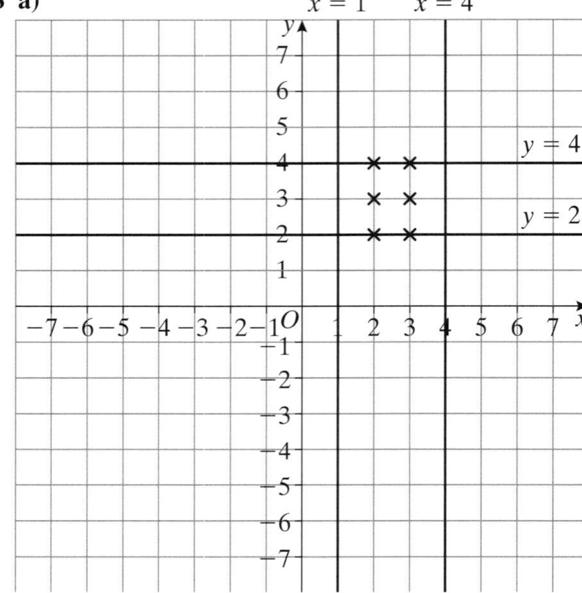

b) $(2, 2)$, $(2, 3)$, $(2, 4)$, $(3, 2)$, $(3, 3)$, $(3, 4)$

9 a)

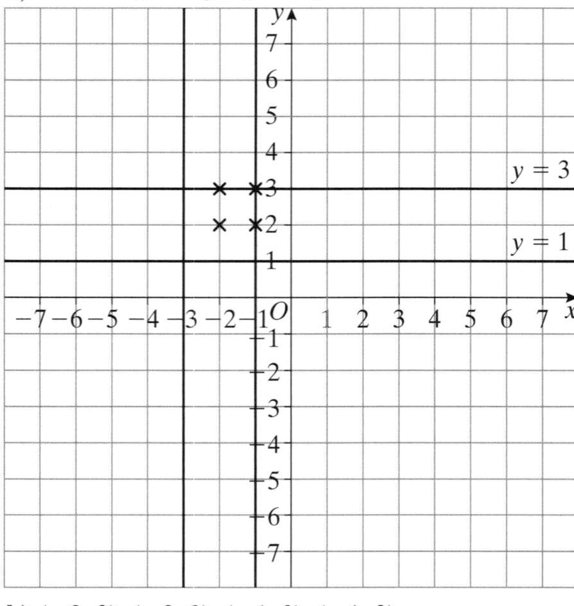

b) $(-2, 2)$, $(-2, 3)$, $(-1, 2)$, $(-1, 3)$

10 a)

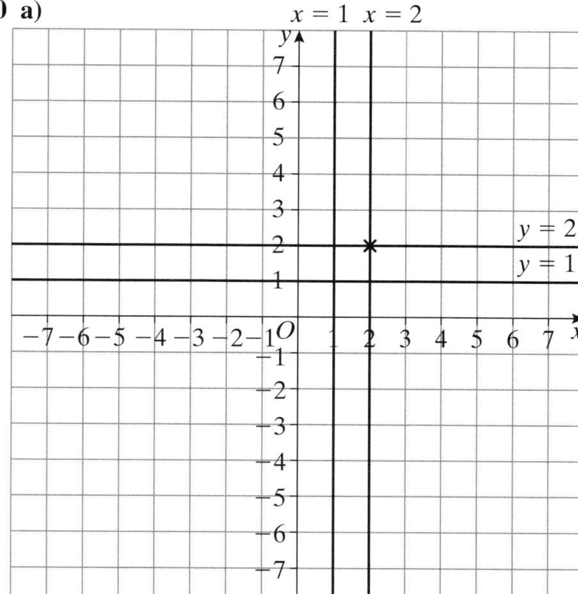

b) $(2, 1)$

Review Exercise 9

1 $0, 1, 2, 3$ **2** $1, 2, 3, 4, 5, 6, 7, 8$

3 $4, 5, 6, 7, 8$ **4** $3, 4, 5, 6, 7, 8, 9$

5 $-1, 0, 1, 2, 3, 4, 5, 6$ **6** $5, 6, 7$

7 $1, 2$ **8** $2, 3, 4, 5$

9 $4, 5, 6$ **10** 2

11 $x \geqslant 1$

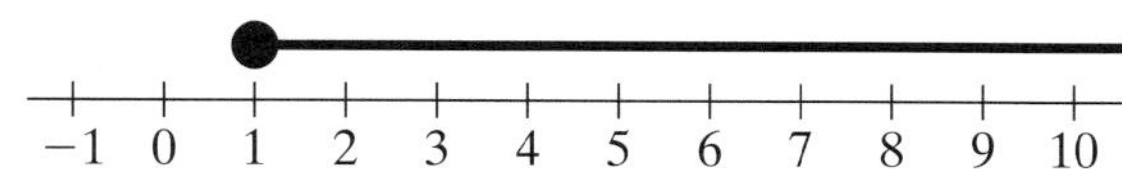

12 $x < -2$

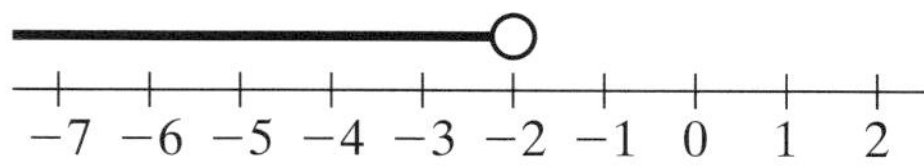

13 $x < 3$

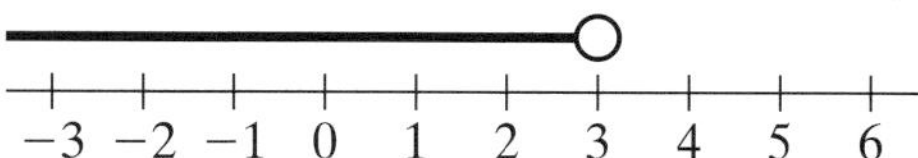

14 $x \leqslant 6$

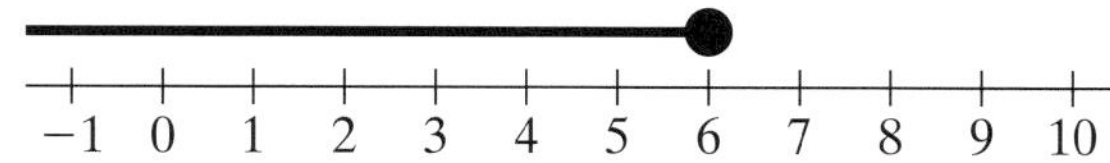

15 $-1 \leqslant x < 1.5$

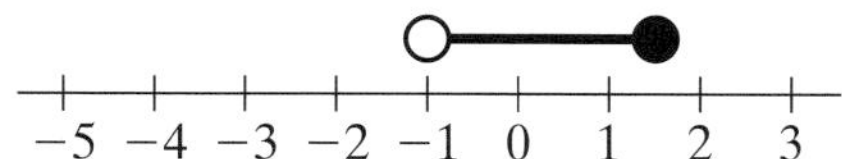

16 $0 < x \leqslant 2$

17 $x < 2$

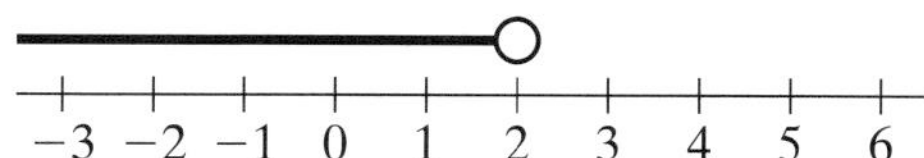

18 $x > 3$

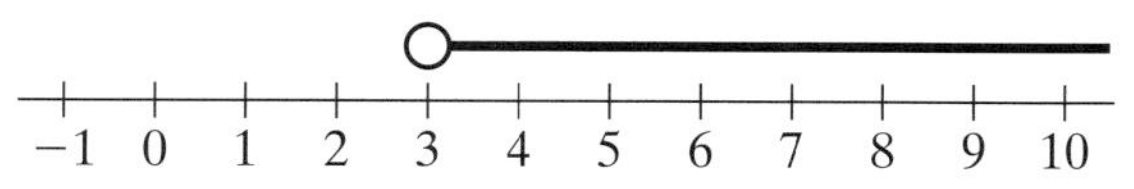

19 $-2 \leqslant x < 7$

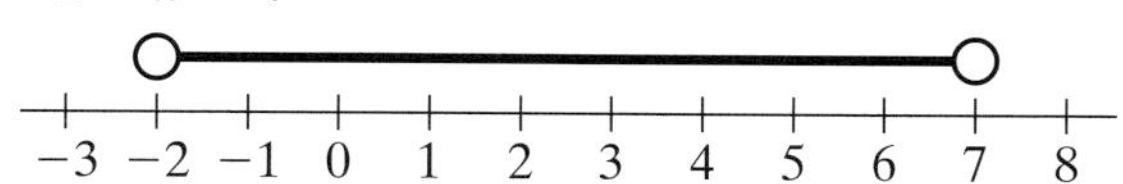

20 $6 < x \leqslant 9$

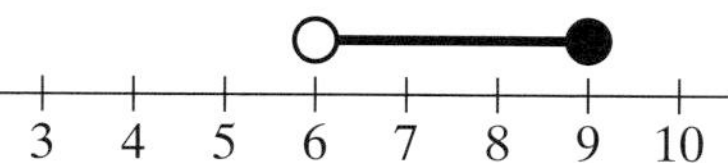

21

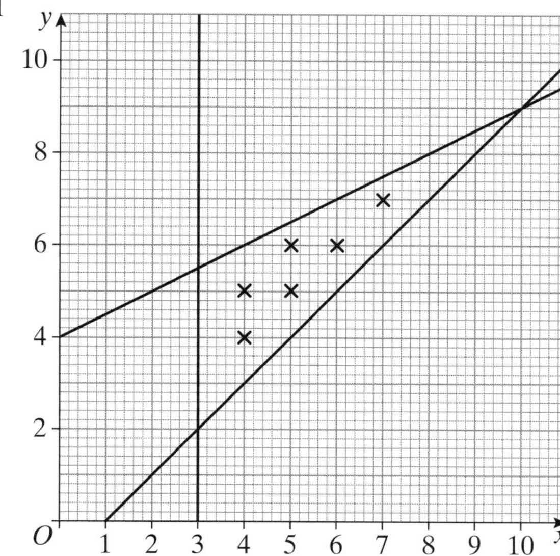

22 a) $-1, 0, 1$

b)

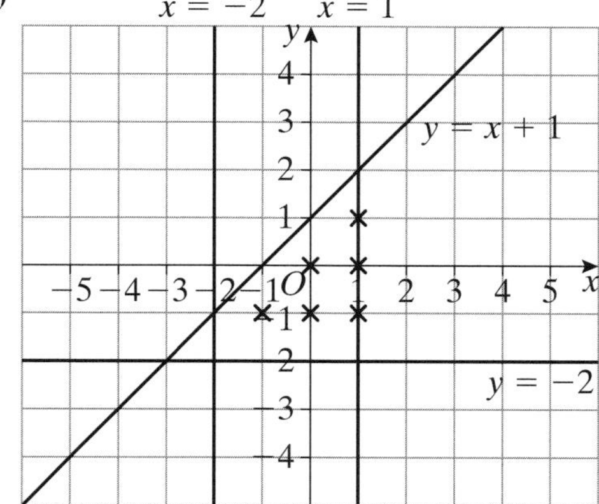

23 $4, 5, 6$

24 a) $y \geqslant -0.5$ **b)** 0

25 a) $y = -\frac{5}{6}x + 2\frac{1}{2}$ **b)** $k = 20$

c) (i)

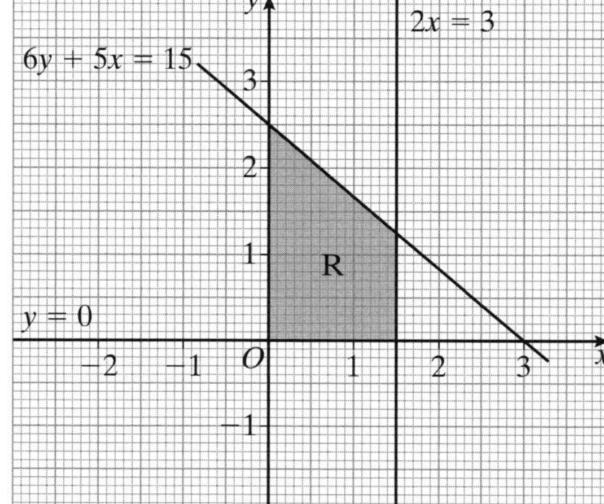

(ii) $(1, 1)$

26 a) (i) $x < 2$

(ii)

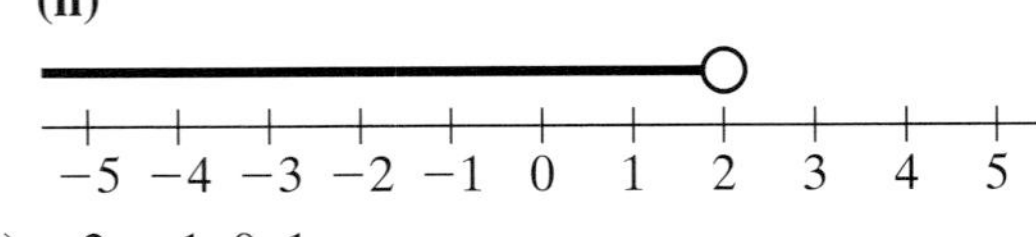

b) $-2, -1, 0, 1$

Internet Challenge 9

ϕ	The Golden Ratio.
Σ	The eighteenth letter of the Greek alphabet, denotes 'the sum of'.
$\div$	This 17th century symbol was formerly used in Europe to indicate subtraction.
∞	A sculpture of this symbol, by Marta Pan, stands on the A6 roadside in France.
i	The (not real) square root of minus one.
$<$	First used in Harriot's *Artis Analyticae Praxis* in 1631.
0	This originated from Hindu mathematics, where it was known as *sunya*.
$=$	Invented by Robert Recorde in 1557.
θ	The eighth letter of the Greek alphabet, used to denote an unknown angle.
$\sqrt{}$	This 16th century symbol may be a corrupted abbreviation for *radix*.

Chapter 10: Number sequences

Starter 10

Task 1: Add 1 each time

Task 2: Add 1, then 2, then 3 (triangular numbers, starting at 0)

Task 3: Seems to double each time

Task 4: Pattern 5 has 5 points, 10 lines, 16 regions
Pattern 6 has 6 points, 15 lines 31 regions
The first two rules seem to work, the third does not.

Exercise 10.1

1 70, 80; add 10; $10n$

2 17, 19; add 2; $2n + 3$

3 63, 65; add 2; $2n + 49$

4 28, 32; add 4; $4n$

5 728, 2186; powers of 3 take 1; $3^n - 1$

6 0.000 01, 0.000 001; divide by 10 each time; $\frac{1}{10^n}$

7 280, 360; triangular numbers $\times$ 10; $10 \times \frac{n(n+1)}{2} = 5n(n+1)$

8 98, 128; double square numbers; $2n^2$

9 a) 13   **b)** $2n - 1$

10 a) n^2 **b)** 900

Worksheet 10.1

1 25, 28   **2** 35, 43

3 68, 65   **4** 48, 96

5 30, 42   **6** 4, 7, 10, 13, 16, 19

7 5, 10, 20, 40, 80, 160   **8** 6, 4, 6, 4, 6, 4, 6, 4, 6, 4

9 a) 3, 8, 13, 18, 23

     b) Yes (goes up in equal steps of 5)

10 a) 3, 8, 15, 24, 35

     b) No (goes up in unequal steps)

11 a) 28   **b)** $2n + 8$

12 a) 61   **b)** $3n + 25$

Exercise 10.2

1 5, 7, 9, 11, 13

2 1, 4, 10, 22

3 a) 7, 15, 23, 31, 39   **b)** 159

4 a) 2, $3\frac{1}{2}$, 5, $6\frac{1}{2}$, 8, $9\frac{1}{2}$ **b)** 35

5 a) Start at 12, go up 3 each time

     b) 39

6 a) 99, 98, 97, 96, 95   **b)** 50

7 a) 3, 9, 27, 81, 243   **b)** Powers of 3

8 a) 10, 17, 24

     b) 73

     c) 150th term

9 a) 1, 3, 6, 10

     b) 465

     c) Either n or $n + 1$ is even

     d) The triangular numbers

10 $6n + 7 = 2770$ gives $n = 460.5$ which is not a whole number, so must be wrong.

Exercise 10.3

1 a) 57   **b)** $5n + 7$

2 a) -6 **b)** $-8n + 66$

3 $3n + 5$

4 $5n - 3$

5 $-n + 11$

6 $5n - 1$

7 $3n + 18$

8 $-2n + 14$

9 a) 19   **b)** $3n + 1$

     c) 3 because 3 sticks are added to form each new square. 1 because 1 stick is needed at the start.

10 a) 7   **b)** 5   **c)** $7n - 2$

Review Exercise 10

1 66, 77, 88; $11n$ **2** 64, 128, 256

3 14, 17, 20; $3n - 1$ **4** 36, 49, 64

5 5, 4, 3; $-n + 11$ **6** 85, 79, 72

7 7, 6, 3, -6

8 a) 2, 5, 9, 14, 20   **b)** No

9 a) 26   **b)** 61   **c)** $5n + 1$

10 a) 15, 9   **b)** $4n - 3$

11 $C = 3n + 2$

12 a) 21, 25   **b)** $4n - 3$

13 $5n + 1$

14 a) $1 + 2 + 3 + 4 = \dfrac{4 \times 5}{2}$

     b) $1 + 2 + 3 + 4 + 5 + 6 + 7 + 8 = \dfrac{8 \times 9}{2}$

     c) $\dfrac{100 \times 101}{2} = 5050$

     d) $\dfrac{n(n + 1)}{2}$

15 a) 20, 27

     b) (i) 65   **(ii)** Add 8, 9, 10, 11

     c) $n + \dfrac{n(n + 1)}{2}$

     d) 5049

3 The ratios get increasingly close to 1.6180

4 $1 - \phi$ and $\dfrac{1}{\phi}$ are equal.

5 Various parts of the Parthenon are rectangles with sides in the Golden Ratio.

6 Leonardo da Vinci   **7** Seurat

8 Born 1170, died 1250

9 Yes, for example Binet's formula
$$\frac{(1 + \sqrt{5})^n - (1 - \sqrt{5})^n}{2^n \sqrt{5}}$$

10 Nautilus

Chapter 11: Travel and other graphs

Starter 11

1 235   **2** 2333   **3** 10.2 seconds   **4** 634 yards

Exercise 11.1

1 a) 450 metres (above his start point)

b) 30 minutes

c)

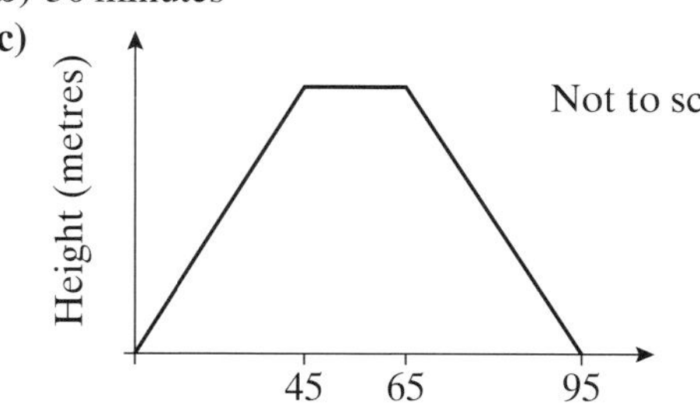

2 B. The initial rate at which the water level rises is fast, since the cross section is small. The rate of increase in depth decreases as the cross section of the bowl gets wider.

3 a) $50 - 3 \times 12 = 14$ litres

b) Not to scale:

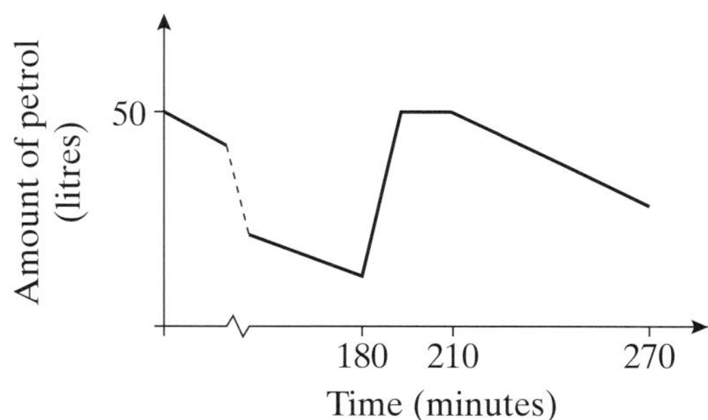

c) 380 litres

4

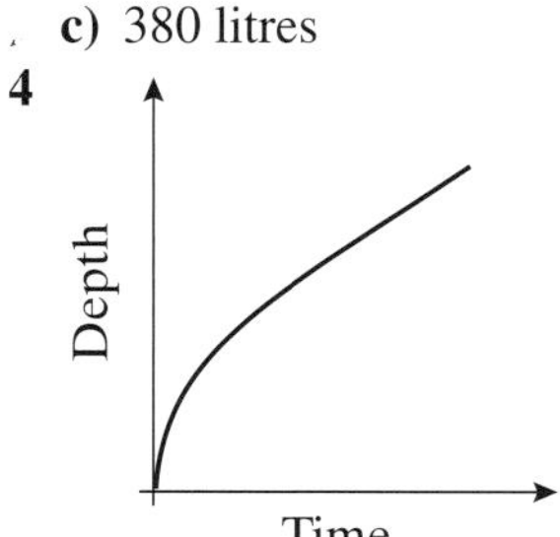

5 a)

No. days (*n*)	Total no. pages written (*T*)
1	15
2	35
3	55
4	75

b) $T = 20n - 5$

c)

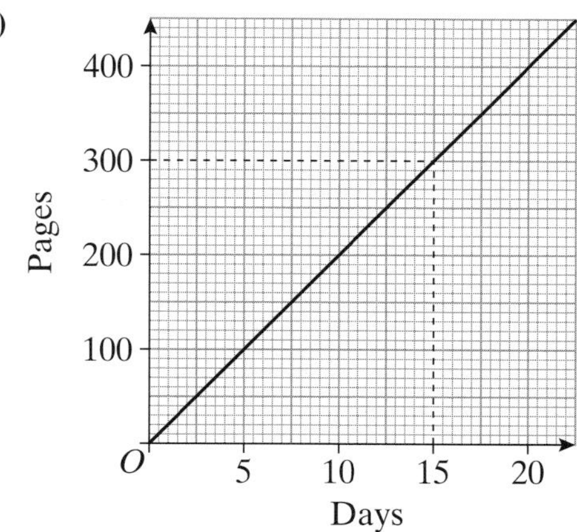

d) 15 days (approx)   **e)** 30 days

6 A. The depth decreases at a constant rate.

Worksheet 11.1

1 a) Filling   **b)** Washing   **c)** Draining

d) Filling   **e)** Rinsing   **f)** Draining

g) Drying

2 a) 3 minutes   **b)** 6 minutes

c) Adds more water   **d)** James

e) 13 minutes   **f)** 16 cm

g) Let out

3 a)

Hours (*h*)	0	2	4	6	8
Cost (*T*)	40	80	120	160	200

b)

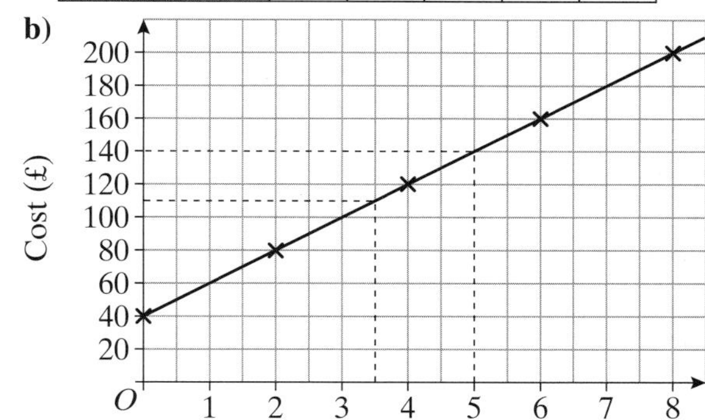

c) (i) £110   **(ii)** 5 hours

4 a) B   **b)** D   **c)** C   **d)** A

Exercise 11.2

1 a) 20 minutes   **b)** 12 km/h   **c)** No

2 a) 70 miles per hour

b) Return journey stops for about 5 minutes half way.

c) Speed when moving is the same in both directions.

d)

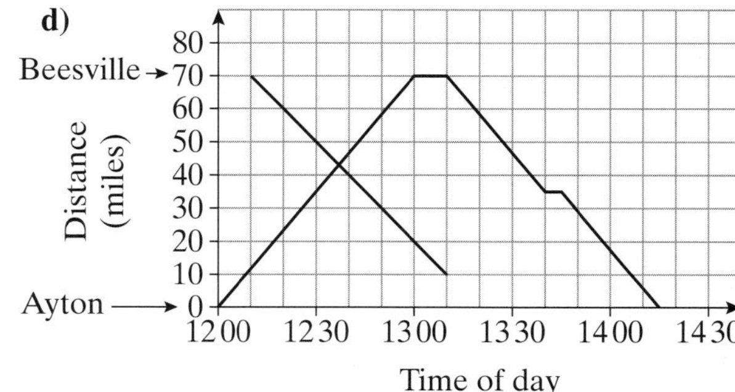

e) About 1237 or 1238

3 a) Day number *n*   3   4

Total distance travelled (*D* km)   42   60

b), c) Tom is back on schedule by end of Day 6.

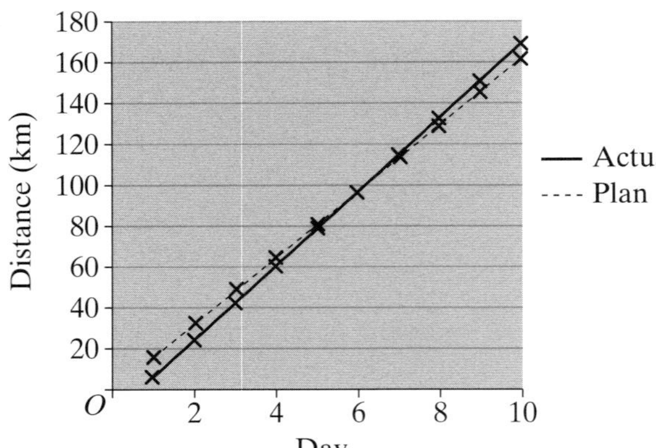

d) $D = 18n - 12$

e) End of Day 14; original schedule end of Day 15.

4 a), b)

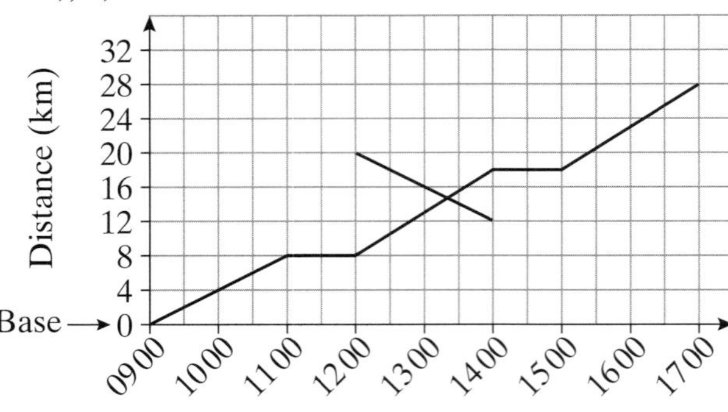

c) About 1320

Worksheet 11.2

1 a) 10.00 am **b)** 10 minutes **c)** 10 km **d)** 11.30 am
e)

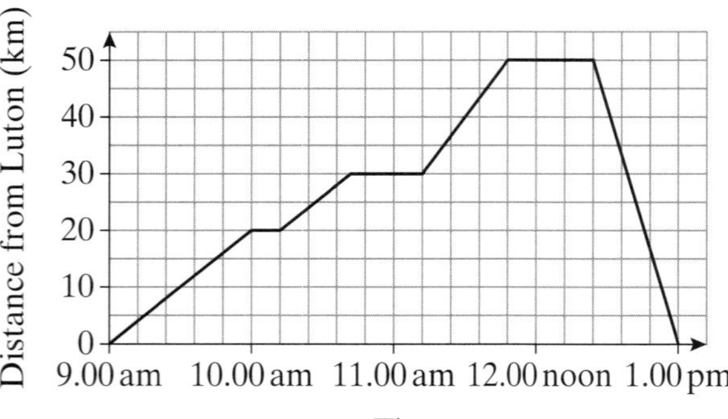

2 a) $1\frac{1}{2}$ hours **b)** 60 km **c)** 40 km/h **d)** $3\frac{1}{2}$ hours
e) 3.30 pm for 30 minutes **f)** 90 km/h

3 a) 8.00 am **b)** He stops for a rest. **c)** 20 miles
d)

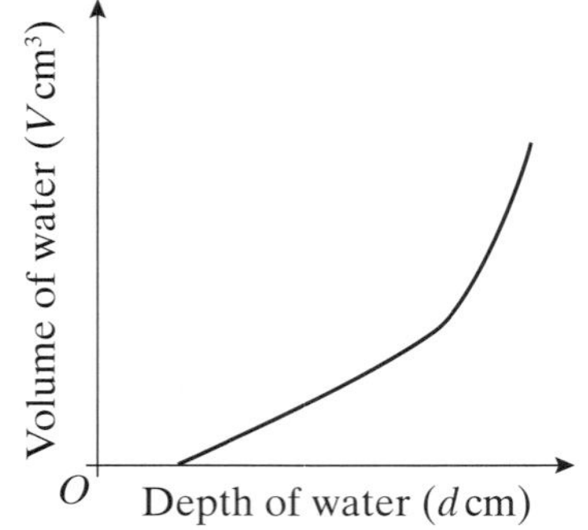

e) 11.25 am **f)** 10.00 am **g)** 30 minutes **h)** Erin

Exercise 11.3

1 a) $3\,\mathrm{m\,s^{-2}}$ **b)** $30\,\mathrm{m\,s^{-1}}$ **c)** 600 m
d) $15\,\mathrm{m\,s^{-1}}$ (which is 34 mph, so the speed limit is
probably 30 mph)

2 a) 120 seconds
b) 22.5 (approx $2g$, i.e. twice the acceleration due to
gravity)

3 a) 3.5 **b)** 49 m **c)** 15.3 seconds

4 a)

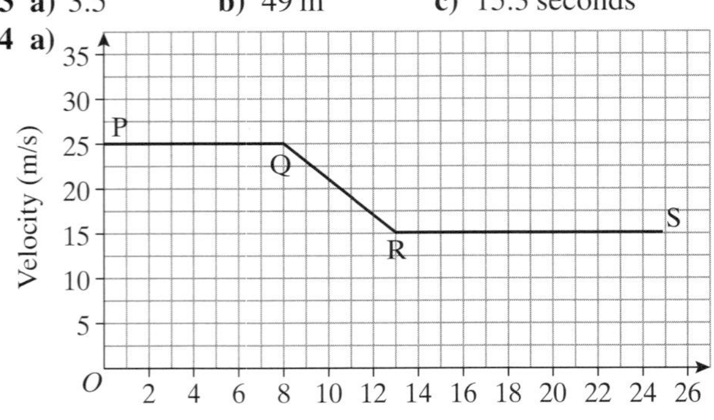

b) 200 m **c)** 180 m

Review Exercise 11

1 a) 0905 **b)** 7 km **c)** 10 minutes **d)** 21 km/h

2 a) 40 km/h
b)

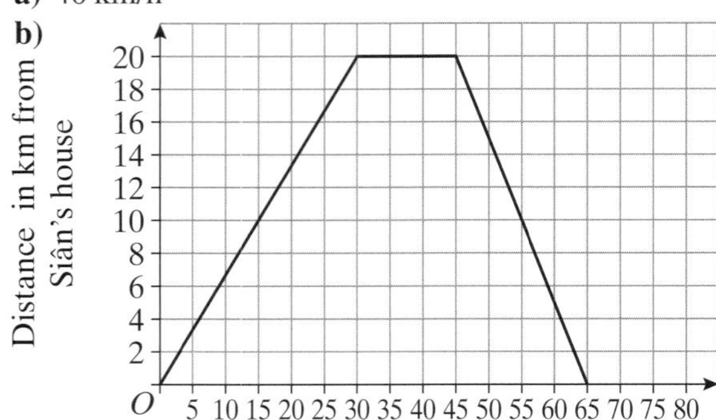

3 a) (i) 1300 **(ii)** 20 km/h
b)

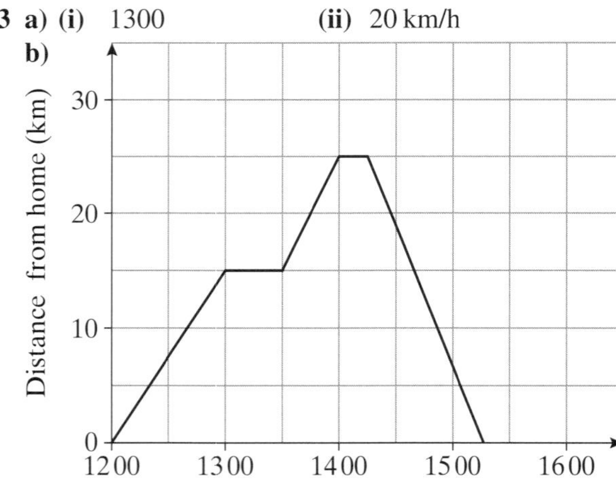

4

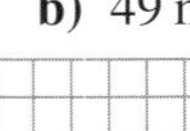

5

Container	Graph
A	R
B	S
C	Q
D	P

6 a) (12 km/h) 7.5 miles per hour   **b)** He stopped.

c)

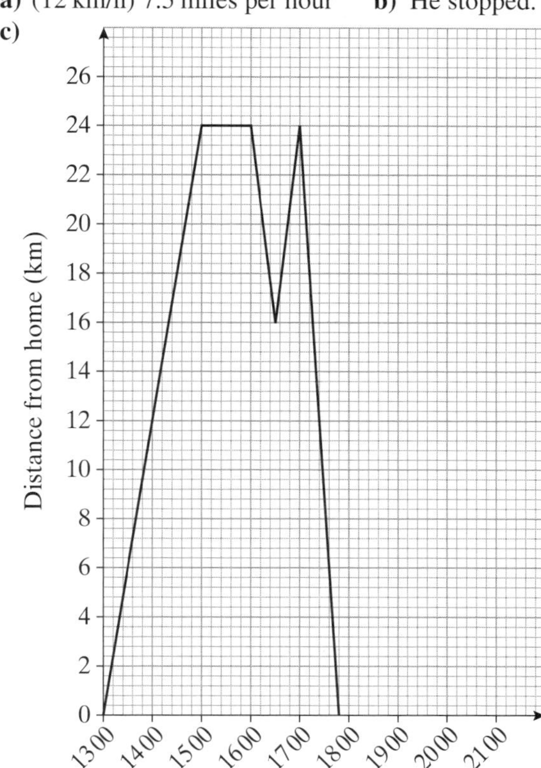

7 a) 270 km   **b)** 180 km/h

c)

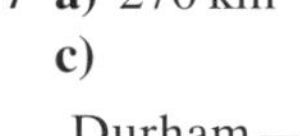

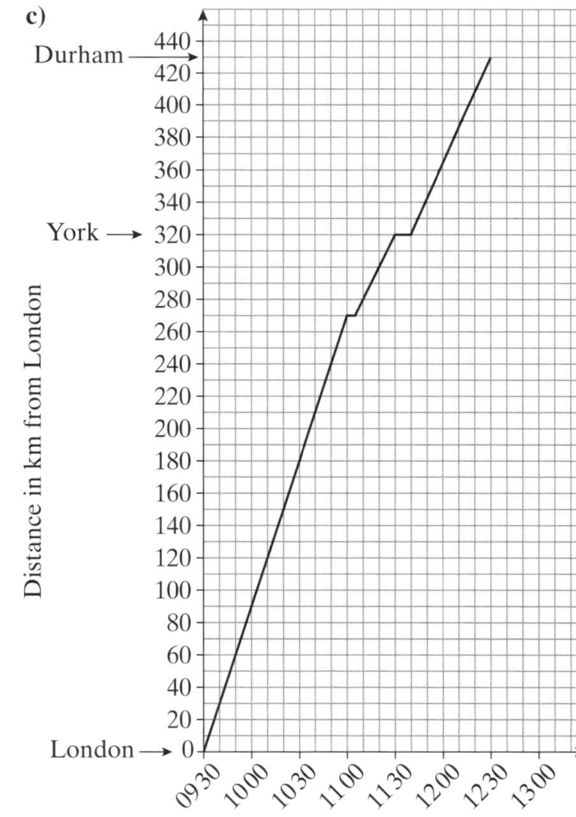

8 a) 1305 (or 1306)   **b)** 1000 and 1100
   c) 50 km/h

9 a), b)

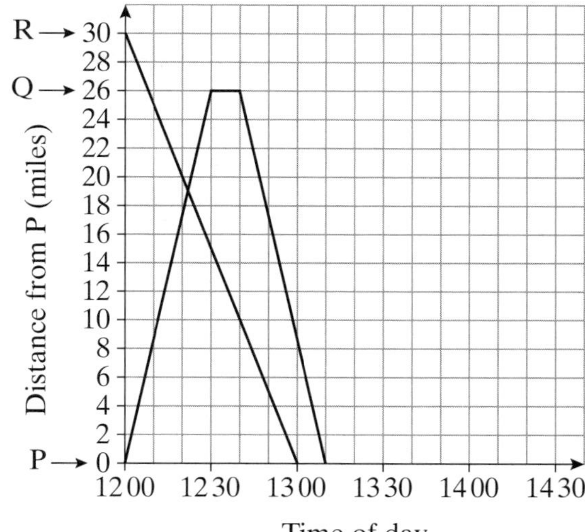

   c) 19 km

10 a) 6 m/s   **b)** 10.7 m/s   **c)** 15 metres

Internet Challenge 11

The tea clipper Cutty Sark	20 mph
Challenger 2 tank	37 mph
Disney's Space Mountain roller coaster (Paris)	43 mph
Intercity 225 train	140 mph
Porsche 911 GT3 RS car	190 mph
Boeing 747-400 passenger jet aircraft	630 mph
Speed of sound (in air)	760 mph
Eurofighter Typhoon jet aircraft	1320 mph
Orbiting Space Shuttle	17 600 mph
Apollo 11 spacecraft	24 500 mph

Chapter 12: Working with shape and space

Starter 12

$a = 40°, b = 140°, c = 40°, d = 55°, e = 60°, f = 36°,$
$g = 65°, h = 30°, i = 60°, j = 60°, k = 120°, l = 120°,$
$m = 36°, n = 18°, o = 54°, p = 63°, q = 58°, r = 45°,$
$s = 66°, t = 15°, u = 45°, v = 30°, w = 140°, x = 58°,$
$y = 14°, z = 166°$

Exercise 12.1

1 $a = 68°, b = 68°, c = 68°$
2 $d = 131°, e = 49°$
3 $f = 61°, g = 61°, h = 119°$
4 $i = 71°, j = 65°, k = 44°$
5 $l = 13°, m = 13°, n = 77°$
6 $o = 52°, p = 52°, q = 90°$
7 $r = 55°$
8 $s = 132°, t = 48°$

Exercise 12.2

1 $a = 25°$
2 $x = 64°$
3 $x = 59°$, largest angle is 71°
4 $8y - 4 = 180$ leading to $y = 23$
   Angles are then 23°, 69°, 88°

5 **a)** $3x + 90 = 180$ **b)** $x = 30°$
  c) $38°, 52°, 90°$
6 **a)** $16c + 4 = 180$ **b)** $c = 11°$
  c) $48°, 48°, 84°$ **d)** Isosceles
7 $85°$ **8** $88°$ **9** $106°$
10 **a)** $10y + 50 = 360$
  b) $y = 31°$ so angles are $82°, 98°, 103°, 77°$
  c) They are parallel
11 $3k + 3 = 180$ so $k = 59$ giving angles of $90°, 115°,$
  $90°, 65°$
12 **a)** $11x + 8 = 360$ **b)** $x = 32°$
  c) $48°, 62°, 105°, 145°$

Worksheet 12.2
1 $a = 140°$, angles on a line
2 $b = 150°$, angles at a point
3 $c = 42°$, angles of a triangle
4 $d = 55°$, angles of an isosceles triangle
5 $e = 58°$, alternate angles
6 $f = 108°$, corresponding angles
7 $g = 30°$, angles on a line
8 $h = 30°, 270° \div 9$
9 $i = 41°$, vertically opposite angles
10 $j = 50°$, angles of a triangle
11 $k = 120°$, alternate angles; $l = 100°$, corresponding
  angles
12 $m = 70°$, alternate angles; $n = 140°$, corresponding
  angles
13 $p = 110°$, angles of a quadrilateral
14 $q = 83°$, alternate angles; $r = 97°$, angles on 2
  straight lines; $s = 87°$, angles of a quadrilateral
15 $t = 72°$, corresponding angles
16 $v = 72°$, angles of a regular polygon; $w = 108°$,
  internal angle of a regular pentagon; $x = 54°$,
  bisected internal angle
17 $y = 25°$, sum of interior angles

Exercise 12.3
1 **a)** $1080°$ **b)** $3240°$
2 $142°$
3 **a)** $60°$ **b)** $24°$
4 **a)** 24 sides
  b) $360 \div 14$ is not a whole number
5 $120°$
6 $5x + 110 = 360$ leading to $x = 50$
  The angles are $110°, 90°, 90°, 130°, 120°$
7 $135°$
8 **a)** $2(a + b + c) = 720$ so $a + b + c = 360$
  b) $c - 20 + c - 10 + c = 360$ leading to $c = 130$
  c) $110°, 120°, 130°, 130°, 120°, 110°$
  d) No. A regular hexagon has all angles equal to $120°$.
9 Check students' diagrams.
10 Check students' diagrams.

Exercise 12.4
1 Perimeter 24 cm, area 24 cm²
2 Perimeter 23.9 cm, area 24.5 cm²
3 Perimeter 30 cm, area 48 cm²
4 Perimeter 40 cm, area 88 cm²
5 Perimeter 42 cm, area 84.7 cm²
6 Perimeter 28 cm, area 42 cm²
7 Perimeter 56 cm, area 84 cm²
8 Perimeter 11.8 cm, area 6.3 cm²

9 Perimeter 26 cm, area 36 cm²
10 Perimeter 34 cm, area 42 cm²
11 Perimeter 34 cm, area 46 cm²
12 Perimeter 32 cm, area 24 cm²
13 Perimeter 148 mm, area 1208 mm²
14 Perimeter 48 cm, area 136 cm²
15 Trapezium
16 No − it could be a rhombus.
17 $x = 5$ cm
18 **a)** $x = 3$ **b)** 25 cm
19 **a)** Equilateral
  b) $4x - 5 = 3x + 1$ leading to $x = 6$
  c) 57 cm
20 **a)** $3x - 10 = x + 6$ leading to $x = 8$
  b) $3y - 1 = 2y + 4$ leading to $y = 5$
  c) 14, 14, 14, 14
  d) Rhombus

Worksheet 12.4
1 **a)** 60 cm²   **b)** 32 cm²   **c)** 64 mm²
2 **a)** 20 cm²   **b)** 12 cm²   **c)** 60%
3 **a)** 100 m²   **b)** 39 m²   **c)** 20 m²
4 $h = 8$ cm
5 **a)** Perimeter $= 2a + 2b$ or $2(a + b)$
  b) Area $= ac$

Exercise 12.5
1 Surface area 600 cm², volume 1000 cm³
2 **a)** 792 cm²   **b)** 1440 cm³
3 **a)** 30 cm²   **b)** 180 cm³   **c)** 240 cm²
4 **a)** 440 cm²   **b)** 35 200 cm³
5 **a)** 1728 cm³   **b)** 864 cm²
6 **a)** 9000 cm³   **b)** 2700 cm²
7 **a)** 8.71 m³   **b)** 20.68 m²
8 **a)** 450 m³   **b)** 450 000 litres
9 **a)** 22 cm along each edge
  b) 2904 cm²
10 **a)** 5 cm by 7 cm by 13 cm
  b) 382 cm²

Review Exercise 12
1 $a = 52°, b = 38°$
2 **a)** $10y + 60 = 360$ leading to $y = 30$
  b) $110°, 70°, 110°, 70°$
  c) Parallelogram
3 **a)** $54°$ **b)** $72°$; isosceles triangle
4 $30°, 150°$ **5** 13 sides   **6** 90 sides
7 **a)** No   **b)** Rectangle
8 **a)** $x = 4$ **b)** 34 cm   **c)** 70 cm²
9 **a)** $31°$ **b)** $135°$
10 20 cm
11 1 cm by 28 cm, 2 cm by 14 cm, 4 cm by 7 cm
12 **a) (i)** $109°$
    (ii) Angles on a line add up to $180°$
  b) (i) $24°$
    (ii) Alternate to angle QSR
13 **a)** 700 cm³   **b)** 13.51 kg
14 **a)** $e = 42°$ **b)** $f = 69°$
15 $30°$
16 **a)** $x = 60°$ **b)** $y = 120°$
17 $x^3 - 8x$
18 8 m²
19 **a)** 20 000 cm³   **b)** 4 minutes

20 a) $6x + 8$ **b)** $x = 7$
21 $140°$
22 a) $11 < 2x + 6 < 20$ so $5 < 2x < 14$
    b) 3, 4, 5, 6
23 a) 16 cm   **b)** 2600 cm³
24 102 cm²   **25** 142°
26 a) 60°   **b)** 120°   **c)** 12 cm²
27 62.96 m² → 25.2 litres of paint → £75.30
    (or 26 × £2.99 = £77.74)
28 a) $4x + 8$ **b)** 15.5 cm
29 24 cm

Internet Challenge 12

1 Francis Guthrie
2 Alfred Kempe
3 Peter Tait
4 1976
5 Kenneth Appel and Wolfgang Haken
6 It was computer-assisted.
7 Four colours (as before)
8 Seven colours
9 Gardner claimed to have found a map that required five colours.
10 A real map-maker might want to colour two separate regions (for example, Alaska and USA) in the same colour to indicate political association. This adds extra restrictions and thus might require extra colours.

Chapter 13: Circles and cylinders

Starter 13

Calculator results for the approximations are:
$3 + \frac{1}{8} = 3.125$
$\frac{22}{7} = 3.142\,857\,143$
$\sqrt{10} = 3.162\,277\,66$
$3 + \frac{8}{60} + \frac{30}{60^2} = 3.141\,666\,667$
$\frac{333}{106} = 3.141\,509\,434$
$\left(\frac{2143}{22}\right)^{\frac{1}{4}} = 3.141\,592\,653$ is the closest
$\frac{88}{\sqrt{785}} = 3.140\,854\,685$
$\sqrt{2} + \sqrt{3} = 3.146\,264\,37$
$\frac{355}{113} = 3.141\,592\,92$
$\left(\frac{4}{3}\right)^{4} = 3.160\,493\,827$
$\pi = 3.141\,592\,654$

Exercise 13.1

1 75.4 cm   **2** 69.1 cm
3 1020 cm²   **4** 104 cm²
5 133.5 cm   **6** 3447 cm²
7 3.927 cm   **8** 0.6504 cm²
9 95.0 cm²   **10** 2.27 mm²
11 785 cm   **12** 6.66 m
13 210 mm   **14** 425 mm²
15 a) 157 m   **b)** 32 laps
16 111 cm²
17 a) 12.6 cm²   **b)** 64.9 cm²
18 a) 356 cm   **b)** 1730 cm²   **c)** 1100 cm²
19 a) 359.4 m   **b)** 334.2 m   **c)** 7.5% longer
20 a) 113 mm²   **b)** 28 mm²   **c)** 85 mm²   **d)** 25%

Exercise 13.2

1 15.4 cm   10.1 cm²
2 31.0 cm   58.7 cm²
3 30.7 cm   58.9 cm²
4 27.8 cm   46.3 cm²
5 95.7 mm   169 mm²
6 35.3 cm   66.0 cm²
7 12.4 cm   9.27 cm²
8 44.3 cm   103 cm²
9 a) 140 cm², 122 cm², 52 cm²   **b)** 30.5 cm
10 a) 72°   **b)** 22.6 in²

Exercise 13.3

1 2.47 cm   **2** 4.07 cm
3 6.18 cm   **4** 7.48 cm
5 0.231 cm   **6** 32.9 cm
7 1.99 cm   **8** 8.46 cm
9 a) 3848.4 cm²   **b)** 35 cm   **c)** 94.4 cm
10 4.77 m

Exercise 13.4

1 8600 cm³
2 62.8 cm²
3 a) 3040 cm³   **b)** 553 cm²
4 509 cm³
5 1810 cm²
6 a) 2 120 000 cm³   **b)** 2120 litres
7 3563 cm³
8 14 cm
9 a) 73.6 cm³   **b)** 73.6 < 75 so not possible.
10 a) Nick is right.
    b) Alan used 14 cm as radius, not diameter.

Exercise 13.5

1 Circumference $= 24\pi$ cm, area $= 144\pi$ cm²
2 Circumference $= 22\pi$ cm, area $= 121\pi$ cm²
3 a) 192π cm²   **b)** 1152π cm³
4 a) 12 cm   **b)** 144π cm²
5 a) 11 cm   **b)** 22π cm
6 3 cm
7 30 cm
8 a) 64π cm²   **b)** $32 + 8\pi$ cm
9 a) 18π cm²   **b)** $144 + 72\pi$ cm²
10 a) $\pi \times 6 \times 8 = 48\pi$ and $\pi \times 8 \times 6 = 48\pi$, so the same.
    b) $\pi \times 3^2 \times 8 = 72\pi$ and $2\pi \times 4^2 \times 6 = 96\pi$, so B has the larger volume.

Review Exercise 13

1 2460 cm²   **2** 283 cm
3 11.94 cm   **4** 3217 mm²
5 a) 8 cm   **b)** 16π cm
6 a) 942 cm²   **b)** 314 cm²   **c)** 1570 cm²
7 a) 2.5 cm   **b)** 19.6 cm²
8 72.7 cm²   **9** 81.7 m²
10 a) 28.3 cm²   **b)** 23.1 cm
11 88.4 cm²   **12** 201 cm
13 218 cm²   **14** 754 cm³
15 7.7 cm   **16** 170 grams
17 58.8 cm
18 a) 737 000 cm³   **b)** 275 grams
19 $18 + 9\pi$ cm
20 a) 12.6 cm   **b)** 240 cm²

Internet Challenge 13

1 The Earth's shadow on the Moon (during a lunar eclipse) is round.
2 The Flat Earth Society.
3 Diameter 12 756 km (7926 miles), circumference 40 074 km (24 900 miles)
4 Diameter 12 714 km (7900 miles), circumference 39 942 km (24 818 miles)
5 A Great Circle is a circle on the surface of a sphere, whose centre coincides with the centre of the sphere. The equator is a Great Circle.
6 Ferdinand Magellan, from August 1519 to September 1522, taking 3 years. (Magellan died during the voyage; the expedition was commanded by Juan Sebastian del Cano thereafter.)
7 Sir Ranulph Fiennes and Charlie Burton, from 1979 to 1982.
8 Round the world yacht races typically take over 50 000 km (over 32 000 miles). They do not complete a Great Circle, but they travel a greater equivalent distance, and cross every line of longitude.
9 Greek *geo* = Earth, *metron* = measure
10 Check students' answers.

Chapter 14: Constructions and loci

Starter 14
Check students' diagrams.

Exercise 14.1
Diagrams are shown to scale but not full size.

1
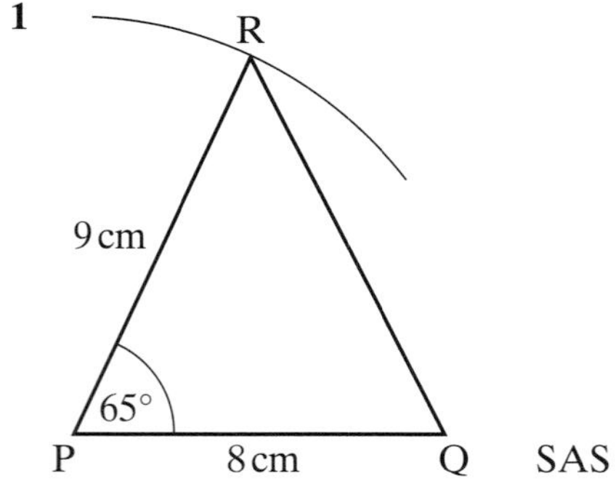

2
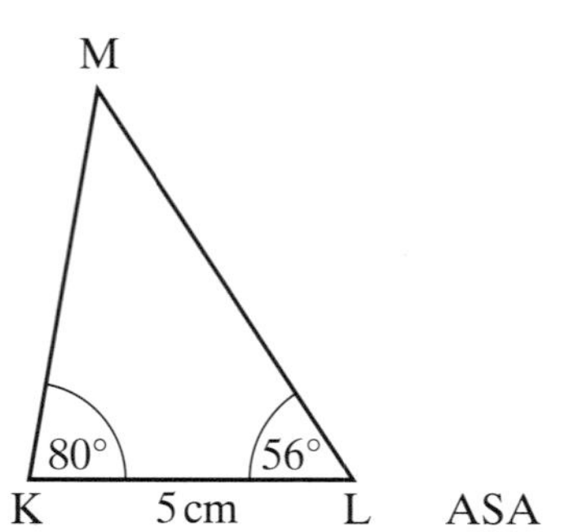

3
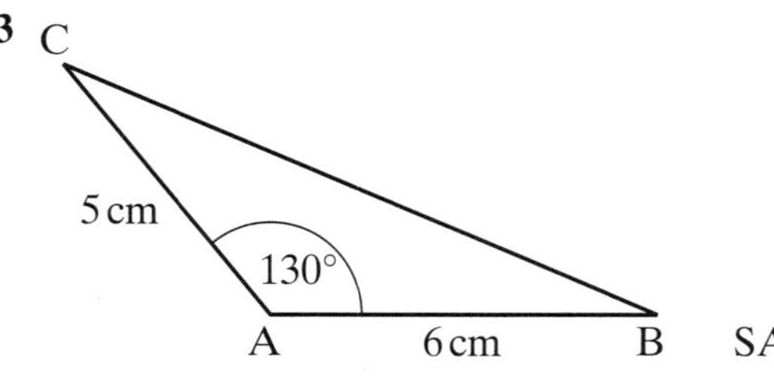

4
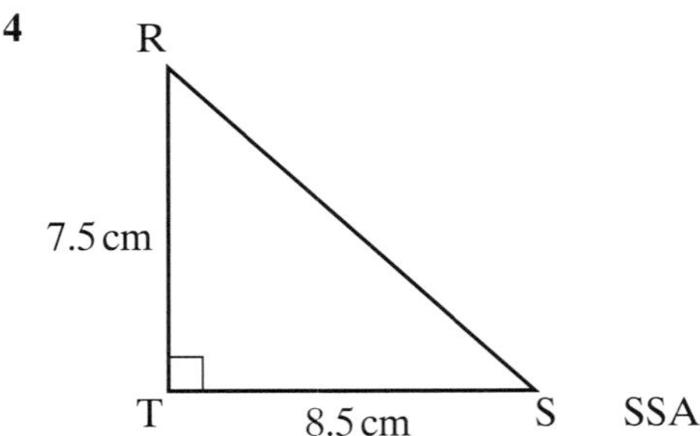

5
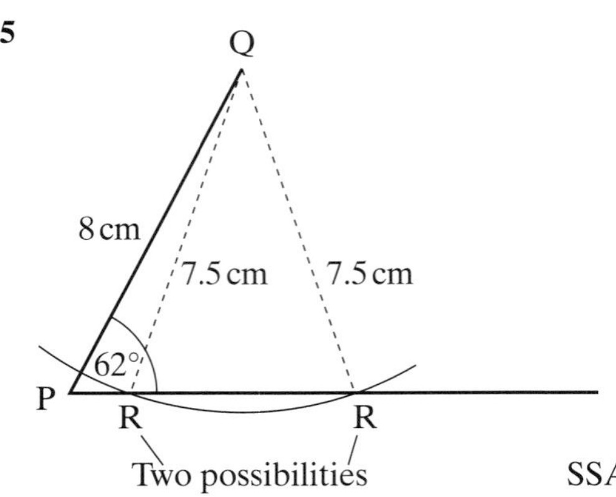

6
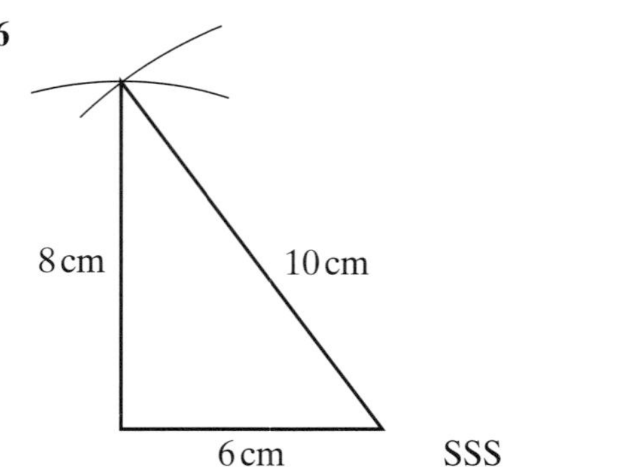

7
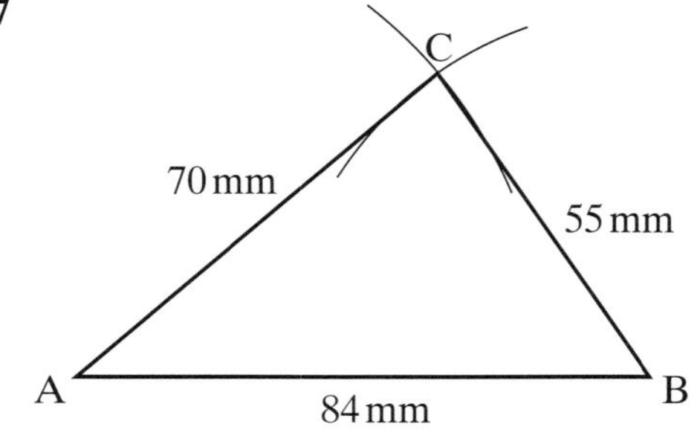

8
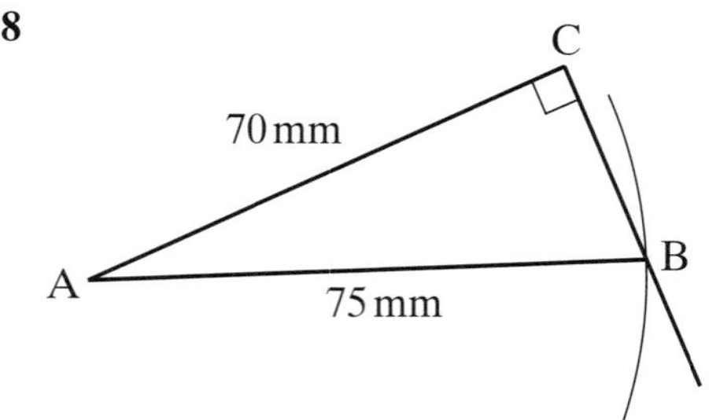

9
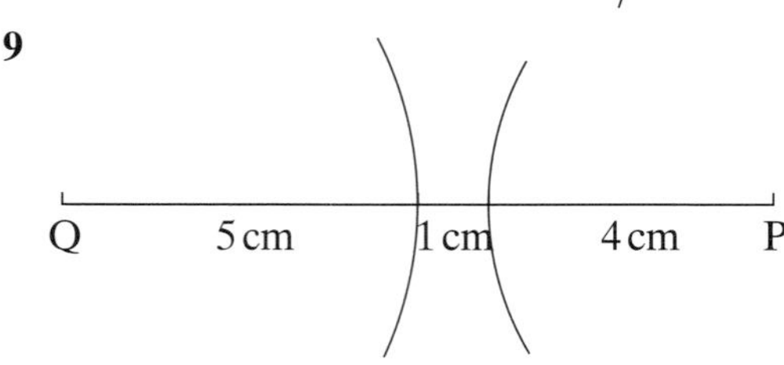

The arcs do not intersect because the sum of the shorter sides is less than the longest side.

10 a)

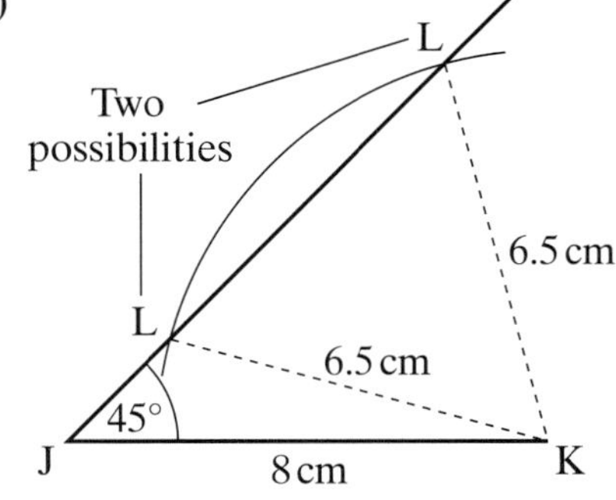

b) There are two possibilities.

Exercise 14.2

Diagrams are shown to scale but not full size.

1

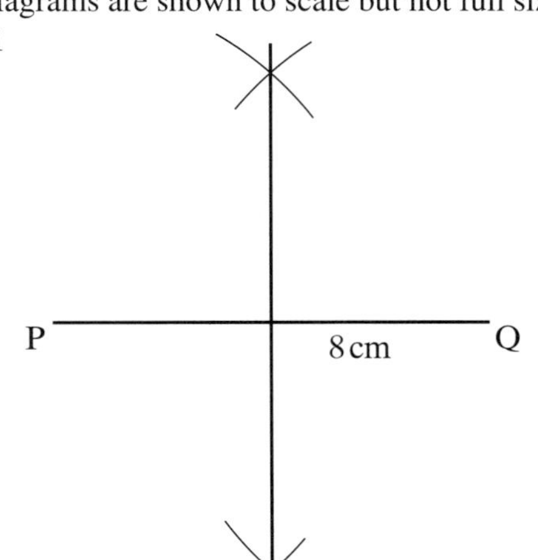

2

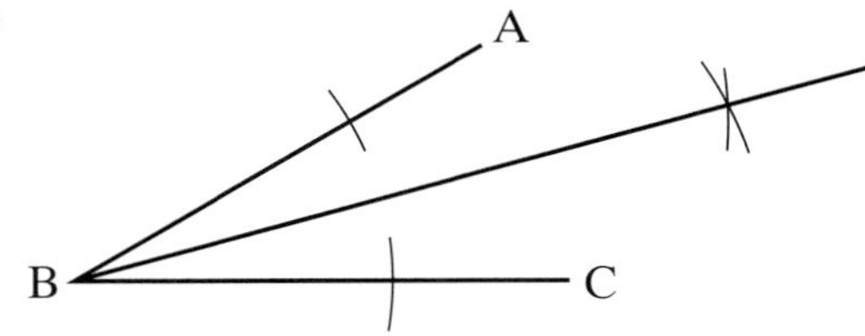

3

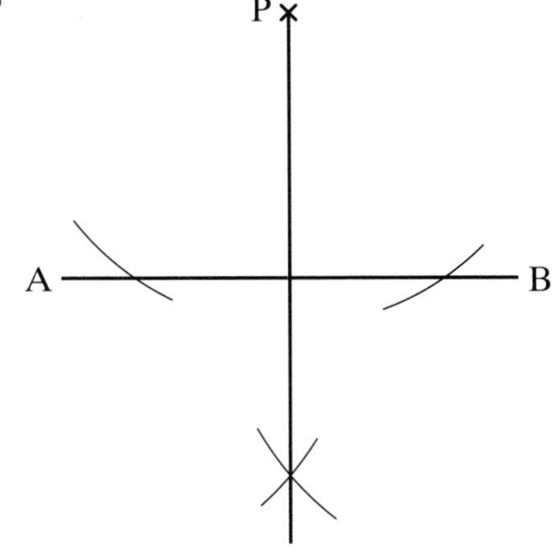

4

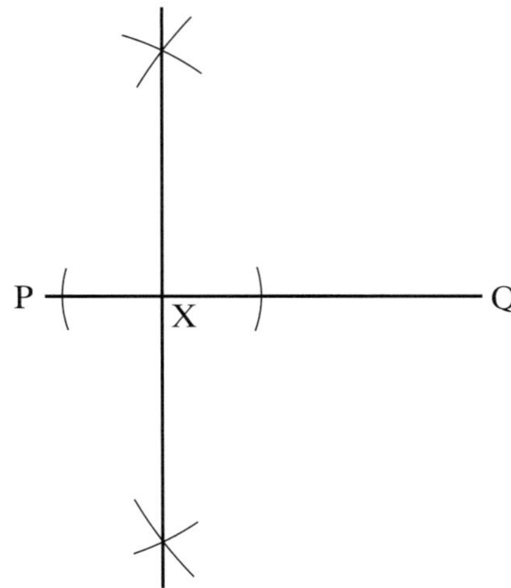

5

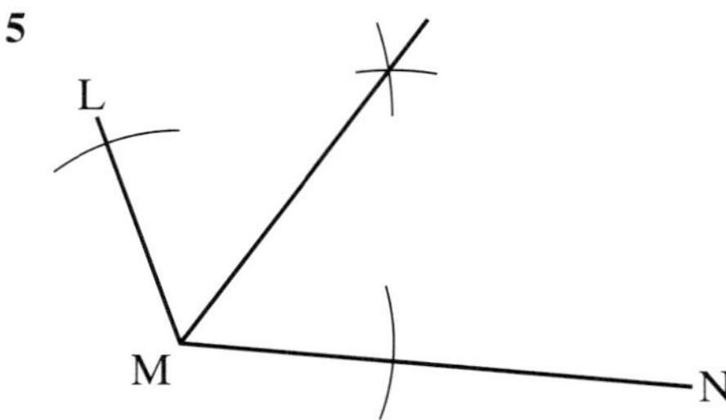

6

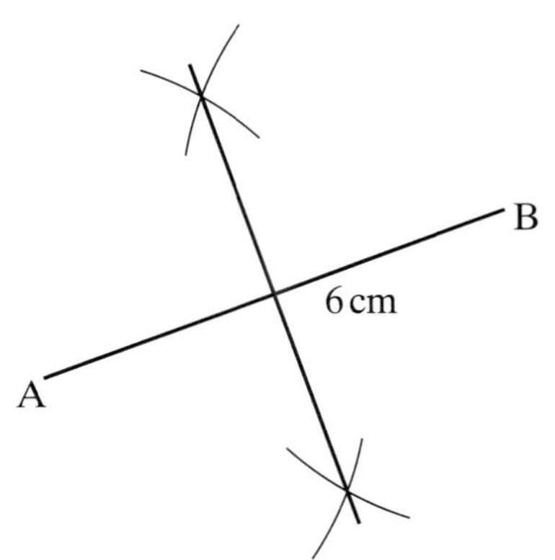

7

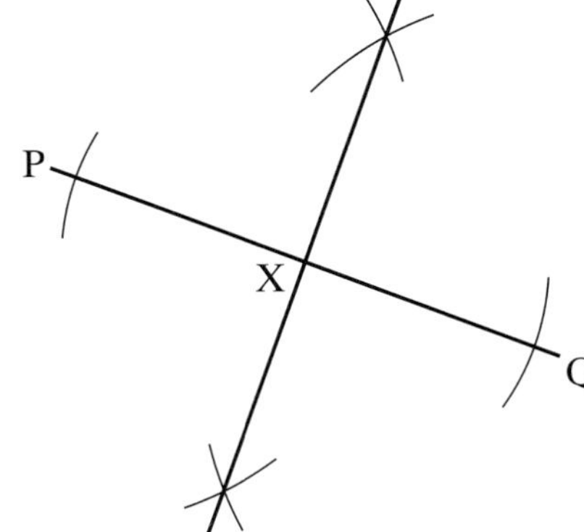

8

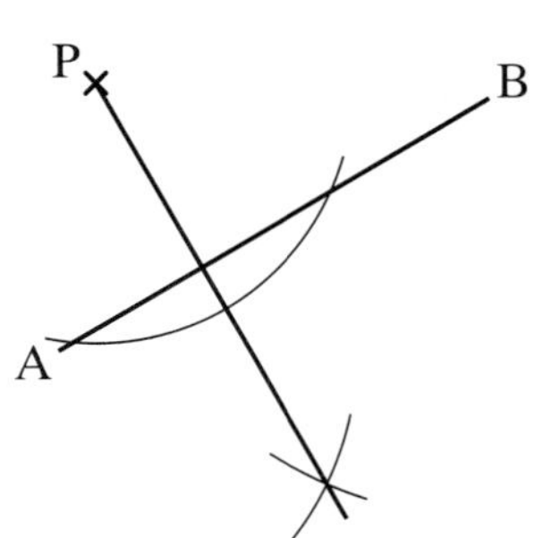

9

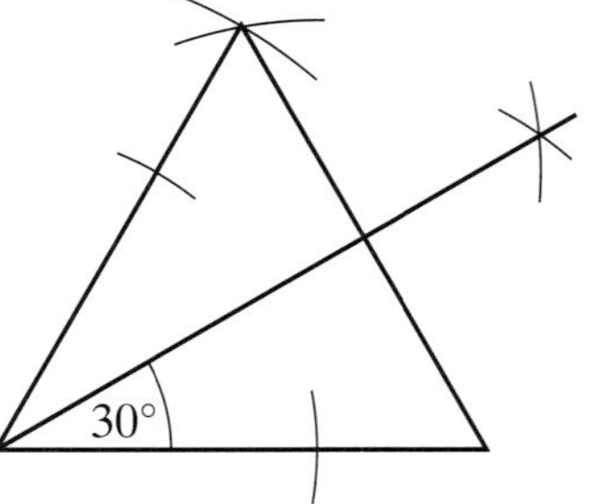

10

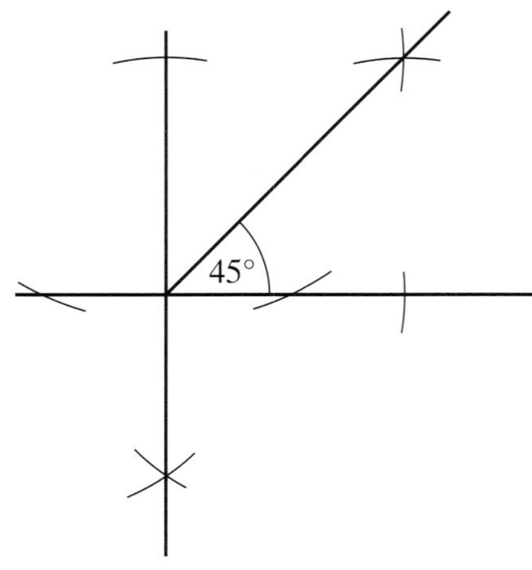

4

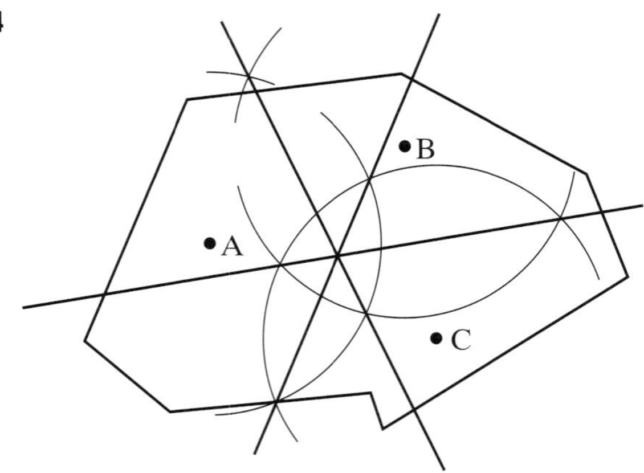

5

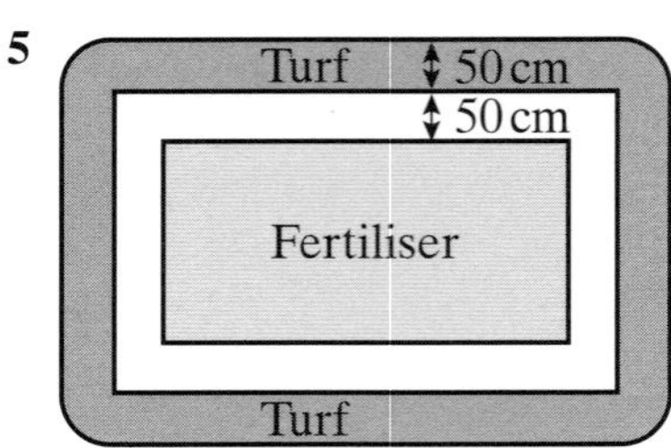

Exercise 14.3
Diagrams are shown to scale but not full size.

1

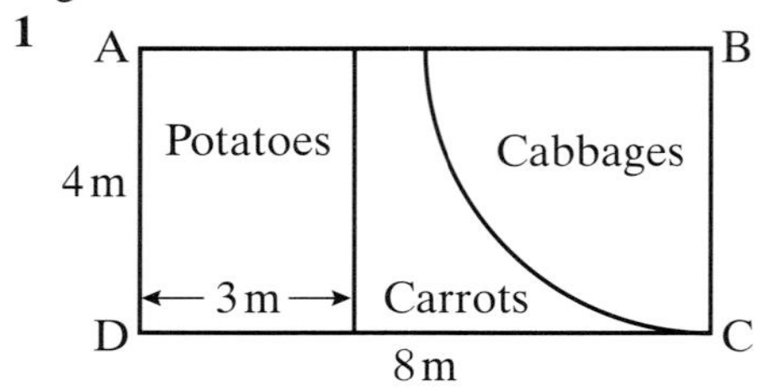

Review Exercise 14
Diagrams are shown to scale but not full size.

1

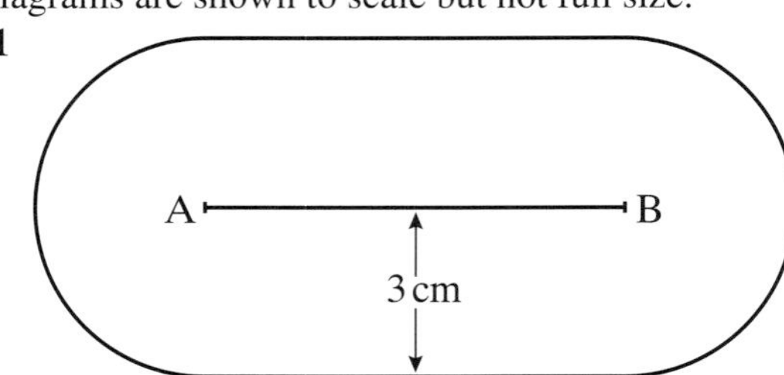

2

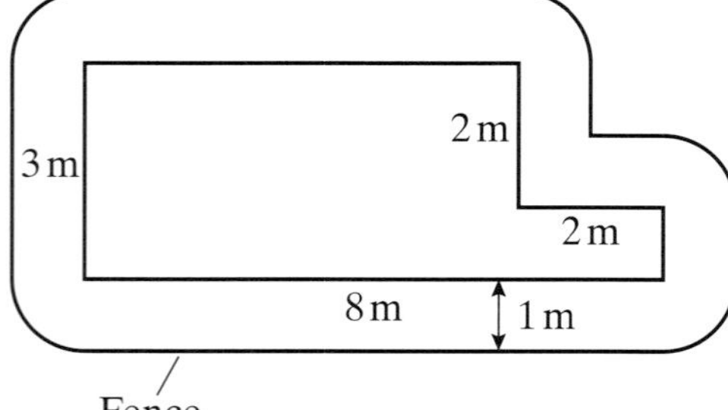

2

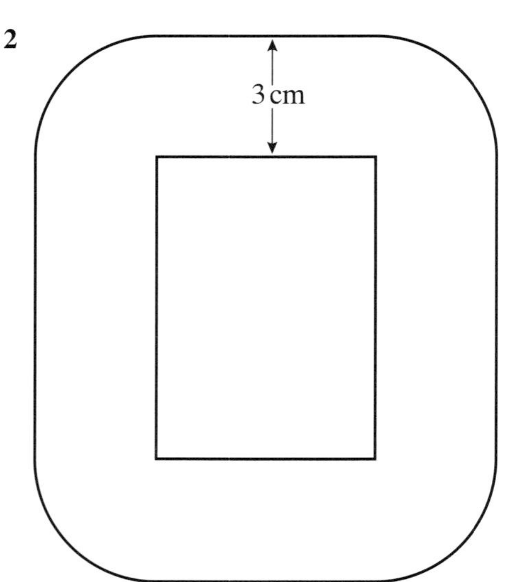

3

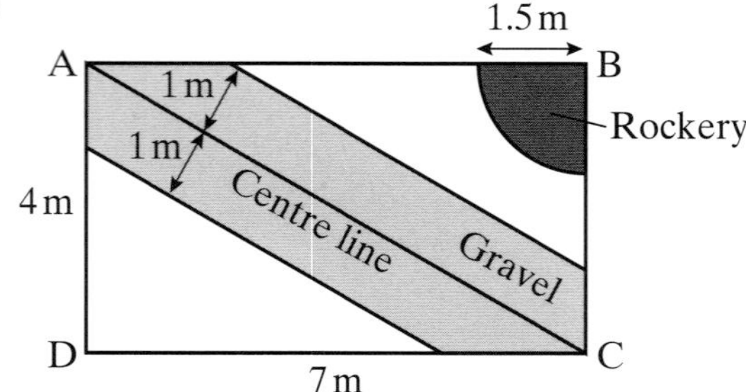

3

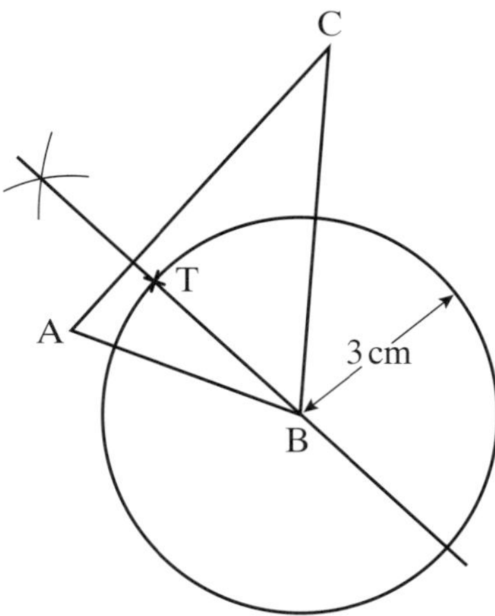

4 a)

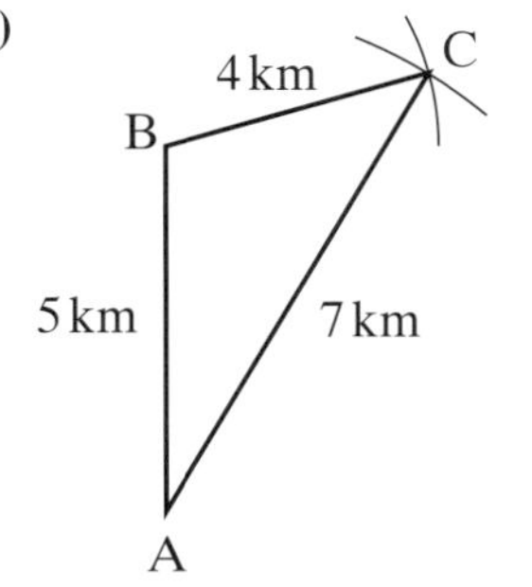

b) 34° **c)** 214°

5

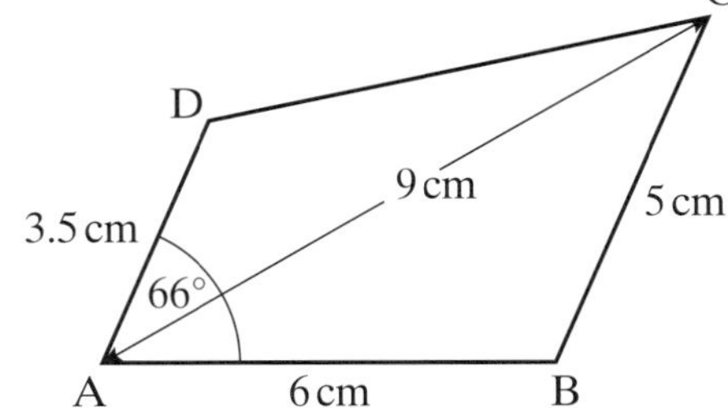

6

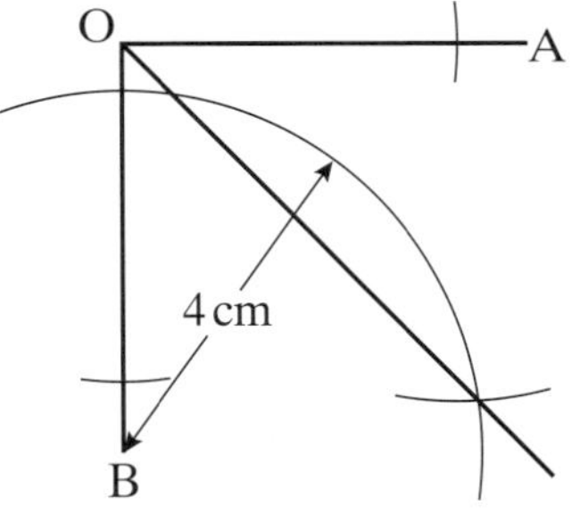

7

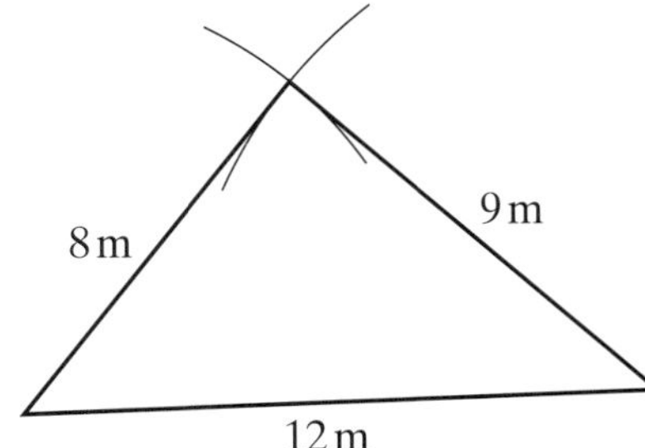

8 a) 3 km
b)

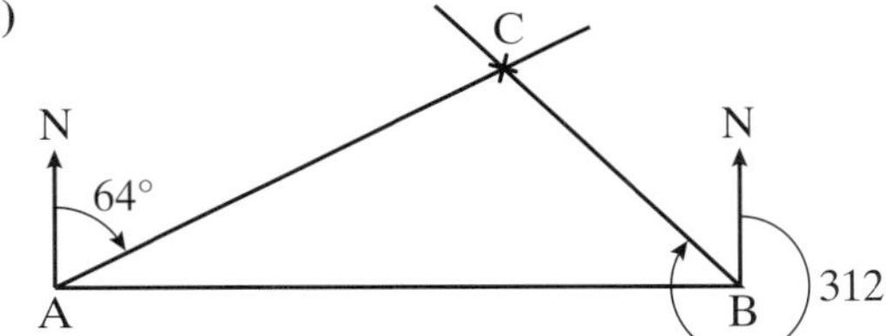

9

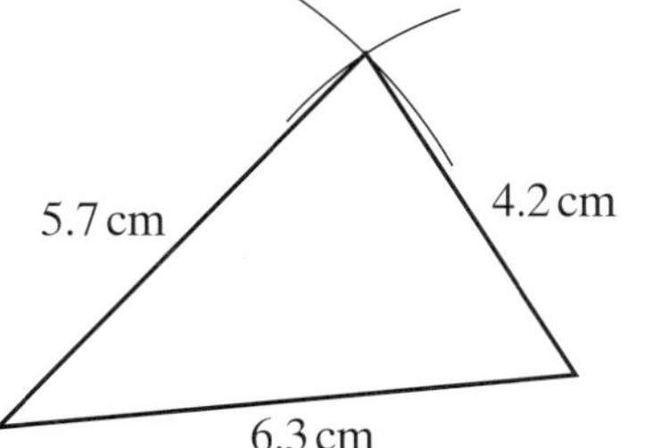

10

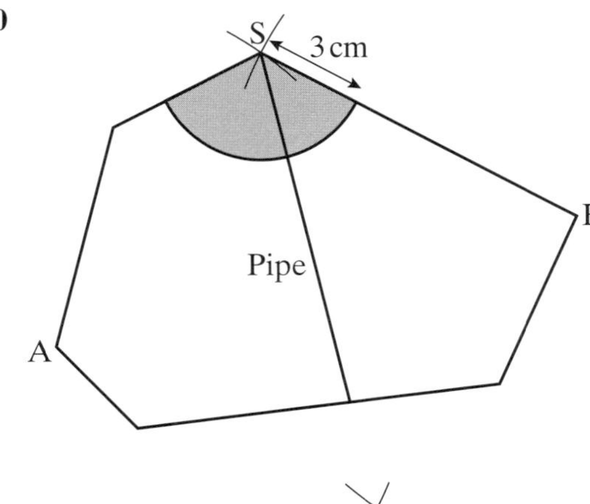

11

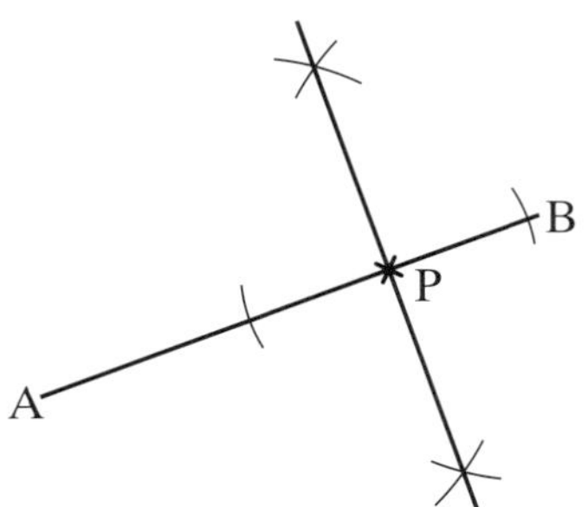

12 a)

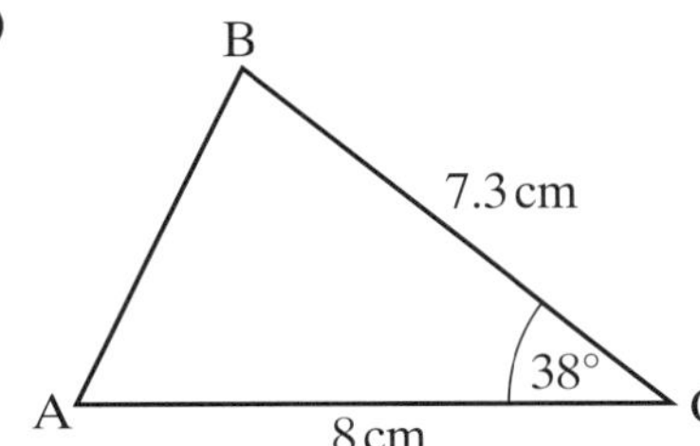

b) 64°

13

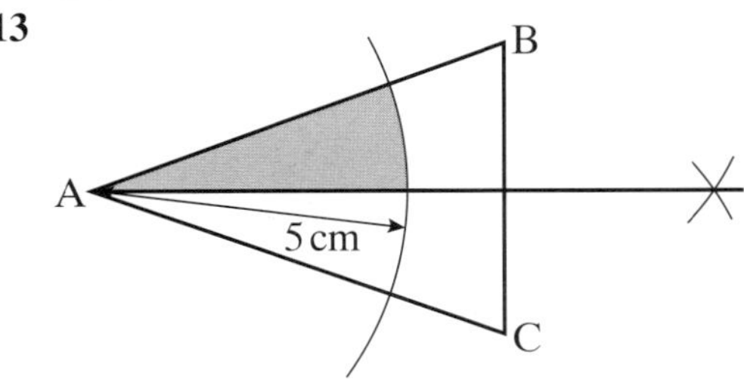

14 081°

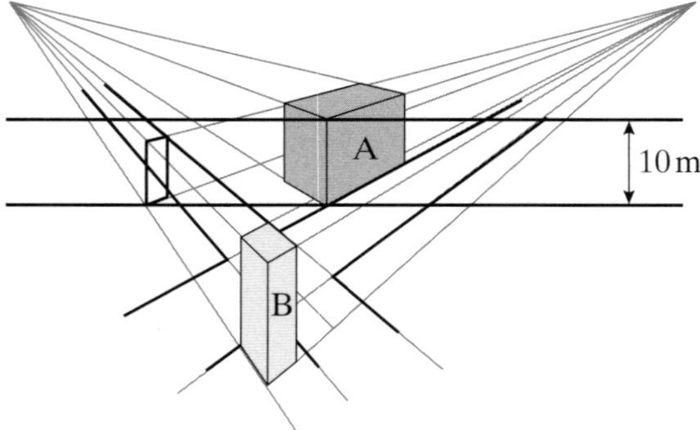

Building A is taller. B is about 7 m tall.

Chapter 15: Transformation and similarity

Starter Exercise

Seven of the nine monkeys are "the same", i.e. congruent; one other is "too long" and one is "too thin". Of the seven congruent monkeys, six are rotationally equivalent and one is flipped – or, if you like, there are six right-handed and one left-handed version of the same picture.

Exercise 15.1

1 a) 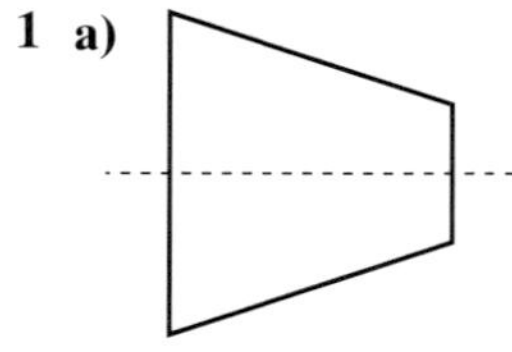   b) 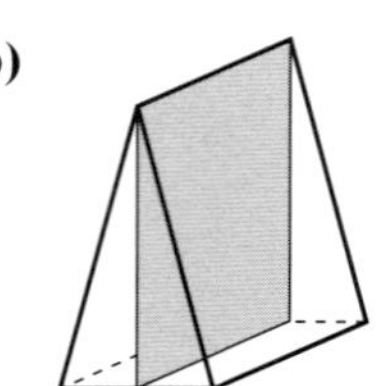

For part b), allow also a plane at right angles to the one shown.

2 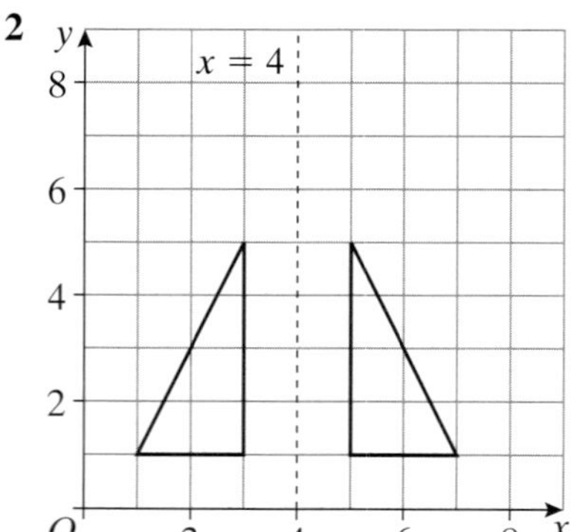

3 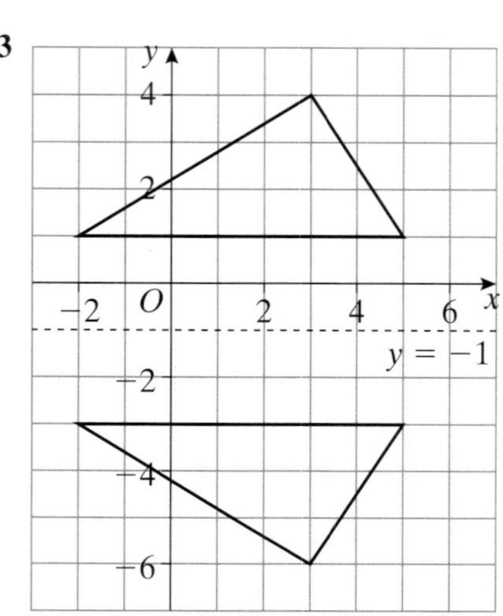

4 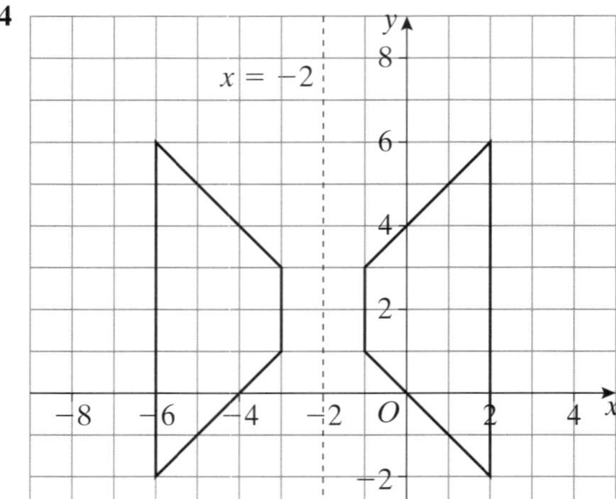

5 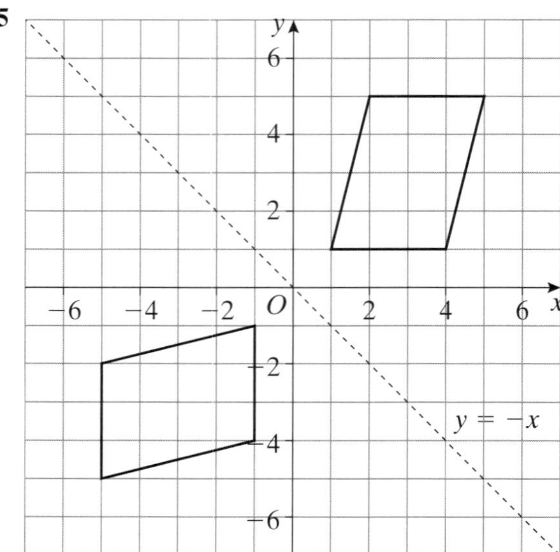

6 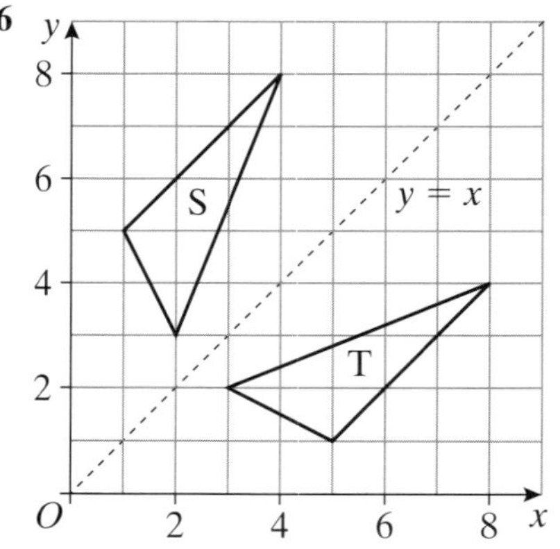

7 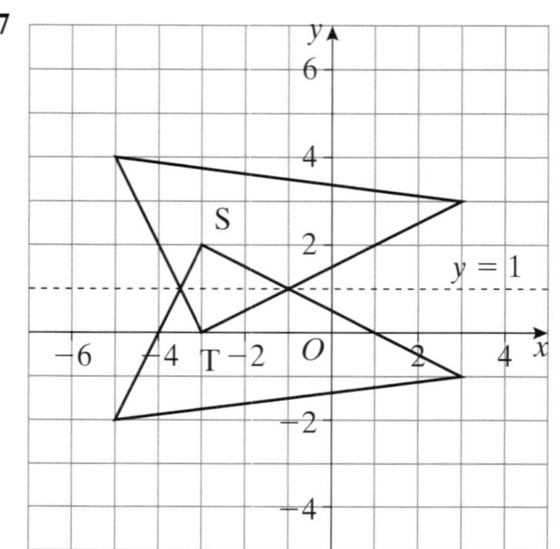

8

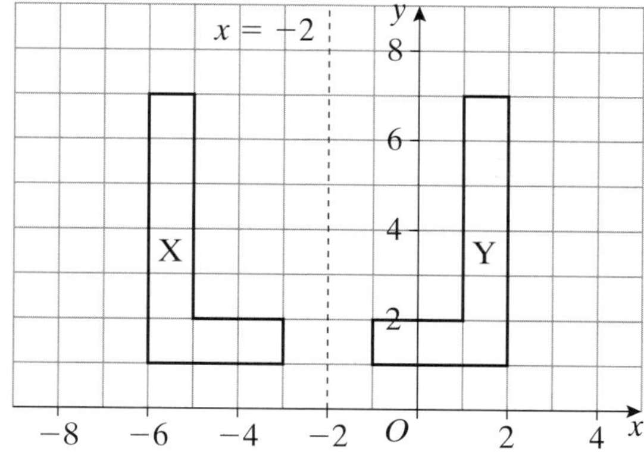

9 a) Two shapes are congruent if they are the same
shape and size.
   b) $y = -1$
   c) B
   d) $y = x$
   e) $y = -x$
10 T and V are coincident (they are, in effect, the same
triangle)

Exercise 15.2

1

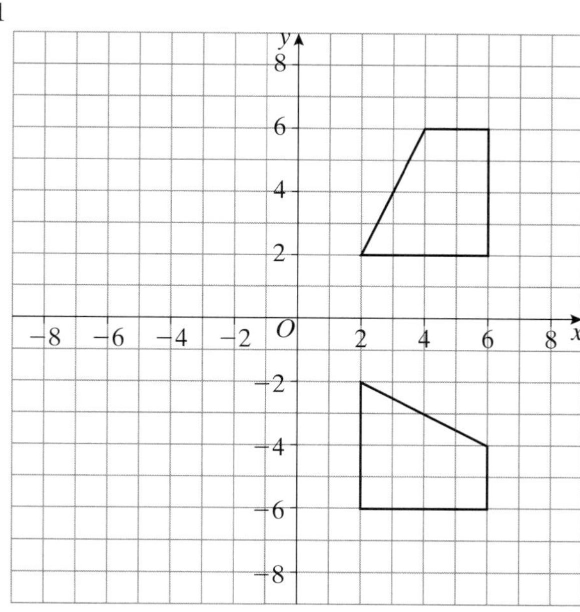

2

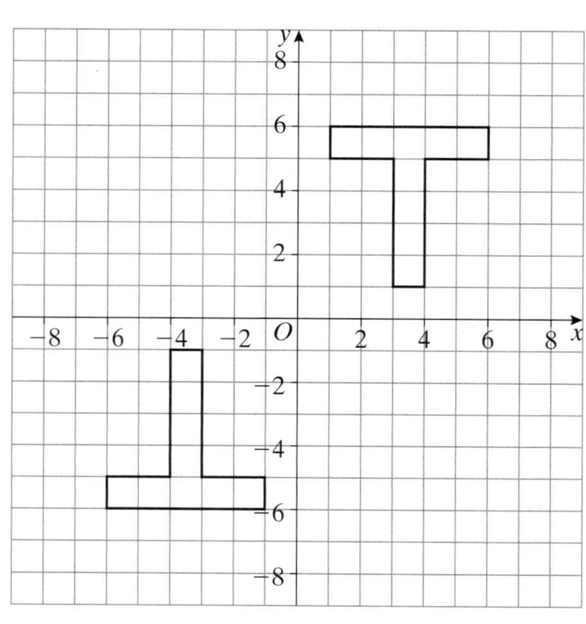

3 a) b)

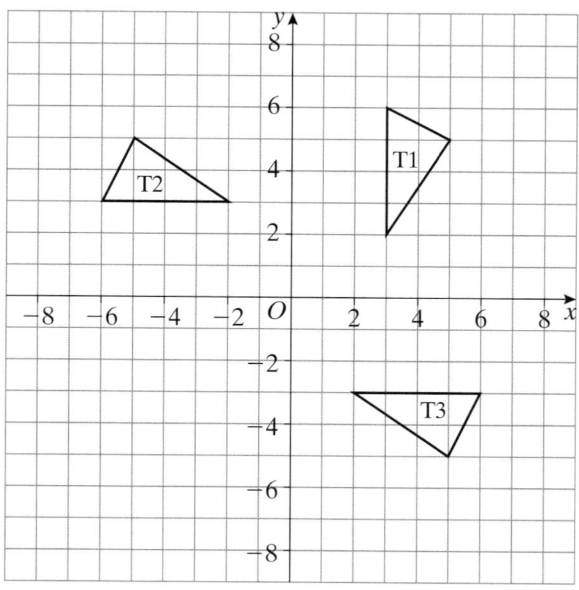

   c) T1 → T3: a rotation of 90° clockwise about O.
4 a) b)

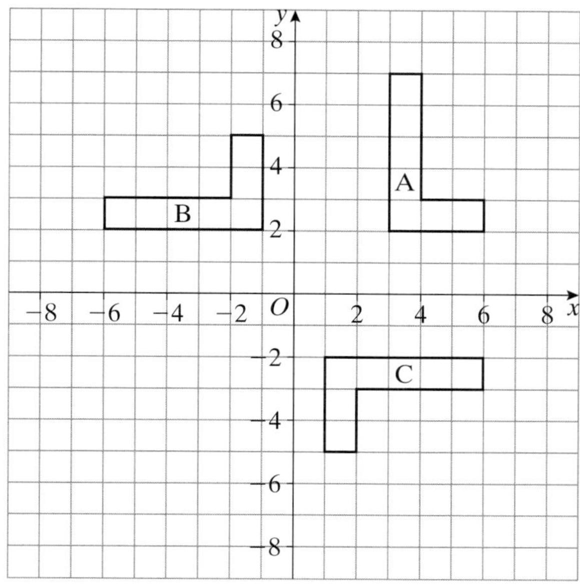

   c) C → A: a rotation of 90° clockwise about (0, 1).
5 a) b)

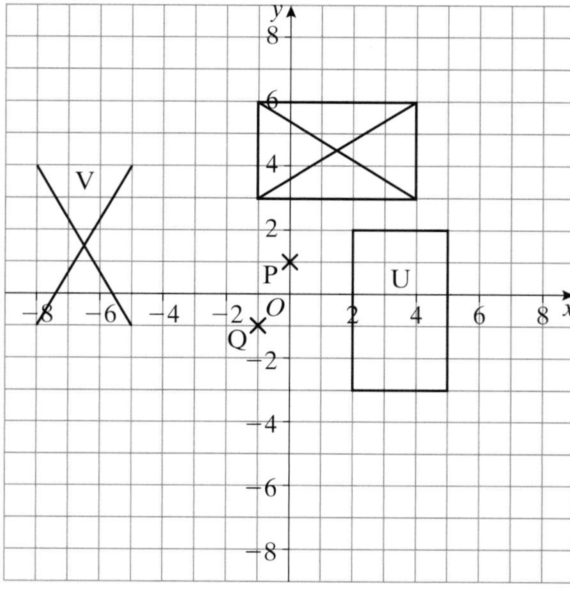

6 a) b)

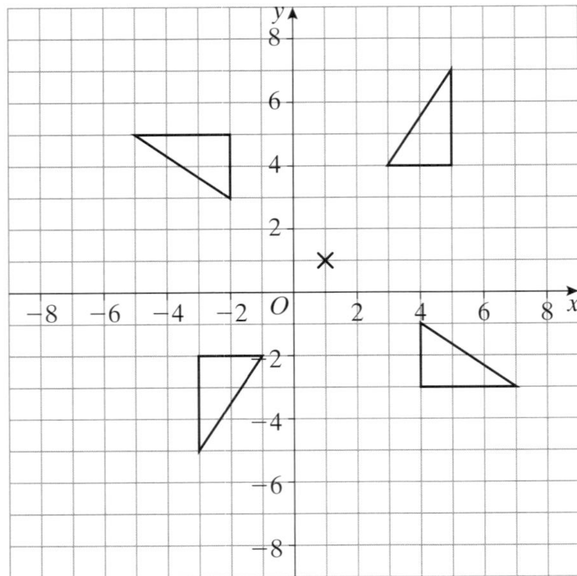

7 a) 90° anticlockwise. **b)** (2, 0)

8 Anita is wrong (for example, 90° anticlockwise and
then 90° clockwise using two different centres is
equivalent to a translation). Bella is (very) wrong.
Therefore Cat is wrong too!

Exercise 15.3

1 a), b)

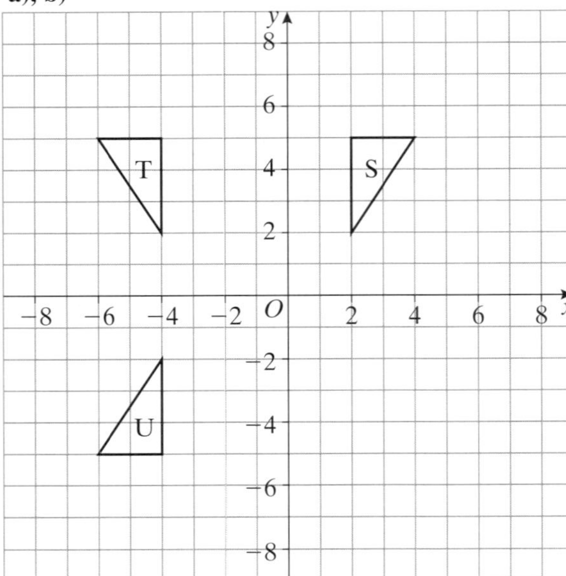

c) Rotation of 180° about (−1, 0)

2 a), b)

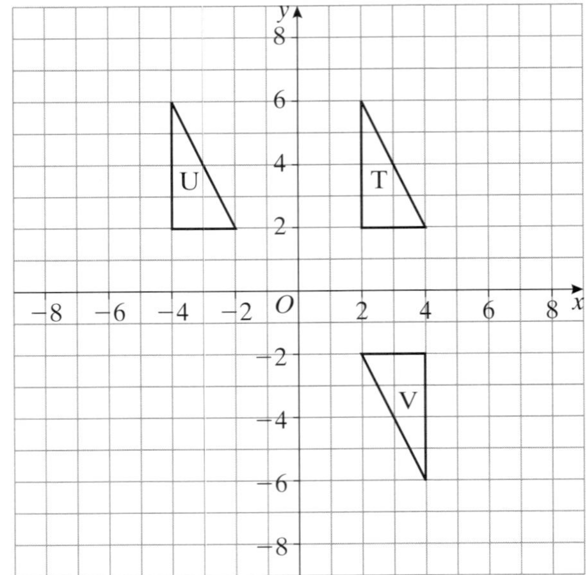

c) Rotation of 180° about (3, 0)

3 a), b)

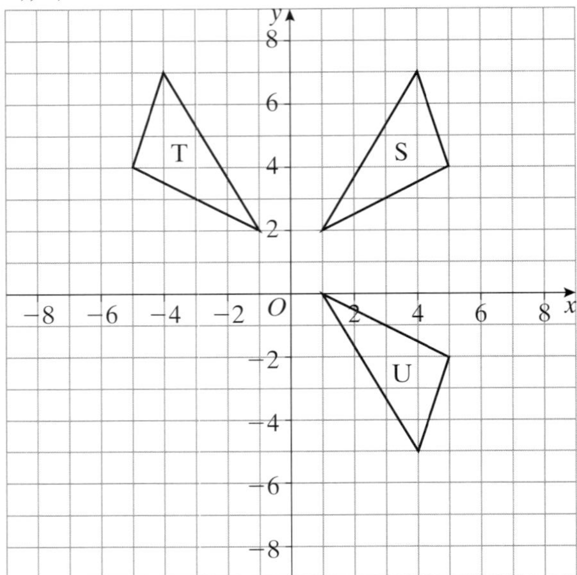

c) Rotation 180° about (0, 1)

4 a), b)

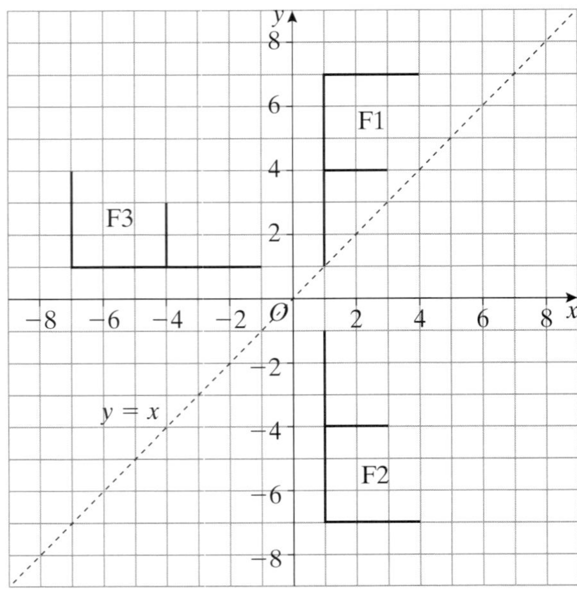

c) Rotation of 90° about (0, 0)

   Answers

5 a), b)

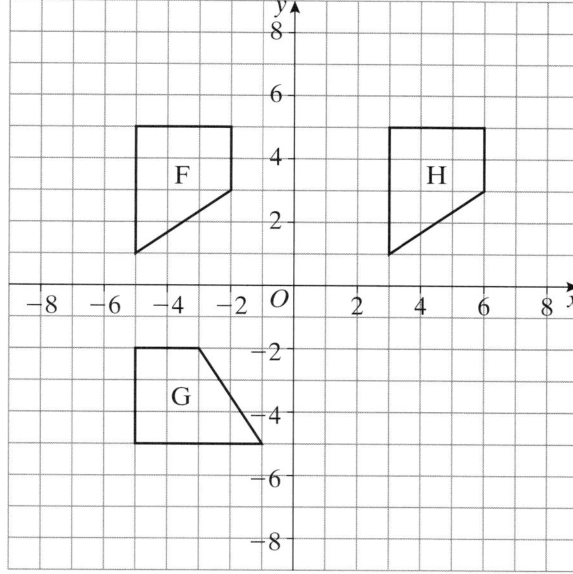

c) Translation by $\begin{pmatrix} 8 \\ 0 \end{pmatrix}$

6 a), b)

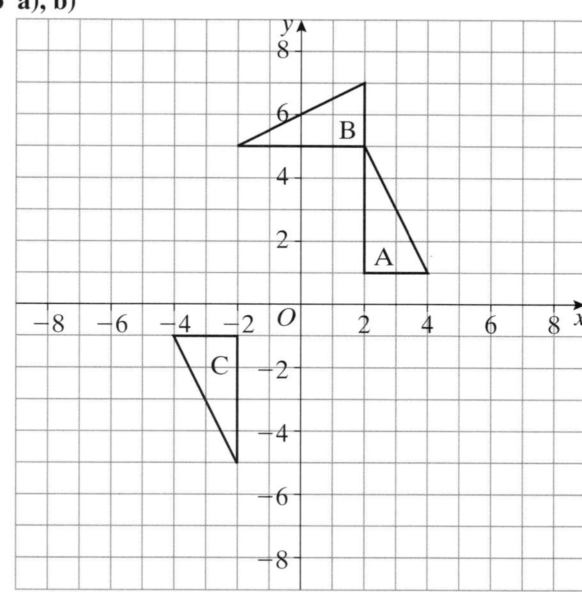

c) Rotation of 90° about (3, 0)

Exercise 15.4

1 a), b)

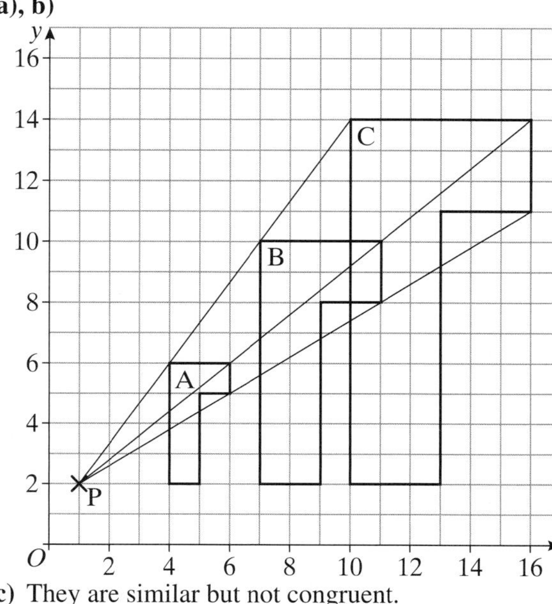

c) They are similar but not congruent.

2

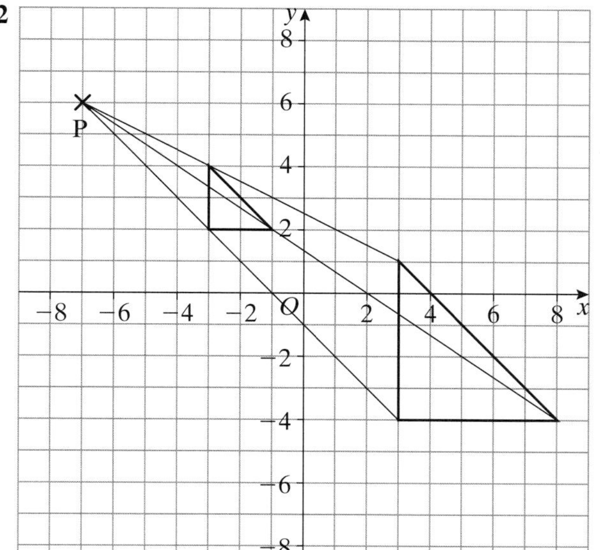

3 a), b)

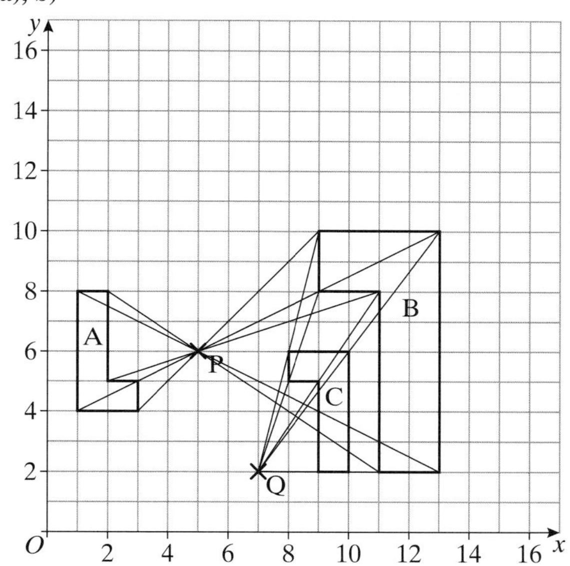

c) A and B are similar but not congruent.

d) A and C are congruent.

4 a) 3   **b)** (2, 5)

5

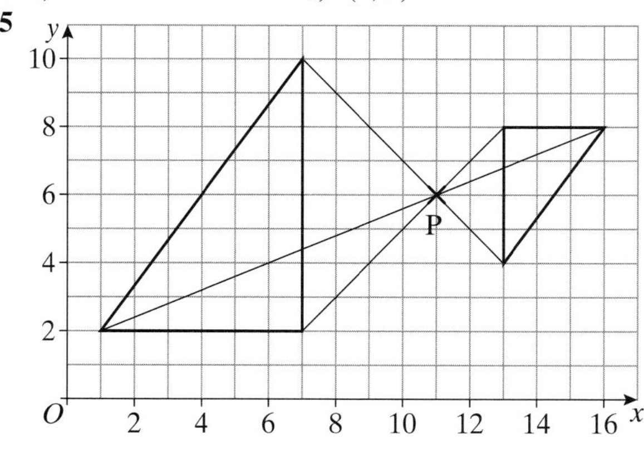

6 a) 1 (or −1)

   b) No, there are two possibilities: 1 and −1.

Exercise 15.5

1 a) 17.6 cm   **b)** 1 : 1.6   **c)** 1 : 2.56

2 54 cm^2

3 8.64 kg

4 a) RT = 11.25 cm, RQ = 20.25 cm

   b) 10.4 cm

5 a) 185 m   **b)** 10 350 tonnes (4 s.f.)

6 $x = 12$ cm, $y = 12$ cm

7 13.0 kg

8 a) Angle RPQ = angle RST (corresponding angles), and angle RQP = angle RTS (same reason), so each angle in one triangle is equal to the corresponding angle in the other one.
   b) 12 cm   **c)** 15 cm

9 a) 1.64 m   **b)** 14.8 m²

10 15.1 cm (3 s.f.)

Review Exercise 15

1 270 cm³

2 a) b)

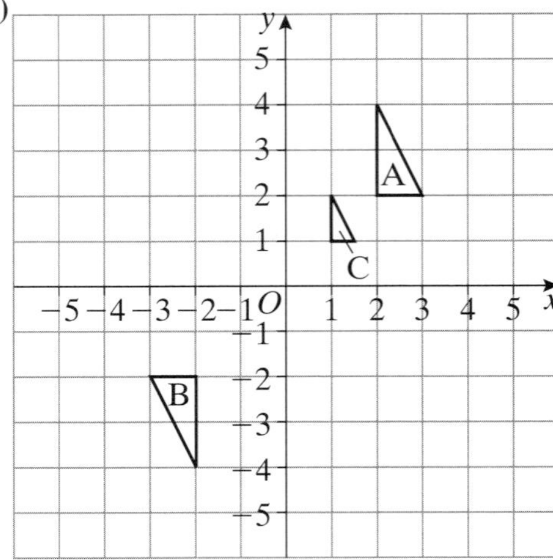

3 6

4 10 248 cm²

5

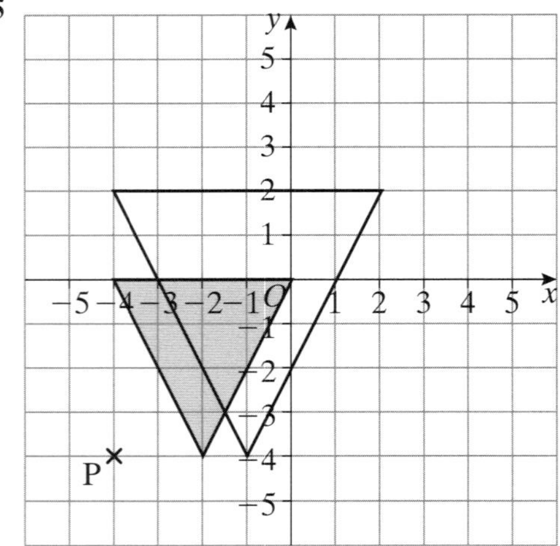

6 a)

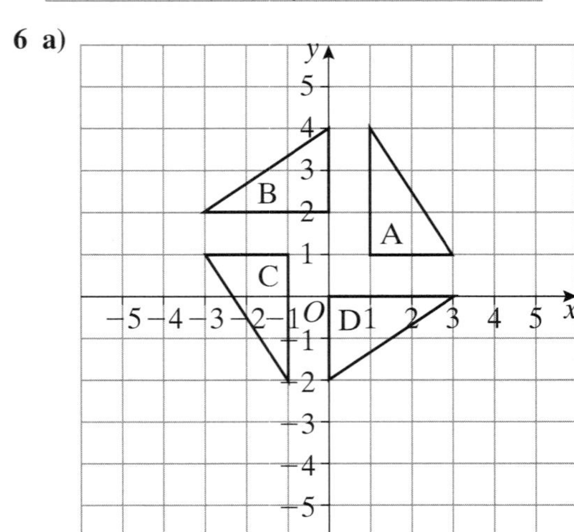

b) Rotation of 180° about (0, 1).

7 a) (i)

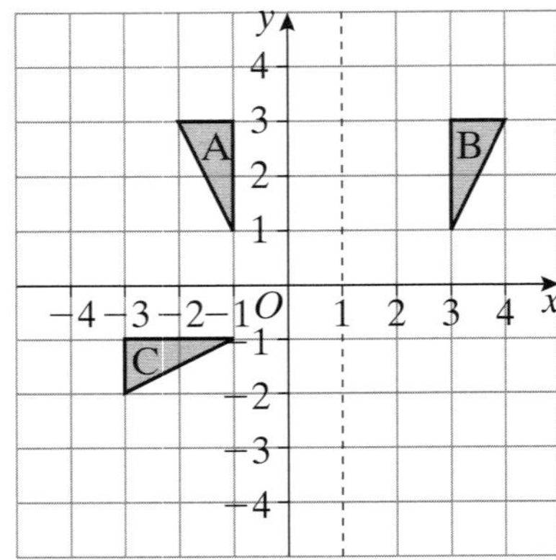

   (ii) $x = 1$
   b) Rotation of 90° anticlockwise about (0, 0)

8 a) Reflection in the y axis
   b) Rotation of 90° clockwise about (0, 0)

9 8.25 cm

10 a) 2
   b), c)

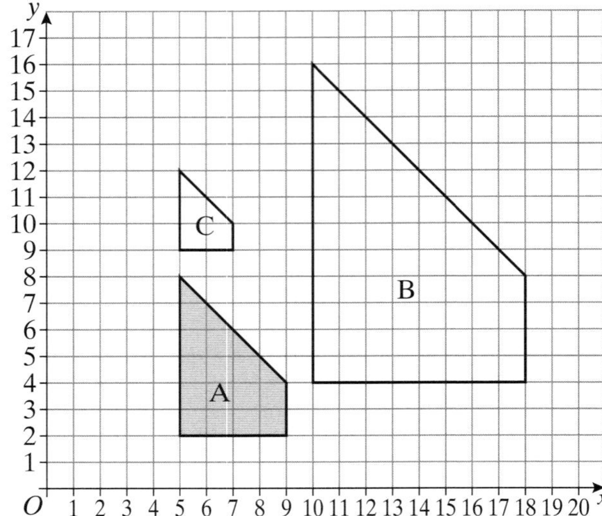

11 a) 13.5 cm   **b)** 2.5 cm

12 a) (i), c)

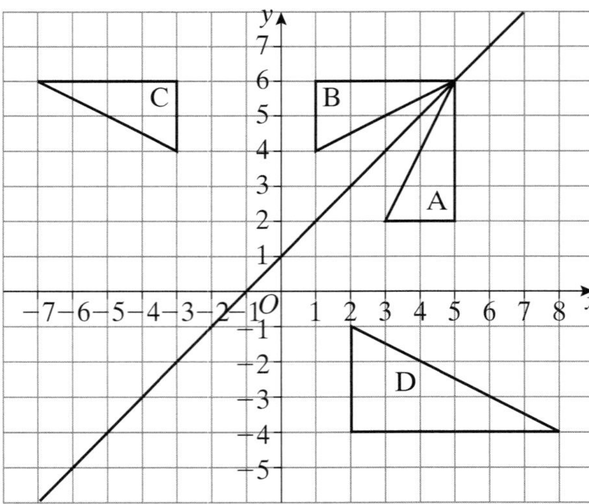

   a) (ii) $y = x + 1$
   b) Rotation of 90° anticlockwise about $(-1, 0)$.

13 317 mm

14 a) 12.5 cm
   b) 7.2 cm

15 23.4 buckets of sand to 3 s.f.

1 Icosahedron   **2** Cylinder
3 Congruent   **4** Similar
5 Alternate   **6** Parallel
7 Torus   **8** Octagon
9 Hemisphere   **10** Apex
11 Tetrahedron   **12** Rhombus
13 Minute of arc   **14** Radian
15 Truncated cone = frustum

Chapter 16: Pythagoras' theorem

Starter 16

Task 1: 16, 49, 6.25, 1.44, 0.64, 169, 36, 256
Task 2: 3.61, 3.16, 4, 4.74, 2.5, 7, 11.0, 11
Task 3: 3: 2.83, 3, 3.16, 3.32, 3.46, 3.61, 3.74, 3.87, 4, 4.12

Some whole numbers like 9 and 16 are the squares of whole numbers, and thus have exact whole number square roots.

Exercise 16.1

1 No   **2** Yes
3 No   **4** No
5 Yes, C   **6** No
7 No   **8** Yes, Q
9 No   **10** No

Exercise 16.2

1 5.39 cm   **2** 2.6 km
3 5.66 cm   **4** 6.71 cm
5 4.22 mm   **6** 7.91 cm
7 2.77 m   **8** 4.72 cm
9 1 km   **10** 7.81 cm
11 2 km   **12** 8.94 mm

Exercise 16.3

1 5.29 cm   **2** 4 km
3 5.03 m   **4** 8.06 cm
5 24 mm   **6** 4.45 cm
7 6.63 cm   **8** 11.5 cm
9 2.24 cm   **10** 7 cm
11 10.6 km   **12** 14.0 cm
13 14.6 cm   **14** 12.0 cm
15 11.8 cm

Exercise 16.4

1 a) 11.4 cm   b) 12.1 cm
2 a) 9.43 cm   b) 9.90 cm
3 b) 16.1 cm
4 Diagonal is 26.4 cm so the rod does fit.
5 a) 20.6 cm   b) 21.0 cm
6 4.5 cm
7 13 cm

Review Exercise 16

1 3.91 cm   **2** 9.93 cm
3 10.1 km   **4** 56.0 cm
5 19.4 mm   **6** 11.2 cm
7 a) 10 cm   b) 6.6 cm
8 a) 5.15 cm, 7.91 cm, 6.96 cm
   b) 8.29 cm
9 3.62 cm

10 a) 86.64 cm^2 b) 45.6 cm
11 13.7 cm
12 a) 21.25 cm^2 b) 9.86 cm
13 9.11 cm

Internet Challenge 16

1 $c = 17$
2 They are multiples of 3, 4, 5 and 5, 12 and 13.
3 3, 4, 5; 5, 12, 13; 6, 8, 10; 7, 24, 25; 8, 15, 17; 9, 12, 15; 12, 16, 20; 15, 20, 25
4 $n^2 - m^2$, $2mn$, $n^2 + m^2$ generates all the irreducible triples, and most of the others.
5 Yes
6 Yes, for example $3^2 + 4^2 + 12^2 = 13^2$
7 Fermat's Last Theorem was proved in 1994/1995 by Wiles (and Taylor).

Chapter 17: Introducing trigonometry

Starter 17

None of the triangles are similar.

The hypotenuse of the fifth triangle is exactly 3 units; this is because their lengths form a progression of $\sqrt{5}$, $\sqrt{6}$, $\sqrt{7}$, $\sqrt{8}$, $\sqrt{9}$.

The angles at the centre of the spiral become progressively smaller.

Exercise 17.1

1 4.23 cm   **2** 3.66 cm   **3** 7.78 cm
4 7.80 cm   **5** 6.10 cm   **6** 7.77 cm
7 5.28 cm   **8** 28.9 mm   **9** 58.9 mm
10 9.54 cm   **11** 7.05 cm   **12** 4.27 cm
13 35.2 mm   **14** 86.4 mm   **15** 42.4 mm

Exercise 17.2

1 9.27 cm   **2** 2.5 cm   **3** 3.05 cm
4 15.9 cm   **5** 10.1 cm   **6** 7.89 cm
7 2.64 cm   **8** 14.8 mm   **9** 53.6 mm
10 13.4 cm   **11** 8.02 cm   **12** 4.91 cm
13 108 mm   **14** 60.4 mm   **15** 57.2 cm

Exercise 17.3

1 7.50 cm   **2** 8.58 cm   **3** 6.64 cm
4 6.29 cm   **5** 25.3 cm   **6** 11.8 cm
7 9.45 cm   **8** 159 mm   **9** 55.4 mm
10 2.27 cm   **11** 19.6 cm   **12** 73.6 cm

Exercise 17.4

1 12.5 cm   **2** 5.91 cm
3 3.20 cm   **4** 2.90 cm
5 7.93 cm   **6** 8.30 cm
7 6.60 cm   **8** 3.32 cm
9 6.59 cm   **10** 8.29 cm

Exercise 17.5

1 36.9°   **2** 41.8°   **3** 17.5°
4 72.5°   **5** 30°   **6** 49.4°
7 47.1°   **8** 47.8°   **9** 41.3°
10 40.9°   **11** 54.5°   **12** 47.2°

Worksheet 17.5A

1 10.1 cm	**2** 1.49 km	**3** 27.2 cm
4 99.7 cm	**5** 22 m	**6** 5.7 m
7 8.49 cm each	**8** 2.62 m	**9** 5.48 km
10 36 m	**11** 5.7 m	**12** 639 m
13 82 m	**14** 63.5 m	**15** 21.3 cm

Worksheet 17.5B

1 031°	**2** 69.4°
3 42.8°	**4** 62.3°
5 67.9°	**6** 41°
7 55.4°	**8** 056°
9 76.08°	**10** 7.11°
11 48.2°, 41.8°	**12** 43.3°
13 7°	**14** 34°, 56°

Worksheet 17.5C

1 2.93 cm	**2** 4.30 cm
3 8.02 cm	**4** 10.85 cm
5 30.4°	**6** 46.2°
7 24.0°	**8** 10.97 cm
9 15.49 cm	**10** 34.18 cm
11 12.20 cm	**12** 15.53 cm
13 9.27 cm	**14** 12.17 cm
15 37.73 cm	**16** 45.4°
17 18.63 cm	**18** 3.15 cm
19 12.81 cm	**20** 14.90 cm

Exercise 17.6

1 a)

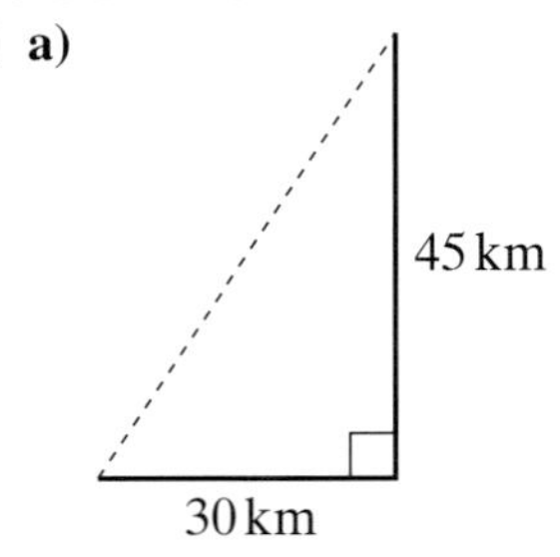

b) 54.1 km

c) 214°

2 a) 12.4 cm

b) 80.9 cm²

c) 38.9°

3 a) Angle PMQ = angle RMQ, so must be $180 \div 2 = 90°$

b) 65.1°

c) 1.72 m (172 cm)

4 a)

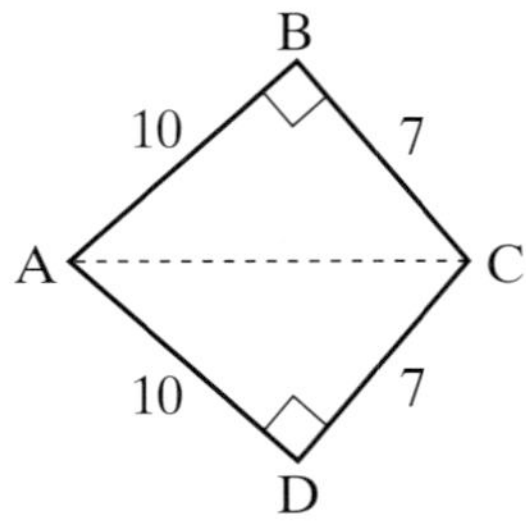

b) AC = 12.2 cm (122 mm)

c) 70°

5 a) 9.44 cm

b) 59.1°

6 a)

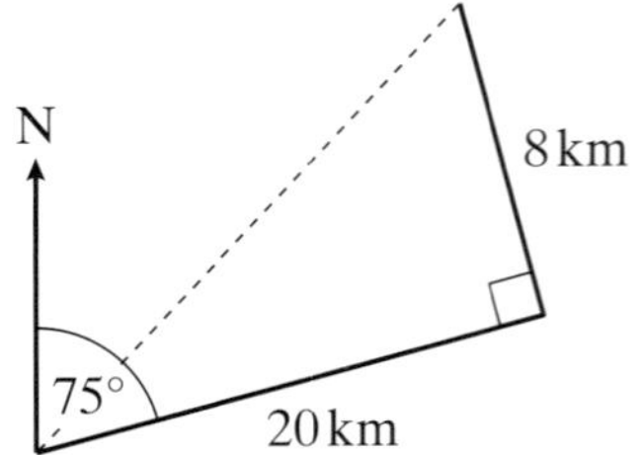

b) 21.5 km

c) $075 - 022 = 053°$

Review Exercise 17

1 $a = 6.74$ cm, $b = 9.18$ cm, $c = 6.10$ cm, $d = 11.0$ cm, $e = 5.54$ cm, $f = 5.91$ cm

2 $a = 44.3°$, $b = 54.3°$, $c = 33.5°$, $d = 53.1°$, $e = 49.9°$, $f = 22.6°$

3 4.28 m

4 a) 5.29 cm

b) 41.4°

c) 12.4 cm

5 8.79 m

6 116 cm

7 a) 20.6°

b) 110.6°

8 50.2°

9 a) 62.0°

b) 19.1 m

10 11.7 m **b)** 36.9°

Internet Challenge 17

1 Euclid	**2** Pythagoras
3 Benoit Mandelbrot	**4** Apollonius of Perga
5 Johann Kepler	**6** Sir Isaac Newton
7 Leonhard Euler	**8** Bernhard Riemann
9 Plato	**10** Felix Klein

Chapter 18: 2-D and 3-D objects

Starter 18

Nets 1, 3, 4, 5 and 8 will make a cube.

Exercise 18.1

1 a)

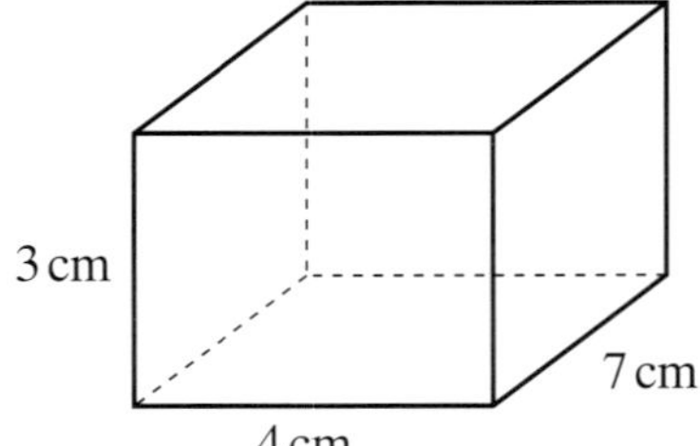

b)

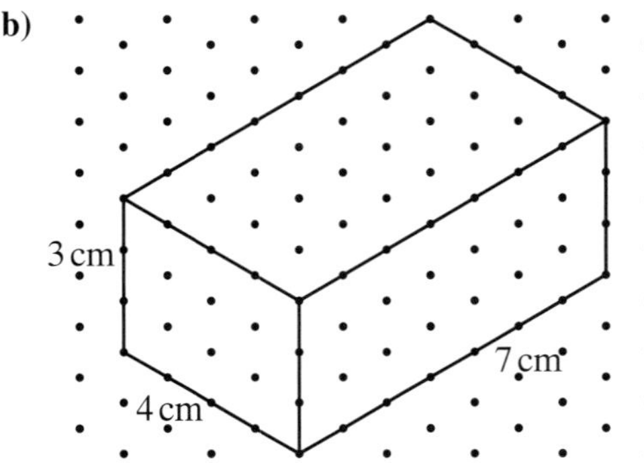

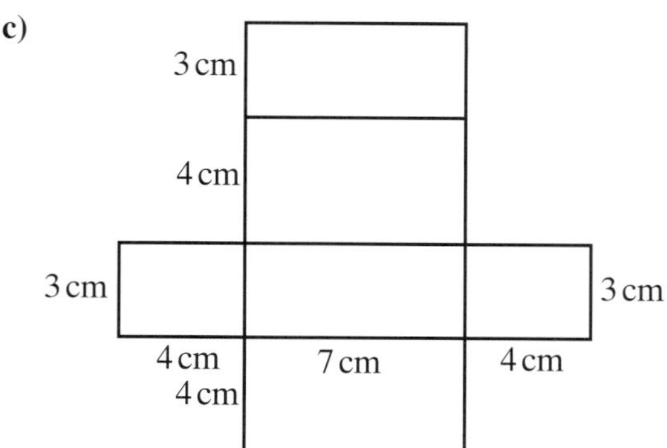

c)

2 a)
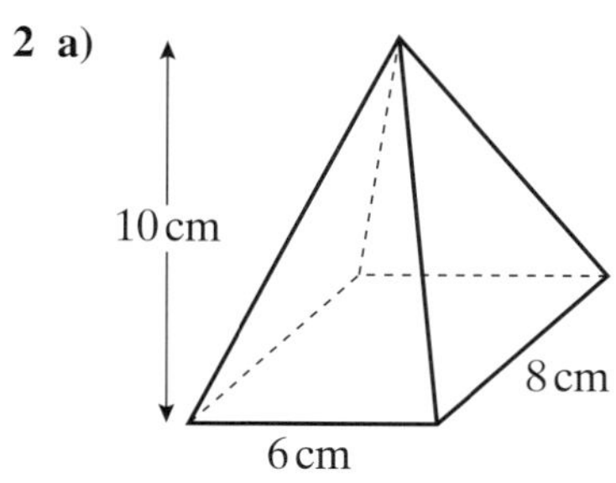

b)
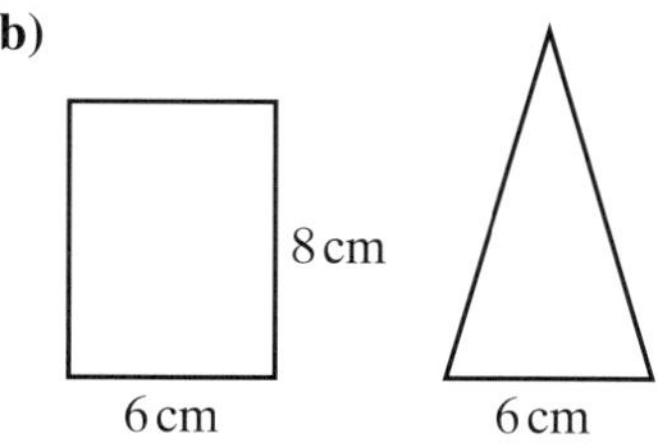

3 a)
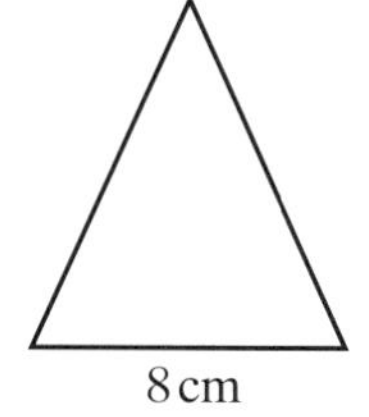

b)
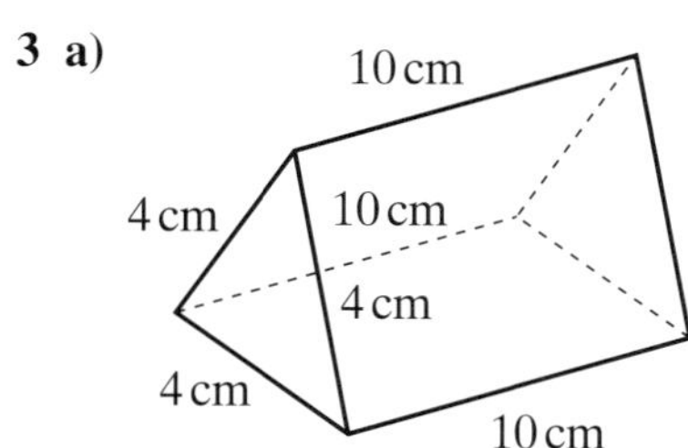

4 a) (i)   **(ii)**   **(iii)**
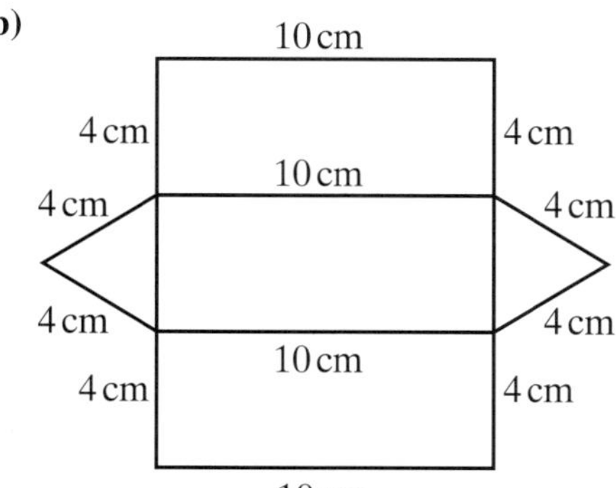

b)
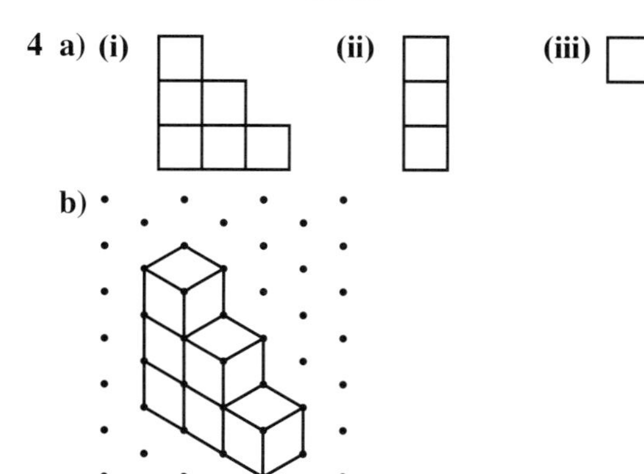

5 a)   **b)**
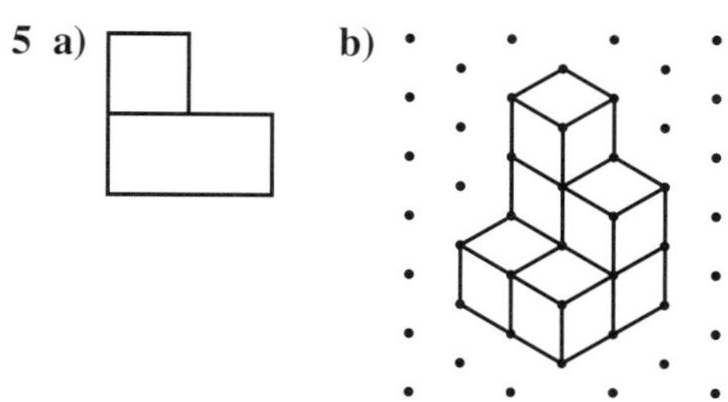

6 a)
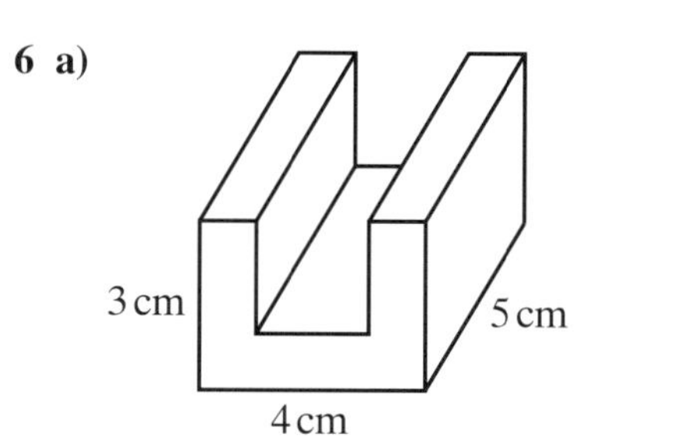

b)
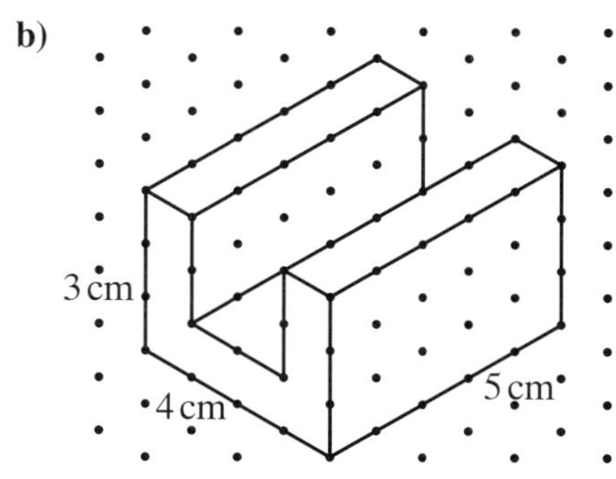

7 Check students' work.

Exercise 18.2

1 a) B (7, 6, 0), C (0, 6, 0), Q (7, 6, 8), R (7, 0, 8), S (0, 0, 8)
  b) M (3.5, 3, 4)
  c) N (3.5, 6, 4)
2 a) A (5, 0, 6), B (5, 0, 0), C (5, 4, 0)
  b) M (5, 2, 0), N (5, 2, 3)
  c) M and N have the same y coordinates.
  d) Triangular-based pyramid
3 a) R (50, 10, 80), S (0, 10, 80)
  b) V (25, 18, 80)
  c) (25, −2, 40)
4 a) M (6, 2, 2)   **b)** N (6, 1, 4)
5 a) T
  b) It is right angled (and isosceles).

Exercise 18.3

1 3050 cm³, 1020 cm²   **2** 172 cm³, 188 cm²
3 286 cm³, 267 cm²   **4** 2304π cm³, 576π cm²
5 144π cm³, 108π cm²   **6** 12π cm³, 24π cm²
7 144π   **8** 8 cm square
9 ($r = 15$) 4500π cm³   **10** ($r = 4$) 16π cm²

Exercise 18.4

1 2 000 000 cm³   **2** 0.5 m²
3 3 km³   **4** 6.6 cm²
5 1 000 000 000 mm³   **6** 0.035 m³
7 24 000 cm²
8 a) 20 000 cm³   **b)** 0.02 m³
9 a) 4.2 m³   **b)** 4 200 000 cm³   **c)** 4200 litres
10 a) 512 000 cm³
  b) 0.512 m³
  c) 0.8 m on each side

Exercise 18.5

1

$2a + 3b$	$3ab$	$a + b + c$	$2a^2c$	$2a^2 + bc$	$ab(b + 2c)$
	A		V	A	V

2

Expression	Length	Area	Volume	None of these
$\pi a^2 b$			✓	
$\pi b^2 + 2h$				✓
$2ah$		✓		

3 πpq, $\dfrac{qr}{2}$ and $r(p + q)$

4

Expression	Length	Area	Volume	None of these
$3rl$		✓		
$\dfrac{2(r + l)^2}{h}$				✓
$\dfrac{4\pi r^4}{3l}$			✓	

5

$\dfrac{ab}{h}$	$2\pi b^2$	$(a + b)ch$	$2\pi a^2$	πab	$2(a^2 + b^2)$	$\pi a^2 b$
	✓			✓	✓	

Review Exercise 18

1

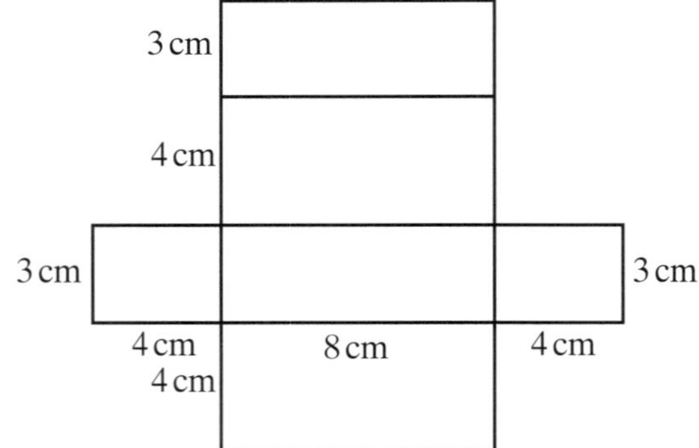

2 a)

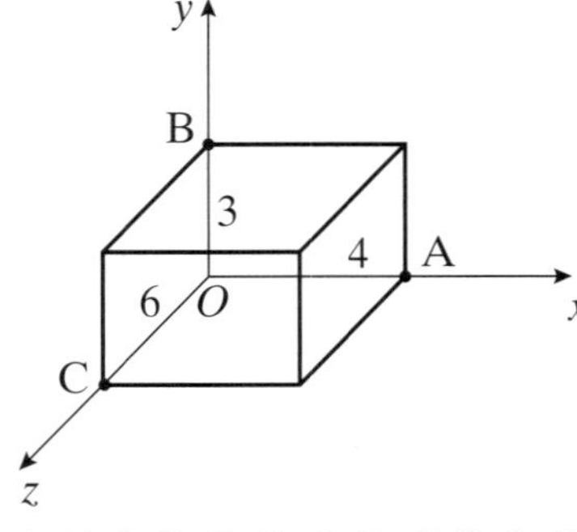

b) A $(4, 0, 0)$, B $(0, 3, 0)$, C $(0, 0, 6)$
c) M $(0, 1.5, 3)$
3 a) $35\,\text{cm}^2$ **b)** $2\,500\,000\,\text{cm}^3$

4 a) $5.58\,\text{m}^3$ **b)** $5\,580\,000\,\text{cm}^3$
c) $15.2\,\text{m}^2$ **d)** $152\,000\,\text{cm}^2$
5 $2.52\,\text{m}$
6 a)

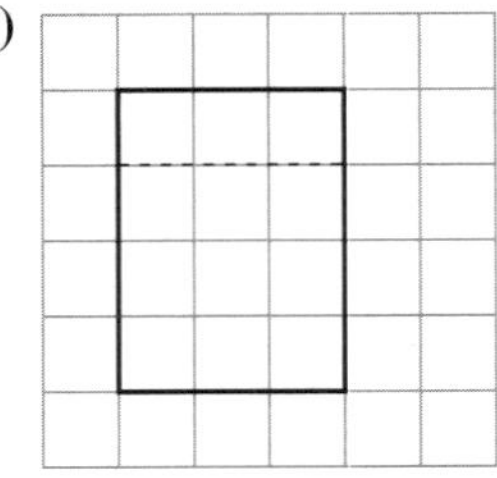

b)

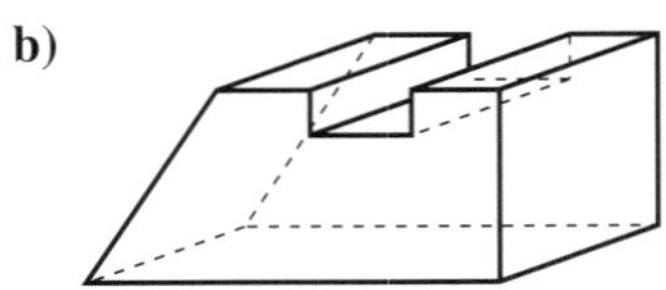

7 a) (i) D **(ii)** C
b) 2
8 a)

$\dfrac{\pi r^2}{x}$	$\pi(r + x)$	$\pi r + r$	$\dfrac{\pi r^3}{x}$	$\pi r^2 + rx$	$\dfrac{r^2}{\pi x}$
			✓	✓	

b)

πr^3	$\pi r^4 + \pi x$	$\dfrac{\pi r^4}{x}$	$\pi r^2 + \pi x$
	✗		

9 $137\,000\,\text{cm}^3$
10 $8.9\,\text{cm}$
11 $18\pi + 15\pi = 33\pi$
12 15π

Internet Challenge 18

1 There are five; we will never find more (Euclid proved there are only five).
2 The Greek mathematician Plato wrote extensively about them.
3 4 Check students' answers.
5 Such a football is not a Platonic solid, since it uses two different polyhedra for the faces.
6 7 Check students' answers.
8 59
9 Double Planet = stellated octahedron, Gravity = stellated dodecahedron
10 Check students' answers.

Chapter 19: Circle theorems

Starter 19

1 Radius **2** Circumference **3** Centre
4 Arc **5** Diameter **6** Chord
7 Tangent **8** Sector **9** Segment
10 a) Arc-en-ciel $\rightarrow$ curve in the sky $\rightarrow$ a good name
b) A 'segment' of an orange is really a sector $\rightarrow$ not a good name

Exercise 19.1

1 a) $x = 90°$ (tangent and radius at right angles)
b) $y = 61°$
2 RT $= 25 - 7 = 18\,\text{cm}$

3 **a)** 136° (radius/tangent at 90° and quadrilateral
angles add up to 360°)
b) Cyclic quadrilateral (opposite angles add up to 180°)
4 **a)** Isosceles   **b)** 65°   **c)** 50°
5 8 cm
6 18.8 cm (3 s.f.)
7 **a)**

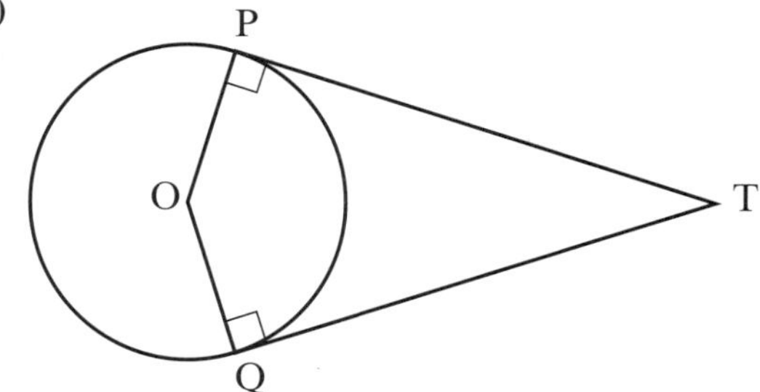

b) Square
8 **a)** Radius bisects a chord at right angles
b) 17 cm
c) 6 cm
d) 19.3 cm

Exercise 19.2
1 $a = 59°, b = 28°$
2 $c = 48°, d = 44°$
3 $e = 42°, f = 45°, g = 51°$
4 $h = 39°$
5 $i = 88°$
6 $j = 58°$
7 $k = 63°, l = 75°$
8 $m = 76°$
9 $n = 51°$
10 $p = 36°$
11 $q = 72°$
12 $r = 44°, s = 46°$
13 $t = 21°$
14 $u = 15°$
15 $v = 30°$

Exercise 19.3
1 $a = 92°, b = 67°$ 2 $c = 105°, d = 95°$
3 $e = 109°, f = 78°$ 4 $g = 126°, h = 27°$
5 $i = 37°$ 6 $j = 29°, k = 61°$
7 $l = 114°$ 8 $m = 54°$
9 $n = 93°$ 10 $p = 162°$
11 $q = 41°$ 12 $r = 87°, s = 83°$
13 $t = 116°$ 14 $u = 150°$
15 $v = 132°, w = 42°$

Review Exercise 19
1 **a)**

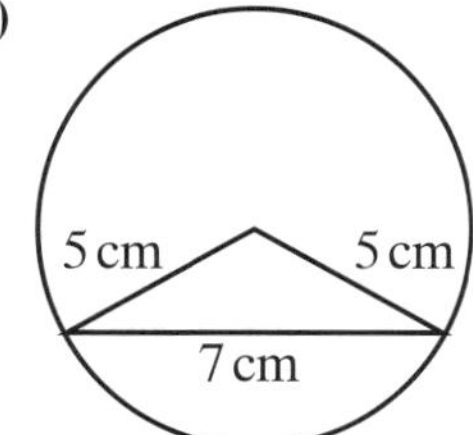

b) 3.57 cm
2 **a)** 36°
b) Yes (opposite angles add up to 180° − cyclic
quadrilateral)
3 **a)** 74°
b) No (74° and 32° do not add up to 180°)

4 **a)** 27° (tangent and radius meet at right angles)
b) 63° (alternate segment theorem)
5 **a)** 90° (angle in a semicircle)
b) 56° (same segment as angle PSQ)
c) 112° (angle at centre = twice angle at
circumference)
6 **a)** 60°
b) 35°
c) Yes (angle DAB = 65° + 25° = 90°)
7 **a)** 55° (angle ABC = 90°, angles in a triangle add up
to 180°)
b) 35°
c) 110°
8 **a)** 48°
b) 18°
c) Angle CDA is not 90°
9 **a)** 40°
b) 12°
c) Angle BAD = 78° so cannot be the angle in a
semicircle
10 **a)** 27° (90° − 36° = 54°, 54° ÷ 2 = 27°)
153° (opposite angles in a cyclic quadrilateral)
11 **a) (i)** 56°   **(ii)** 112°
b) (i) 23°   **(ii)** 65°

Worksheet 19R
1 $a = 108°$ 2 $b = 48°$
3 $c = 40°$ 4 $d = 44°$
5 $e = 19°, f = 51°$ 6 $g = 55°$
7 $h = 50°, i = 98°$ 8 $j = 32°, k = 58°$
9 $l = 60°, m = 30°$ 10 $n = 114°$
11 $p = 90°, q = 39°$ 12 $r = 90°$
13 $s = 124°$ 14 $t = 43°$
15 $u = 108°$ 16 $v = 70°, w = 110°$
17 $x = 33°, y = 57°$ 18 $z = 71°$
19 $a = 60°, b = 120°$ 20 $c = 36°, d = 29°$

Internet Challenge 19
The line segment AP1 is perpendicular to BC, BP2 is
perpendicular to CA and CP3 is perpendicular to AB.

The three line segments AP1, BP2, CP3 intersect at M.

P4 is the midpoint of AM, P5 is the midpoint of BM and
P6 is the midpoint of CM.

P7 is the midpoint of BC, P8 is the midpoint of AC, P9 is
the midpoint of AB.

Check students' constructions.

This construction was probably first made by Feuerbach,
Poncelet or Brianchon around 1820.

Chapter 20: Collecting data
Starter 20
Enrico's method is too coarse: 0 to 50 has a frequency of
0, so there are only two groups containing any data at all:
51−100 (9) and 101−150 (11).

Similarly, Kwame's sheet is too fine; most of the
frequencies are 0.

Sunita's table is the best; there are 8 groups, with
frequencies of 2, 2, 5, 4, 2, 2, 1, 2 respectively.

Exercise 20.1

1 a) Biased sample – those who like school are liable to arrive early!

b) Select 50 names at random from a school roll of names.

2 Andy's method is probably better. Neither method produces a random sample, but if the classes are all the same size then Andy's method produces a representative sample. It is very unlikely that the numbers of students whose names begin with A, B, C etc. will be even approximately equal, so this is not a fair process at all.

3 This is not a good sampling method – Janice should buy the sweets from a selection of different shops. (Those at one shop may all have come from the same production run.)

4 Frequencies of 10, 10, 11, 8, 11 (Note that the last two have, necessarily, been rounded to the nearest integer.)

5 Each continent contains a different number of countries, so Georgie's method is better because each continent is represented in the same proportion.

6 a) Secondary

b) Primary

c) Primary

d) Primary

e) Primary (if you measure them), secondary (if you read the values off the sleeve)

f) Primary (if surveyed), secondary (if from e.g. internet)

Exercise 20.2

1 *First question:*

What gender are you?   Boy ☐   Girl ☐

Second question:

What is your favourite type of music?

Classical   Jazz   Rock   Hip hop   Other (specify)
☐   ☐   ☐   ☐   ☐

2 a) It is not possible to see which responses are from boys and which from girls.

b)

Boys	Hockey	Soccer	Tennis	Other

Girls	Hockey	Soccer	Tennis	Other

3 People may be reluctant to answer question 1. It should be rewritten with tick boxes for ages e.g. 11−20, 21−30, etc.

Question 2 is too vague. It should be rewritten with tick boxes with the names of different newspapers.

Question 3 is a leading question, and should be rewritten as follows:

What do you think of newspaper prices?

Too low   About right   Too high
☐   ☐   ☐

4 Cassie has noticed that the local shop is mostly preferred by men, whereas the supermarket is mostly preferred by women.

5 a)

	German	Spanish	Italian	Total
Male	(25)	5	10	(40)
Female	22	25	(13)	60
Total	47	(30)	(23)	(100)

b) 53 students do not study German.

6

	Liked coffee	Did not like coffee	Total
Boys	6	4	10
Girls	8	12	20
Total	14	16	30

7

Name of club	Tally	Frequency
United		
Rovers		
City		
Other		

Review Exercise 20

1

Type of video	Tally	Frequency
Comedy		
Romance		
Horror		
Other		

2 The question should be more specific: Which of these foods do you eat each week? The question should include response boxes (tick boxes)

Chips   Salad   Meat   Eggs   Other
☐   ☐   ☐   ☐   ☐

3 a) Nigel could obtain a list of all the pupils in the school, and number them 1, 2, 3, …, 1000. He could then use random numbers to select the required number of students.

b) (i) 11   **(ii)** 8

4 What gender are you?   Boy ☐   Girl ☐

Do you have a part-time job?   Yes ☐   No ☐

5 a) No − the classes are not all of the same size.

b) 11

6

Type of vehicle	Tally	Frequency
Car		
Lorry		
Motor cycle		
Other		

7 a)

	School dinners	Sandwiches	Other	Total
Boys	(12)	(3)	1	(16)
Girls	(8)	4	(2)	14
Total	20	7	3	30

b) 4

8 a) The question does not allow negative responses.

b) (i) The question is too vague.

(ii) How much do you typically spend per day in the canteen?

Less than £1   ☐
Over £1 but less than £3   ☐
Over £3   ☐

9 a) 8   **b)** 85

10 a) What is your favourite type of food?

Italian   ☐
Indian   ☐
Chinese   ☐
Greek   ☐
Other (specify) ☐ __________

b) It is a leading question (don't you agree that…).
The sample is biased (members of his own family).

Internet Challenge 20

1 1841   **6** William (the Conqueror)
2 2011   **7 c)** 336 000
3 Fine of £1000   **8 b)** 22 million
4 1086   **9 a)** 15%
5 The Domesday survey   **10 a)** 52 million

Chapter 21: Working with data

Starter 21

Medical research: The y axis should be drawn all the way down to zero, then the recent growth may be seen to be much more gradual.

Milk bottles: The bottle for Sykes Farm is 60% taller and 60% wider, making it look much more than just 60% more volume.

Staff cars: The yellow sector looks much bigger because it is at the front of the pseudo-3-D diagram.

Exercise 21.1

1
```
2 | 1  4  9          Key: 2|1 = 21
3 | 0  5  7
4 | 1  3  7  8
5 | 2  5  6  7  9  0
6 | 2
7 |
8 | 1
```

2
```
0 | 1  5  6          Key: 3|6 = 36
1 | 0  1  3  4  8
2 | 1  2  4  5
3 | 6  8  9
```

3 a)
```
1 | 9                Key: 1|9 = 19
2 | 5  9
3 | 0  1  2  5  6  6
4 | 1  4  7  8
5 | 0
```

b) Range = 50 − 19 = 31

4
```
4 | 9                Key: 4|9 = 49
5 |
6 | 1  3  8
7 | 2  3  4  6  9
8 | 1  4  5  6  8  8
9 | 1  2
```

5
```
 8 | 8               Key: 9|3 = 93
 9 | 3  4  6  7
10 | 0  0  2  5  6
11 | 1  3
12 |
```

Exercise 21.2

1 a) 40   **b)** 2.55
2 a) 35.2   **b)** 35 to 39
c) Largest possible range is 44 − 25 = 19
3 24.7 cm
4 40.2 years
5 a) 17.25 minutes
b) Exact raw values are not recorded.

Exercise 21.3

1 a)

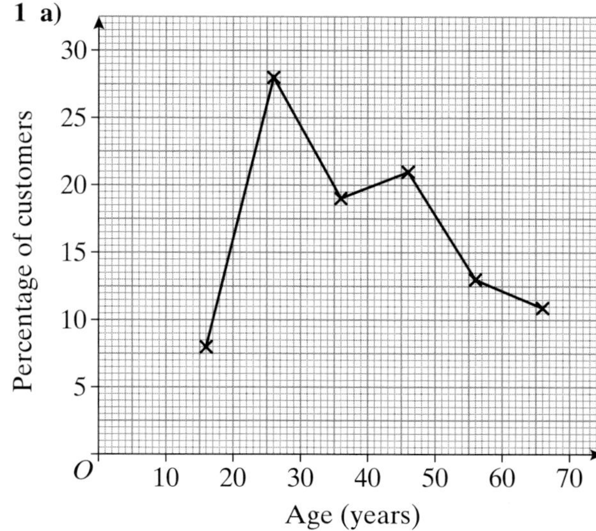

b) 31 − 40

2 a)

b) 65 grams
c) $60 \leqslant w < 80$

3 a)

Height (h cm)	Frequency
$140 \leqslant h < 150$	15
$150 \leqslant h < 160$	35
$160 \leqslant h < 165$	20
$165 \leqslant h < 170$	18
$170 \leqslant h < 180$	22
$180 \leqslant h < 190$	12
$190 \leqslant h < 210$	12

b)

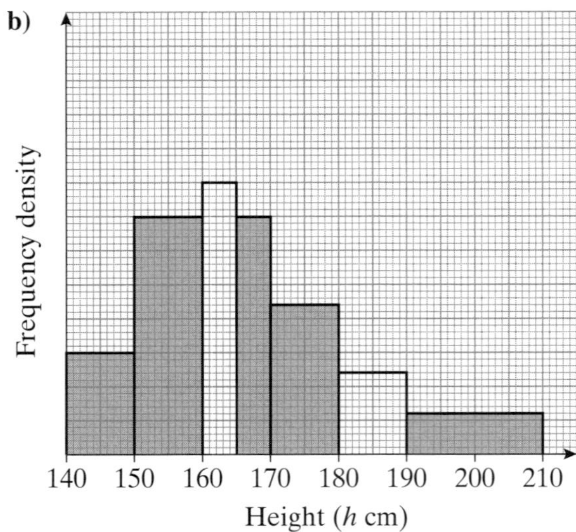

4 a)

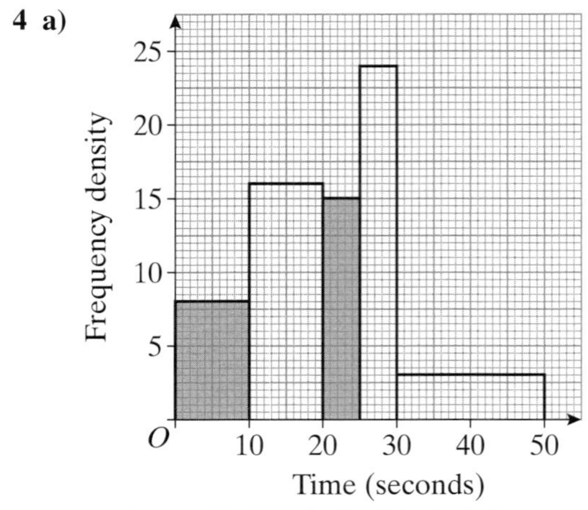

b)

Time (t seconds)	Frequency
$0 < t \leqslant 10$	10
$10 < t \leqslant 20$	18
$20 < t \leqslant 25$	14
$25 < t \leqslant 30$	10
$30 < t \leqslant 50$	8

5 a)

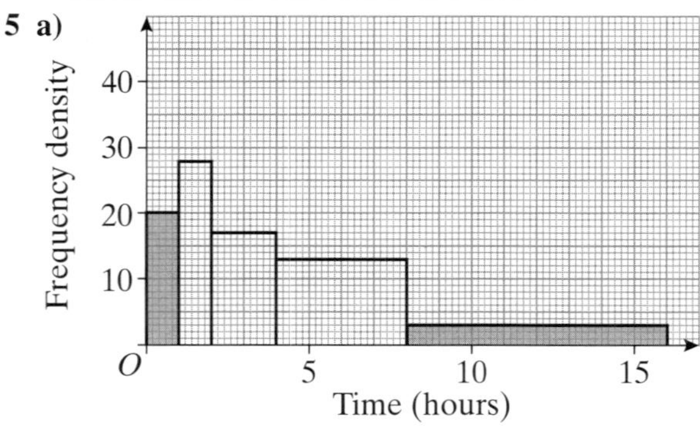

b)

Time (t seconds)	Frequency
$0 < t \leqslant 1$	20
$1 < t \leqslant 2$	28
$2 < t \leqslant 4$	34
$4 < t \leqslant 8$	52
$8 < t \leqslant 16$	24

Exercise 21.4

1 a) 70   **b)** 18 seconds   **c)** 21% approx

2 a) 1200

b)

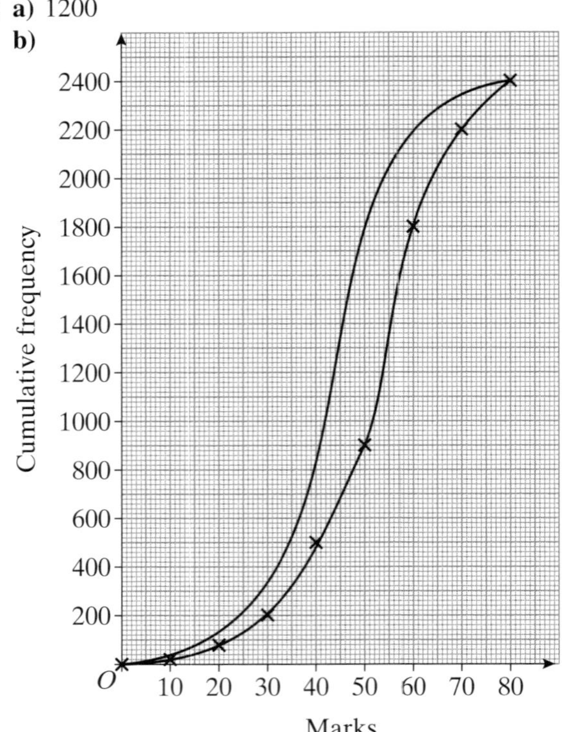

c) 53 approx

3 a)

Waiting time (t seconds)	Cumulative frequency
$0 \leqslant t < 50$	4
$0 \leqslant t < 100$	11
$0 \leqslant t < 150$	21
$0 \leqslant t < 200$	37
$0 \leqslant t < 250$	67
$0 \leqslant t < 300$	80

b)

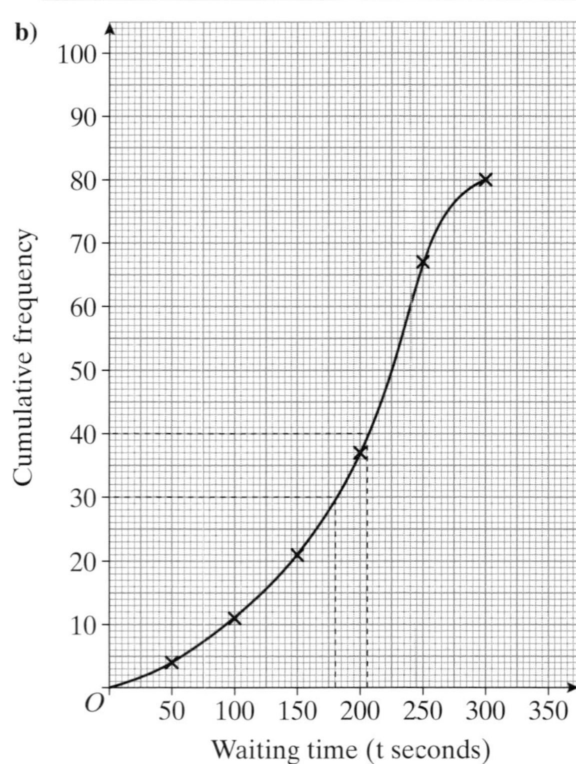

Answers

(i) 205 seconds
(ii) $80 - 30 = 50$ people

4 a) £172.72 ÷ 35 = £4.92
  b) (i)

Hourly rate of pay (£x)	Cumulative frequency
$3.00 < x \leqslant 3.50$	1
$3.00 < x \leqslant 4.00$	3
$3.00 < x \leqslant 4.50$	7
$3.00 < x \leqslant 5.00$	14
$3.00 < x \leqslant 5.50$	33
$3.00 < x \leqslant 6.00$	35

(ii)

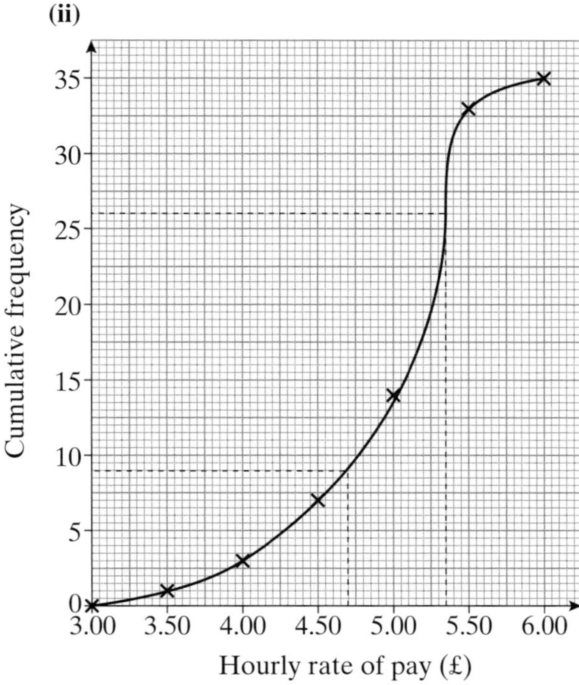

  c) £0.65 approx
  d) $35 - 21 = 14$ approx
5 A and Q, B and P, C and S, D and R, E and T

Exercise 21.5

1 a), c)

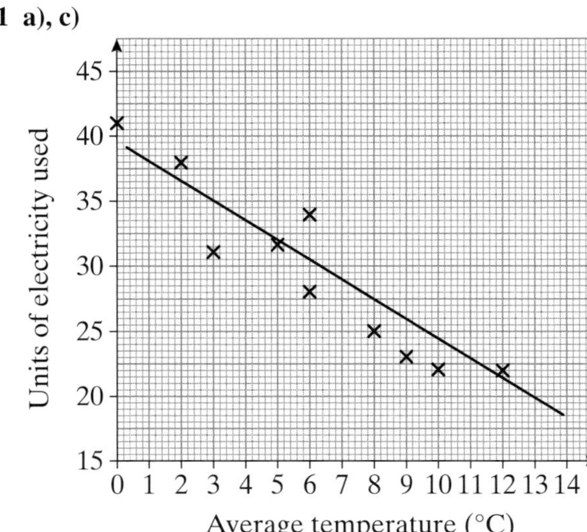

  b) Negative correlation
  d) (i) 3°C   **(ii)** 29 units

2 a)

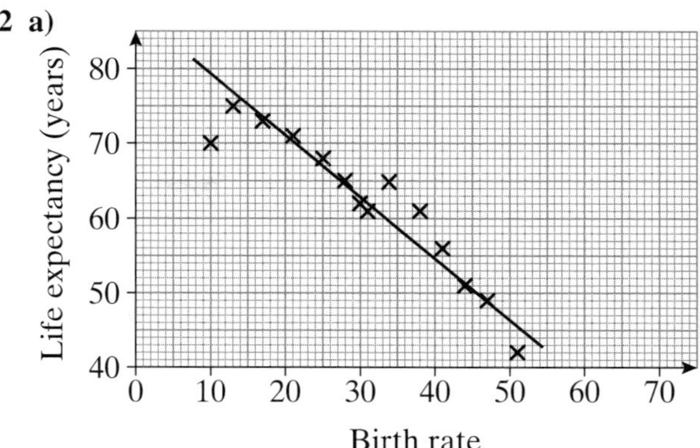

  b) 53 years approx
  c) 26 approx
3 a) (i) No correlation   **(ii)** Length of hair
  b) (i) Positive correlation
    (ii) Waist measurement (or distance round neck)
4 a), c)

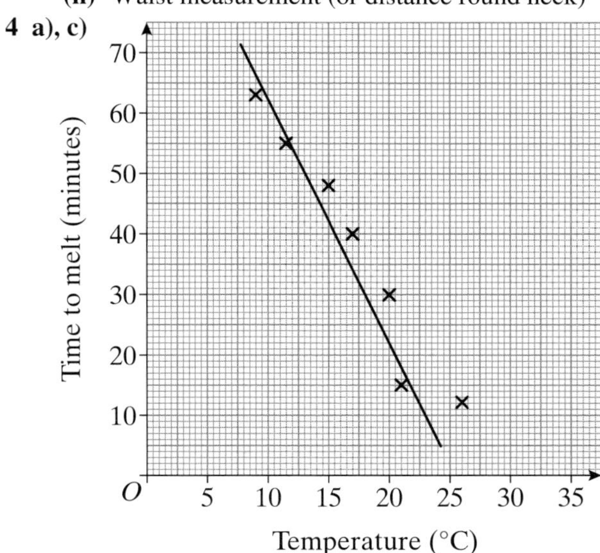

  b) Negative correlation
  d) 50 minutes
  e) 23°C
  f) 35°C is outside the range of the data.
5 a), c)

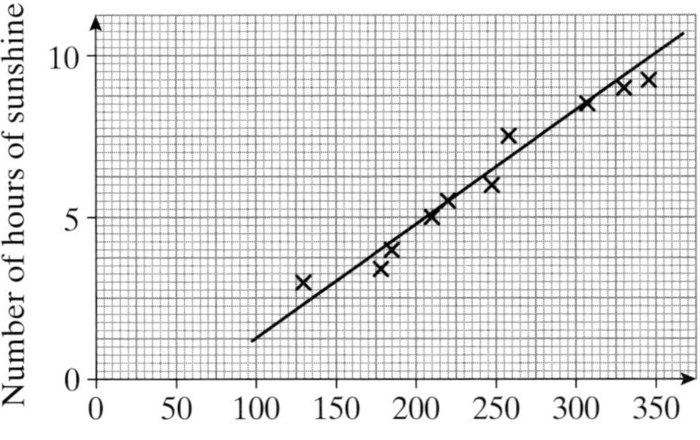

  b) Positive correlation
  d) (i) 265 approx   **(ii)** 4.75 hours approx

Review Exercise 21

1

Stem	Leaf	
14	7 7 7 8	Key: 14 | 7 = 147
15	1 2 3 5 8 8	
16	4 5 6 6 7	
17	1 2 8	
18	9 9	

2

Stem	Leaf
0	5 7 8 8
1	0 0 0 0 2 5 5 5 6
2	0 0 0 4 5 3 3 5
3	3 5

Key: $1\,|\,2 = 12$

3 a) 11 to 15   **b)** $1230 \div 75 = 16.4$

4 $£53\,000 \div 50 = £1060$

5 $988 \div 10 = 98.8$ grams

6

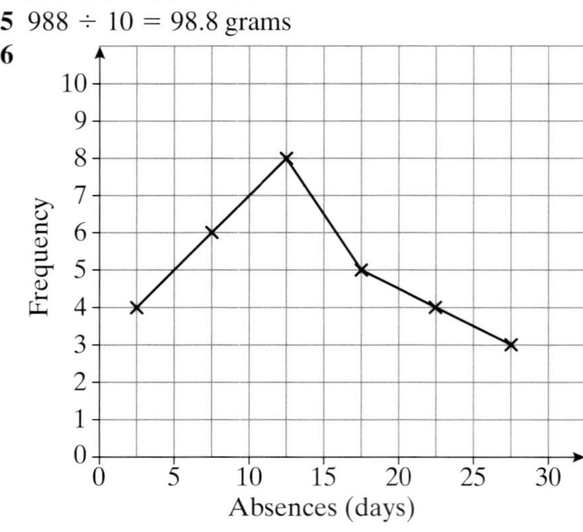

7

Time (t minutes)	Frequency
$0 < t \leqslant 10$	20
$10 < t \leqslant 15$	18
$15 < t \leqslant 30$	45
$30 < t \leqslant 50$	52
Total	135

8 a)

Revision time (t minutes)	Frequency
$20 \leqslant t < 25$	20
$25 \leqslant t < 40$	36
$40 \leqslant t < 60$	29
$60 \leqslant t < 85$	30
$85 \leqslant t < 95$	32

b)

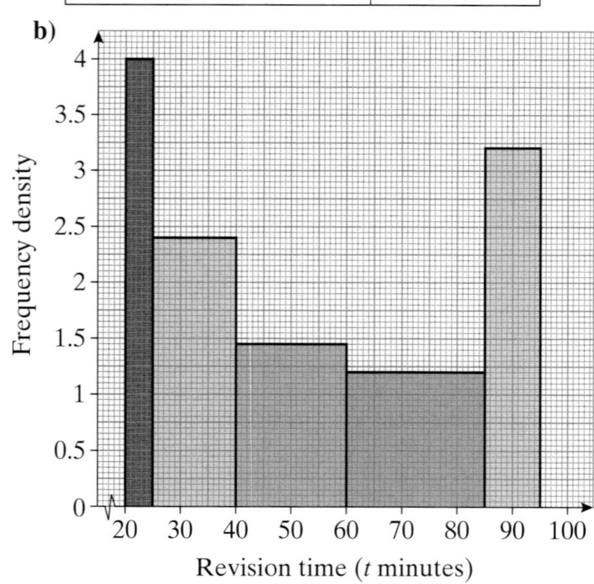

c) 6 students

9 a)

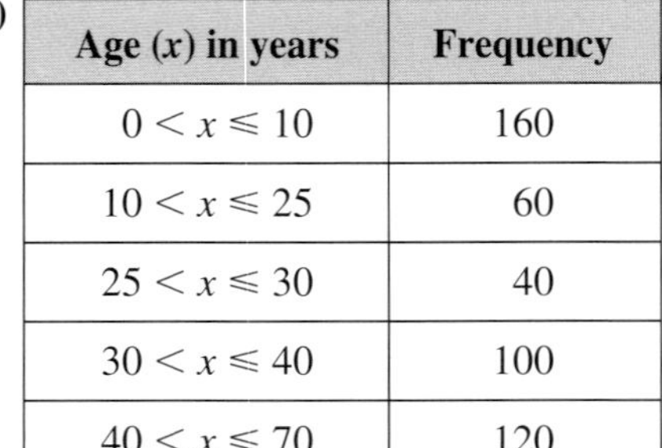

Age (x) in years	Frequency
$0 < x \leqslant 10$	160
$10 < x \leqslant 25$	60
$25 < x \leqslant 30$	40
$30 < x \leqslant 40$	100
$40 < x \leqslant 70$	120

b)

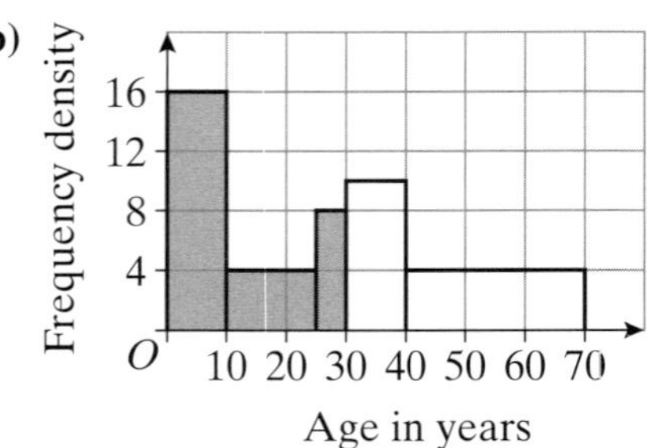

10 a) $12\,150 \div 200 = 60.75$ hours

b)

Number of hours worked (t)	Cumulative Frequency
$0 < t \leqslant 30$	0
$0 < t \leqslant 40$	4
$0 < t \leqslant 50$	22
$0 < t \leqslant 60$	90
$0 < t \leqslant 70$	169
$0 < t \leqslant 80$	200

c)

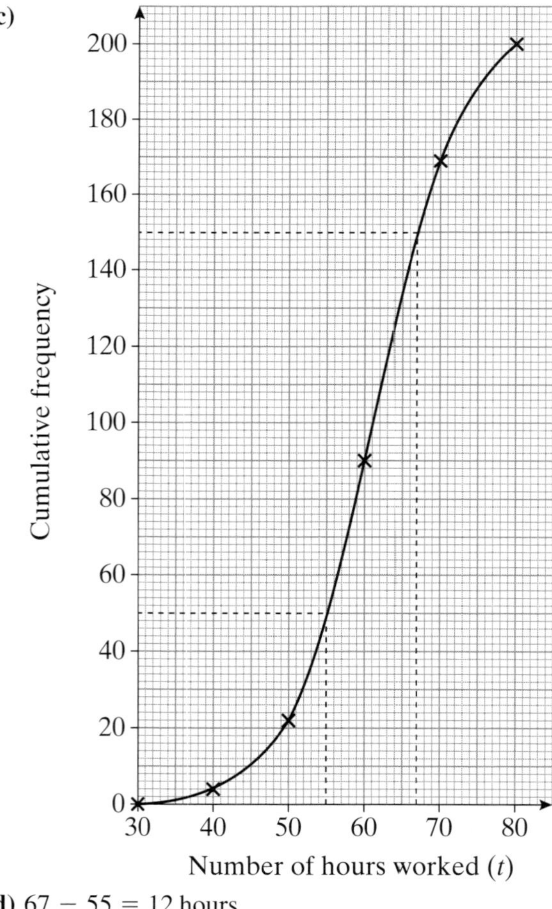

d) $67 - 55 = 12$ hours

11 a) 32 seconds

b)

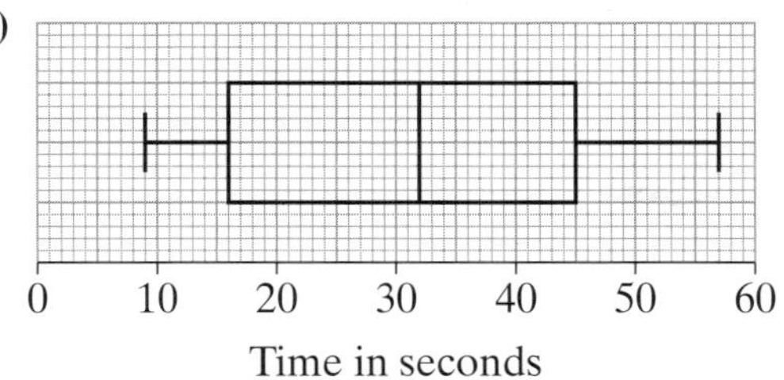

c) The average times for boys and girls are very similar.
There is more variation in the boys' times than the girls' times.

12 a), c)

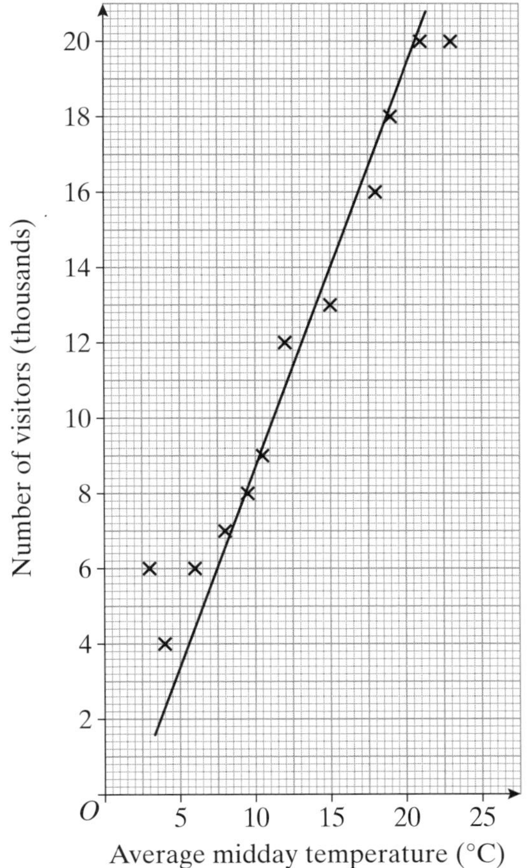

b) Positive correlation

d) 13 000 visitors approx

13 a), c)

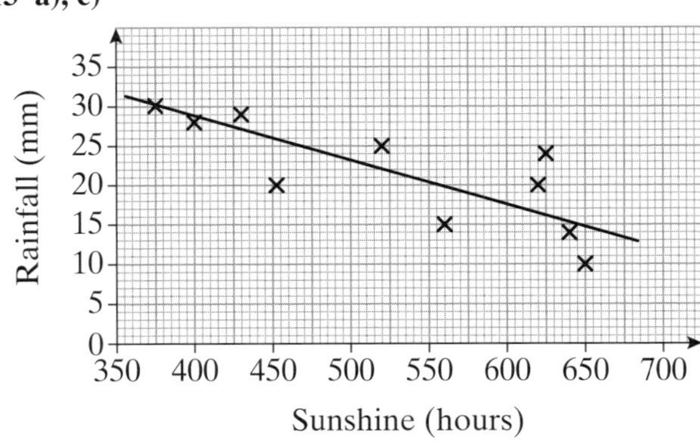

b) Negative correlation

d) (i) 26 mm approx

(ii) 590 hours approx

Internet Challenge 21

Values may differ, depending on your source.

1

		Population (millions)	Area (millions of sq km)
1	China	1306	9.60
2	India	1080	3.29
3	USA	296	9.63
4	Indonesia	242	1.92
5	Brazil	186	8.51
6	Pakistan	162	0.80
7	Bangladesh	144	0.14
8	Russia	143	17.10
9	Nigeria	129	0.92
10	Japan	127	0.38
11	Mexico	106	1.97
12	Philippines	88	0.30

2

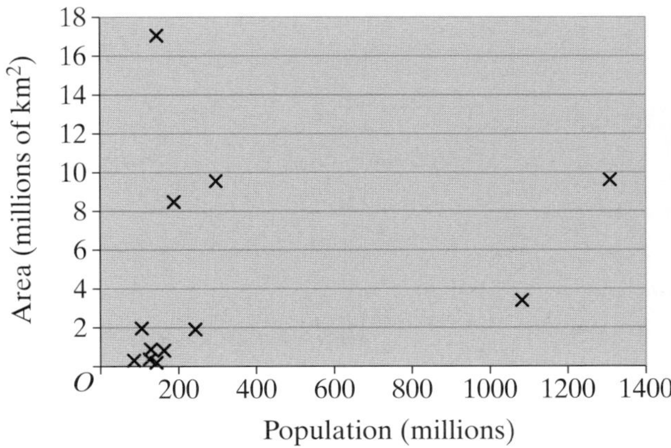

There is possibly a hint of positive correlation, but very weak at best.

3 pmcc = 0.237 (no evidence of linear correlation, since 5% critical value would be 0.4973)

4 Spearman = 0.531 (evidence of rank correlation, since 5% critical value would be 0.5035)

The Spearman is preferred for this data set, since the underlying requirements for a pmcc do not seem to apply (no elliptical distribution of points).

Chapter 22: Probability

Starter 22

One dice: All six outcomes should occur with similar frequencies.

Two dice: 7 should occur most often, with 6 and 8 also having high frequencies.

Exercise 22.1

1 a) $\frac{15}{40} = \frac{3}{8}$ **b)** $\frac{16}{40} = \frac{2}{5}$

2 a) $\frac{1}{7}$ **b)** $\frac{6}{7}$ **c)** 0

3 0.78

4 a) $\frac{12}{50} = \frac{6}{25}$　　　**b)** 96

5 a) $\frac{10}{30} = \frac{1}{3}$　　**b)** $\frac{20}{30} = \frac{2}{3}$　　**c)** $\frac{5}{30} = \frac{1}{6}$

6 a)

	Science	History	Total
Hardback	10	18	28
Paperback	20	32	52
Total	30	50	80

　　b) $\frac{10}{30} = \frac{1}{3}$　　**c)** $\frac{32}{52} = \frac{8}{13}$

7 a) $\frac{46}{50} = \frac{23}{15}$　　**b)** 28

8 a) $\frac{3}{50}$　　**b)** $\frac{21}{29}$

　　c) Disagree – data from one day, so not representative (may be biased).

9 a)

	Tea	Coffee	Other	Total
Morning	78	32	0	110
Afternoon	22	48	20	90
Total	100	80	20	200

　　b) $\frac{32}{110} = \frac{16}{55}$　　**c)** $\frac{22}{90} = \frac{11}{45}$　　**d)** 520

10 a) $\frac{18}{30} = \frac{3}{5}$　　**b)** 288

　　c) Class 3G might not be representative of the school as a whole.

Worksheet 22.1

1 a) $\frac{3}{7}$　　**b)** $\frac{4}{7}$　　**c)** 0　　**d)** 1

2 a) $\frac{1}{4}$　　**b)** $\frac{3}{4}$

3 a) $\frac{1}{6}$　　**b)** $\frac{2}{6} = \frac{1}{3}$　　**c)** $\frac{3}{6} = \frac{1}{2}$　　**d)** $\frac{2}{6} = \frac{1}{3}$　　**e)** $\frac{5}{6}$

4 a) $\frac{4}{5}$　　**b)** $\frac{2}{5}$　　**c)** $\frac{2}{5}$　　**d)** $\frac{1}{5}$　　**e)** $\frac{3}{5}$

5 a) $\frac{13}{15}$　　**b)** 40

6 a) 4　　**b)** $\frac{5}{12}$

7 a) 0.55　　**b)** 27

8 a) $\frac{10}{20} = \frac{1}{2}$　**b)** $\frac{8}{20} = \frac{2}{5}$　**c)** $\frac{2}{20} = \frac{1}{10}$　**d)** $\frac{3}{20}$　**e)** $\frac{7}{20}$

9 a) $\frac{3}{9} = \frac{1}{3}$　　**b)** $\frac{2}{9}$　　**c)** $\frac{1}{9}$　　**d)** 0

10 a) 0.2　　**b) (i)** 9　　**(ii)** 4

Exercise 22.2

1 a) 0.2　　**b)** 0.9　　**c)** 0.2　　**d)** 0.4

2 a)

Activity	Cinema	Pizza	Stay in
Frequency	0.25	0.45	0.3

　　b) Pizza

　　c) 0.75

3 a)

Type of bird	Blackbird	Sparrow	Starling	Robin
Frequency	0.35	0.25	0.3	0.1

　　b) Blackbird

　　c) 0.65

　　d) 0.55

4 a) $\frac{4}{7}$　　**b)** $\frac{6}{7}$

5 a) 0.3　　**b)** 0.4

6 a) $\frac{1}{3}$　　**b)** $\frac{2}{3}$

7 a) 0.7　　**b)** 0.3　　**c)** 0.15

8 a) Fred has added the two probabilities together.

　　b) Julie thinks the two events might not be mutually exclusive; a car can have unsafe lights and unsafe tyres.

9 a) 0.31　　**b)** 144

10 a) 0.2　　**b)** 20

　　c) $\frac{7}{25}, \frac{11}{25}, \frac{2}{25}, \frac{5}{25}$　　**d)** 25

Worksheet 22.2

1 0.75

2 a) $\frac{3}{10}$　　**b)** $\frac{2}{10} = \frac{1}{5}$　　**c)** $\frac{5}{10} = \frac{1}{2}$　　**d)** $\frac{8}{10} = \frac{4}{5}$

3 a) 0.25　　**b)** 0.75　　**c)** 7

4 a) $\frac{17}{30}$　　**b)** $\frac{21}{30} = \frac{7}{10}$　　**c)** $\frac{9}{30} = \frac{3}{10}$

5 a) 0.46　　**b)** 0.67　　**c)** 20

6 a) 0.06　　**b)** 0.52

　　c) 2 and 4　　**d)** 0.8

7 a) 0.93　　**b)** 0.25　　**c)** 0.5

8 16

9 a) $\frac{1}{2}$　　**b)** $\frac{1}{4}$　　**c)** $\frac{1}{2}$　　**d)** $\frac{3}{4}$　　**e)** $\frac{1}{2}$

10 a) $\frac{5}{12}$　　**b)** $\frac{17}{60}$　　**c)** $\frac{5}{6}$　　**d)** 120 discs

Exercise 22.3

1 a) 0.09　　**b)** 0.11

2 a)

+	1	2	3	4
1	2	3	4	5
2	3	4	5	6
3	4	5	6	7
4	5	6	7	8

　　b) $\frac{2}{16} = \frac{1}{8}$　　**c)** $\frac{1}{2}$

3 a) $\frac{2}{5}$　　**b)** $\frac{4}{25}$

4 a)

+	1	2	3	4
5	6	7	8	9
6	7	8	9	10

　　b) $\frac{2}{8} = \frac{1}{4}$　　**c)** 5

5 a) $\frac{1}{6}$　　**b)** $\frac{5}{36}$　　**c)** $\frac{1}{36}$

6 a) 0.7　　**b)** 0.09　　**c)** 0.063

7 a) $\frac{11}{144}$　　**b)** $\frac{121}{144}$

8 a) 0.24　　**b)** 0.01　　**c)** 15 days

Exercise 22.4

1 a)

0.2 — Correct — 0.2 Correct / 0.8 Incorrect

0.8 — Incorrect — 0.2 Correct / 0.8 Incorrect

　　b) 0.04

　　c) 0.32

2 a)

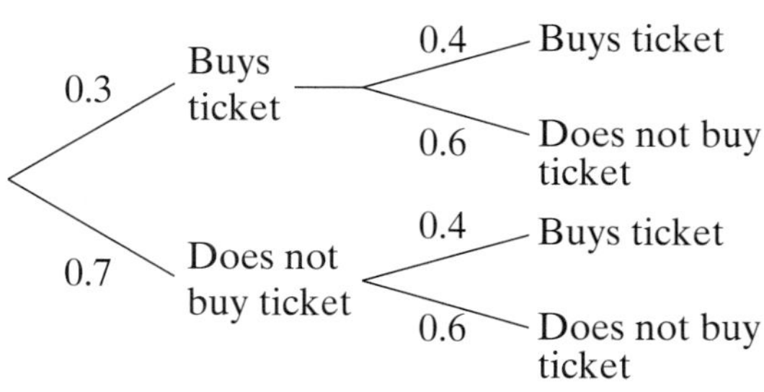

b) $\frac{4}{9}$ **c)** $\frac{2}{9}$ **d)** $\frac{4}{9}$

3 a)

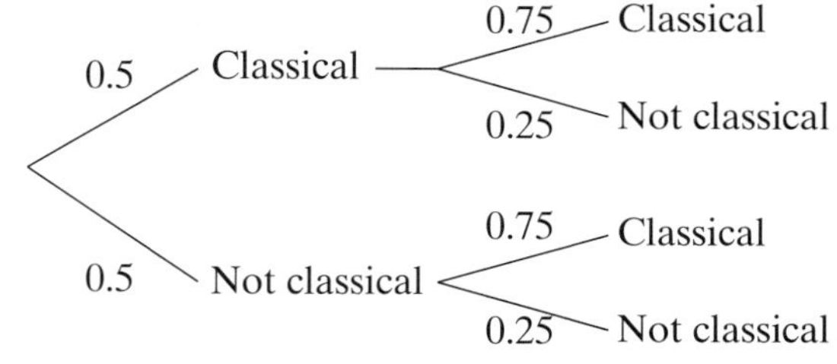

b) 0.46   **c)** 0.42

4 a) 0.5   **b)** 0.75

c)

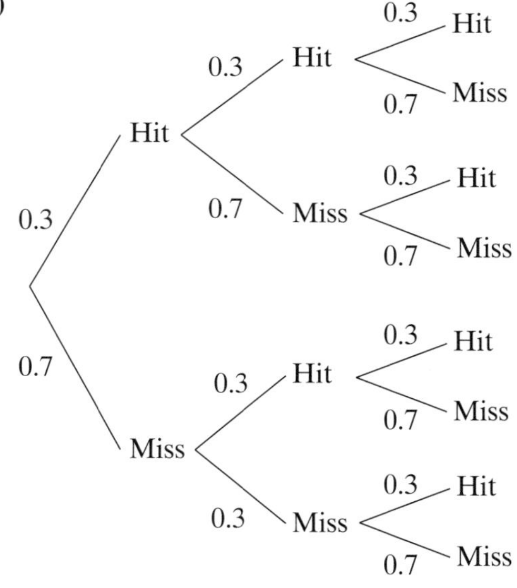

d) 0.5

5 a)

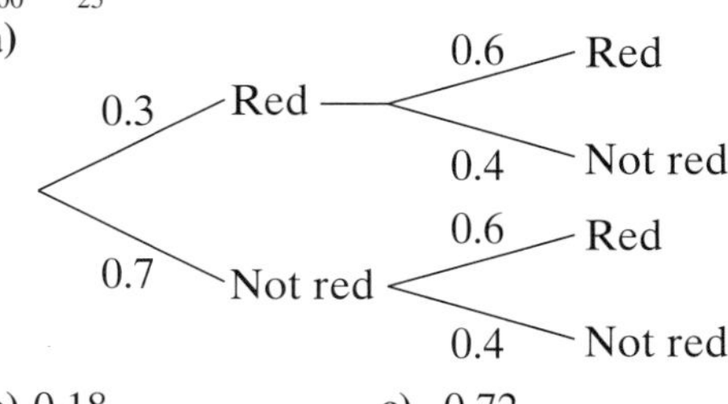

b) 0.027   **c)** 0.441

Worksheet 22.4

1 $\frac{52}{100} = \frac{13}{25}$

2 a)

b) 0.18   **c)** 0.72

3 a) $\frac{1}{4}$ **b)** $\frac{1}{4}$

4 0.18

5 0.4875

6 a) $\frac{25}{144}$ **b)** $\frac{16}{144} = \frac{1}{9}$ **c)** $\frac{9}{144} = \frac{1}{16}$

7 a) $\frac{24}{100} = \frac{6}{25}$ **b)** $\frac{36}{100} = \frac{9}{25}$ **c)** $\frac{52}{100} = \frac{13}{25}$

8 a) $\frac{1}{9}$ **b)** $\frac{2}{9}$ **c)** $\frac{4}{9}$ **d)** $\frac{5}{9}$

9 a) (i) $\frac{7}{12}$ **(ii)** $\frac{5}{12}$

   b) (i) $\frac{49}{144}$ **(ii)** $\frac{25}{144}$

10 a) 0.35   **b)** 0.55   **c)** 0.1925

Review Exercise 22

1 a) 0.5   **b)** 0.35

2 $\frac{3}{8}$

3 a) (i) 0.2   **(ii)** 0

   b) 40 times

4 a) 0.48   **b)** 0.2

5 a)

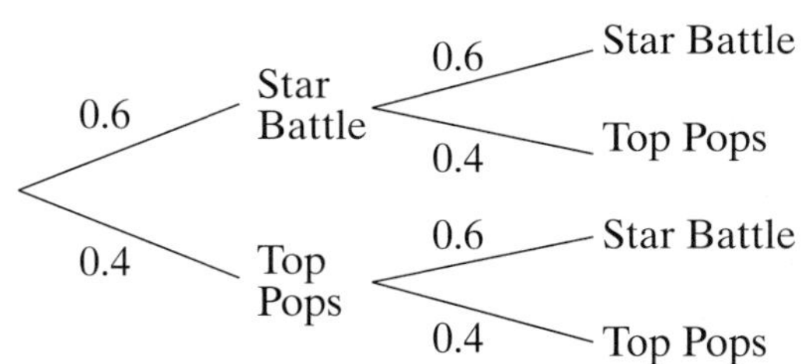

b) (i) $\frac{25}{144}$ **(ii)** $\frac{35}{72}$

6 a) (1, Heads), (2, Heads), (3, Heads), (4, Heads), (5, Heads)
     (1, Tails), (2, Tails), (3, Tails), (4, Tails), (5, Tails)

   b) (i) 0.14   **(ii)** 0   **(iii)** 1   **(iv)** 0.25

   c) 0.125

7 a)

   b) 0.36

   c) 0.48

8 a) 0.16   **b)** 4000   **c)** 212

9 a) $\frac{4}{24} = \frac{1}{6}$ **b)** $\frac{6}{24} = \frac{1}{4}$

10 a) 0.91   **b)** 3

   c) (i) 0.0081   **(ii)** 0.1638

11 a)

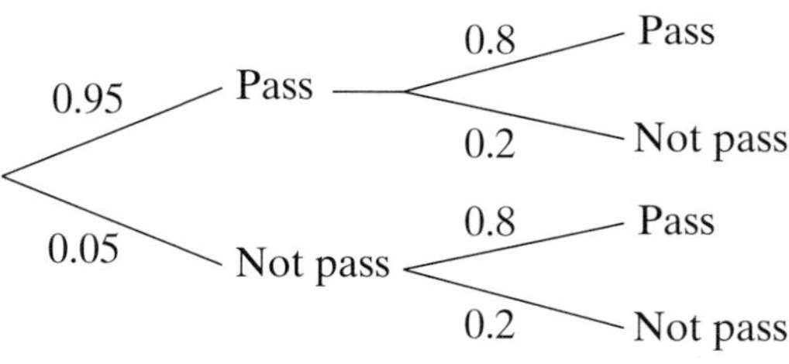

   b) 0.76

   c) 0.23

12 a)

<table>
<tr><td></td><td colspan="5" align="center">Spinner B</td></tr>
<tr><td></td><td>×</td><td>1</td><td>2</td><td>3</td><td>4</td></tr>
<tr><td rowspan="3">Spinner A</td></tr>
<tr><td>1</td><td>1</td><td>2</td><td>3</td><td>4</td></tr>
<tr><td>2</td><td>2</td><td>4</td><td>6</td><td>8</td></tr>
<tr><td>3</td><td>3</td><td>6</td><td>9</td><td>12</td></tr>
</table>

b) $\frac{2}{12} = \frac{1}{6}$ **c)** $\frac{4}{12} = \frac{1}{3}$

Internet Challenge 22

The probability of initially selecting the right door is $\frac{1}{3}$, and remains at this value if the contestant does not switch. Thus it is better to switch, since the probability of winning increases to $1 - \frac{1}{3} = \frac{2}{3}$, i.e. is doubled.

Chapter 23: Using a calculator efficiently

Starter 23

3 to the power of 4^{21} has about 2×10^{12} digits!

Exercise 23.1

1 20 736
2 3375
3 1.32×10^{14}
4 6.98×10^{13}
5 5.66
6 3.30
7 2.85
8 8.94
9 0.0294
10 0.4
11 5
12 0.111
13 3.56
14 2.29
15 41.6
16 0.267
17 8.66
18 2.63
19 16
20 9

Exercise 23.2

1 24.058
2 24.01
3 5.089 204 26; 5.09
4 11
5 0.292 279 412; 0.292
6 4.929 503 018; 4.93
7 2.342 857 143; 2.34
8 1.161 290 323; 1.16
9 17.576
10 5.88
11 4.301 162 634; 4.30
12 1.16
13 2.150 326 797; 2.15
14 5.816 356 248; 5.82
15 1.666 666 667; 1.67
16 14.534 883 72; 14.5
17 4.6×10^{12}
18 9.42×10^{13}
19 2.0×10^{3}
20 3.25×10^{10}

Exercise 23.3

1 $\frac{50}{77}$
2 $\frac{10}{39}$
3 $\frac{11}{39}$
4 $\frac{3}{7}$
5 $11\frac{13}{70}$
6 $3\frac{13}{20}$
7 $3\frac{1}{5}$
8 $\frac{11}{18}$
9 $2\frac{1}{4}$
10 $2\frac{1}{2}$
11 $8.75, 8\frac{3}{4}$
12 $11.352, 11\frac{44}{125}$
13 $8.85, 8\frac{17}{20}$
14 $1.35, 1\frac{7}{20}$
15 $8.26, 8\frac{13}{50}$
16 $24.45, 24\frac{9}{20}$
17 $10.08, 10\frac{2}{25}$
18 $20.52, 20\frac{13}{25}$
19 $0.65, \frac{13}{20}$
20 $5.8, 5\frac{4}{5}$

Exercise 23.4

1 a) £212 b) £224.72 c) £479.31
2 a) £13 500 b) £10 125 c) £1014
3 12 years
4 4 years
5 a) £260 b) £273 c) £838.53
6 a) 3, 4, 7 b) 581 130 736

Exercise 23.5

1 a) 50 cm, 46 cm b) 156.25 cm^2, 132.25 cm^2
2 a) 34 cm b) 52.25 cm^2
3 a) 29.25 cm^2 b) 0.538
4 a) 60 mph b) 56.86 mph c) 63.27 mph
5 28.5 cm^2

Review Exercise 23

1 41.2
2 a) 46.416 376 42 b) 46
3 1.7×10^{12}
4 a) 0.787 965 006 b) 0.79
5 a) 17.9867 b) $(1.6 + 3.8 \times 2.4) \times 4.2$
6 2.56
7 a) 53.898 666 67 b) 53.9
8 1.865
9 18
10 0.0205

Internet Challenge 23

<table>
<tr><td></td><td></td><td></td><td></td><td></td><td></td><td>¹T</td><td>U</td><td>R</td><td>I</td><td>N</td><td>G</td><td></td></tr>
<tr><td></td><td>²P</td><td>E</td><td>R</td><td>F</td><td>E</td><td>C</td><td>T</td><td></td><td></td><td></td><td></td><td></td></tr>
<tr><td></td><td></td><td></td><td>³B</td><td>R</td><td>A</td><td>C</td><td>K</td><td>E</td><td>T</td><td>S</td><td></td><td></td></tr>
<tr><td></td><td>⁴Q</td><td>U</td><td>O</td><td>T</td><td>I</td><td>E</td><td>N</td><td>T</td><td></td><td></td><td></td><td></td></tr>
<tr><td></td><td></td><td></td><td></td><td></td><td>⁵J</td><td>P</td><td>E</td><td>G</td><td></td><td></td><td></td><td></td></tr>
<tr><td></td><td></td><td></td><td></td><td></td><td>⁶E</td><td>R</td><td>N</td><td>I</td><td>E</td><td></td><td></td><td></td></tr>
<tr><td></td><td></td><td></td><td>⁷G</td><td>O</td><td>O</td><td>G</td><td>O</td><td>L</td><td></td><td></td><td></td><td></td></tr>
<tr><td>⁸B</td><td>A</td><td>S</td><td>I</td><td>C</td><td></td><td></td><td></td><td></td><td></td><td></td><td></td><td></td></tr>
<tr><td></td><td></td><td></td><td></td><td>⁹S</td><td>T</td><td>A</td><td>N</td><td>D</td><td>A</td><td>R</td><td>D</td><td></td></tr>
<tr><td></td><td></td><td></td><td></td><td>¹⁰S</td><td>I</td><td>L</td><td>I</td><td>C</td><td>O</td><td>N</td><td></td><td></td></tr>
<tr><td></td><td></td><td>¹¹H</td><td>A</td><td>C</td><td>K</td><td>E</td><td>R</td><td></td><td></td><td></td><td></td><td></td></tr>
<tr><td></td><td>¹²I</td><td>N</td><td>D</td><td>E</td><td>X</td><td></td><td></td><td></td><td></td><td></td><td></td><td></td></tr>
<tr><td></td><td></td><td></td><td>¹³C</td><td>R</td><td>Y</td><td>S</td><td>T</td><td>A</td><td>L</td><td></td><td></td><td></td></tr>
</table>

Chapter 24: Direct and inverse proportion

Starter 24

1 49 mm – but only assuming the plant continues to grow at the same rate
2 $8 \times 12 \div 6 = 16$ nights
3 $9 \times 15 \times 20 \div (18 \times 6) = 25$ minutes
4 1
5 10 minutes

Exercise 24.1

1 $y = 3x; y = 39$
2 $y = 2.5x; x = 16$
3 a) $y = 2x; 6$ b) $y = 4x; 4, 4.5$
  c) $y = \frac{x}{6}; 2, 42, 17$ d) $y = \frac{5x}{8}; 1.25, 56$

Answers

4 a) $y = \dfrac{2x}{3}$

  b) (i) 40   **(ii)** 37.5

5 a) $y = \dfrac{x^2}{5}$

  b) (i) 180   **(ii)** ± 25

6 $y = 4x^2$; $y = 36$

7 $y = 0.4x^3$; $y = 204.8$

8 a) $T = \dfrac{c^2}{6.4}$ **b)** 22.5 minutes   **c)** 24 cities

9 a) $t = \dfrac{n^2}{50\,000\,000}$

  b) 20 000 seconds

  c) 173 000

10 a) £4.80   **b)** 30 cm

Exercise 24.2

1 25

2 3

3 a) 5, 2   **b)** 8, 36   **c)** 0.2, 0.25   **d)** 0.8, 0.4

4 a) $r = \dfrac{294}{t}$

  b) (i) 21   **(ii)** 6

5 a) $p = \dfrac{180}{s}$

  b) (i) 20   **(ii)** 3

6 36

7 2

8 117; 7.75 m

9 $v = \dfrac{8.424}{m}$; 58.5 km/h

10 a) 2401   **b)** 9603

Exercise 24.3

1 Neither   **2** Inverse proportion

3 Neither   **4** Neither

5 Neither   **6** Direct proportion

Review Exercise 24

1 39

2 2

3 9

4 320

5 a) $T = \dfrac{2x}{3}$ **b)** 24   **c)** 72

6 a) $T = 0.2\sqrt{l}$ **b)** 1.18 seconds

7 a) $d = \dfrac{L^3}{168\,750}$ **b)** 136

8 a) $y = \dfrac{48}{x^2}$ **b)** 1.92

9 a) $y = 9x^2$ **b)** $c = 18$ and $n = -\dfrac{1}{2}$

10 a) $y = \dfrac{72}{x}$

  b) (i) 12   **(ii)** 15

11 a) $d = 5t^2$ **b)** 245   **c)** 3

12 a) $F = \dfrac{36}{x^2}$ **b)** 9   **c)** 0.75

13 a) 41.4 m   **b)** $d = 0.1V^2 + 0.5V$

14 a) $S = \dfrac{8000}{f^2}$ **b)** 500

Internet Challenge 24

1–4

Planet	Mean distance, d, from Sun	Orbital period, T
Mercury	0.387	88 days
Venus	0.723	225 days
Earth	1	1 year
Mars	1.524	1.88 years
Jupiter	5.203	11.86 years
Saturn	9.529	29.41 years
Uranus	19.19	84.0 years
Neptune	30.06	164.8 years
Pluto	39.53	248.5 years

5 Planets move in orbits that are ellipses. Planets move such that the line between the Sun and the planet sweeps out the same area in the same time, no matter where in the orbit.

6 11 500 years

7 Scientists are unsure of how to classify them.

Chapter 25: Quadratic equations

Starter 25

1 $x = 3$ **2** $x = -1, x = 1$

3 $x = 2, x = 5$ **4** $x = 4$

5 $x = -1, x = 1$ **6** $x = 3$

7 $x = -2, x = 1$ **8** $x = -6, x = 6$

9 $x = -5, x = 5$ **10** $x = 1, x = 2, x = 3$

Yes, they do.

Exercise 25.1

1 $-1, -2$ **2** $-1, -5$

3 $1, -8$ **4** $1, -2$

5 $2, -4$ **6** $2, -6$

7 $3, 4$ **8** $3, 5$

9 $4, -2$ **10** 2 (twice)

11 $-\frac{1}{2}, -1$ **12** $\frac{1}{2}, -3$

13 $-2, -\frac{1}{3}$ **14** $1, -1\frac{1}{2}$

15 $-2, -\frac{2}{3}$ **16** $3, 1\frac{1}{2}$

17 $-1, -1\frac{2}{3}$ **18** $2, 2\frac{1}{2}$

19 $-5, -\frac{1}{5}$ **20** $-\frac{1}{2}$ (twice)

21 $\frac{1}{3}, -\frac{1}{2}$ **22** $0, \frac{1}{5}$

23 $\frac{1}{2}, -\frac{1}{2}$ **24** $0, 1$

25 $\frac{1}{3}, \frac{1}{4}$ **26** $0, \frac{1}{10}$

27 $\frac{1}{2}, \frac{3}{4}$ **28** $1, \frac{3}{8}$

29 $-1\frac{1}{2}$ (twice)   **30** $-1\frac{1}{2}, 1\frac{1}{2}$

31 $x^2 - 6x - 7 = 0$; $7, -1$

32 $x^2 - 13x + 40 = 0$; $5, 8$

33 $x^2 + 13x + 30 = 0$; $-3, -10$

34 $x^2 + 7x - 44 = 0$; $4, -11$

35 $2x^2 - 11x - 6 = 0$; $6, -\frac{1}{2}$

36 $3x^2 + 23x - 8 = 0$; $-8, \frac{1}{3}$

37 $3x^2 + x - 2 = 0$; $-1, \frac{2}{3}$

38 $4x^2 - 8x + 3 = 0$; $1\frac{1}{2}, \frac{1}{2}$

39 $6x^2 + 5x - 6 = 0$; $\frac{2}{3}, -1\frac{1}{2}$

40 $4x^2 - 25 = 0$; $-2\frac{1}{2}, 2\frac{1}{2}$

Worksheet 25.1

1	$4, 2$	**2**	$-3, -2$
3	$-1, 5$	**4**	$2, 3$
5	$-8, 7$	**6**	$-7, 5$
7	$-5, -3$	**8**	$-8, 5$
9	$-1, 2$	**10**	$4, 5$
11	$-1\frac{1}{2}, -1$	**12**	$\frac{1}{2}, 5$
13	$-1\frac{1}{3}, 1$	**14**	$-\frac{1}{2}, \frac{1}{3}$
15	$-\frac{2}{5}, 3$	**16**	$-1\frac{1}{2}, \frac{2}{3}$
17	$0, 3$	**18**	$-6, 6$
19	$-1\frac{1}{2}, 0$	**20**	$-5, 5$

Exercise 25.2

1 $-0.438, -4.562$

2 $-0.757, -9.243$

3 $5.898, 1.102$

4 $0.422, -5.922$

5 $6.854, 0.146$

6 $0.721, -1.387$

7 $0.193, -5.193$

8 $2.351, -0.851$

9 $0.558, -0.358$

10 $0.212, -4.712$

11 $x^2 + 5x - 7 = 0$; $1.14, -6.14$

12 $2x^2 - 3x - 1 = 0$; $1.78, -0.281$

13 $3x^2 - 4x - 5 = 0$; $2.12, -0.786$

14 $x^2 + 10x - 2 = 0$; $0.196, -10.2$

15 $2x^2 + 11x - 1 = 0$; $0.0895, -5.59$

16 $3x^2 - 12x - 1 = 0$; $4.08, -0.0817$

17 $5x^2 - 2x - 4 = 0$; $1.12, -0.717$

18 $7x^2 - 21x - 1 = 0$; $3.05, -0.0469$

19 $6x^2 + 17x + 4 = 0$; $-0.259, -2.57$

20 $9x^2 - x - 2 = 0$; $0.530, -0.419$

Exercise 25.3

1 a) $x(x + 7) = 144$
c) 9 and 16 or -16 and -9

2 a) $x(2x - 5) = 3000$
c) $x = 40$. The field is 40 m by 75 m

3 a) $x(x + 3) = 180$
c) 12 and 15

4 a) $x^2 + x(2x + 1) = 114$
c) 6

5 a) $(3x + 1)(2x + 5) - 2x^2 = 55$
c) $x = 2$. The rectangle is 7 cm by 9 cm

6 a) $x(2x + 3)$ and $(x + 3)(x + 4)$
b) $x(2x + 3) = (x + 3)(x + 4)$ which becomes
$x^2 - 4x - 12 = 0$
c) $x = 6$. The rectangles are 6 cm by 15 cm and 9 cm by 10 cm

Exercise 25.4

1 a) $(x + 2)^2 + 11$　　**b)** $(x + 5)^2 - 23$

2 $(x - 3)^2 + 1$; $p = 3, q = 1$

3 $(x - 7)^2 + 1$; $f = 7, g = 1$

4 $(x - 6)^2 - 6$; $a = -6, b = -6$

5 a) $(x + 8)^2 - 25$; $a = 8, b = -25$
b)

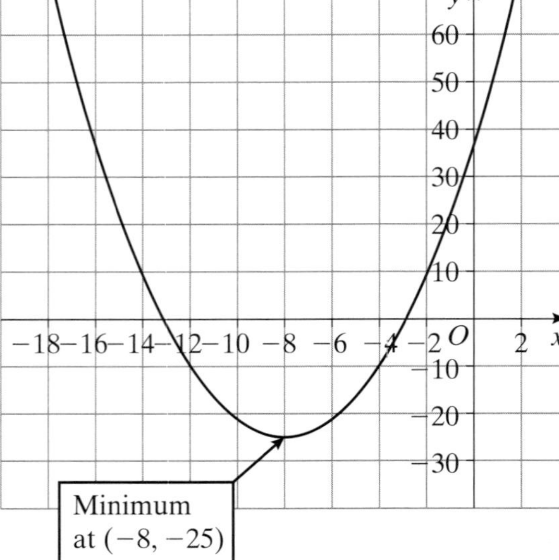

6 a) $(x + 10)^2 - 10$;　$a = 10, b = -10$
b)

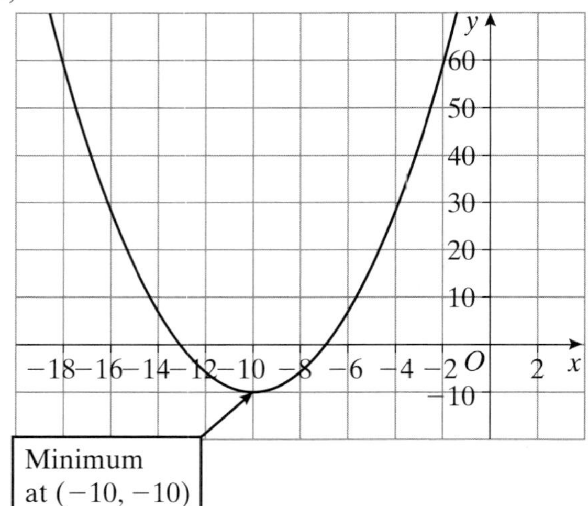

7 a) $(x + 1)^2 + 4$; $a = 1, b = 4$
b)

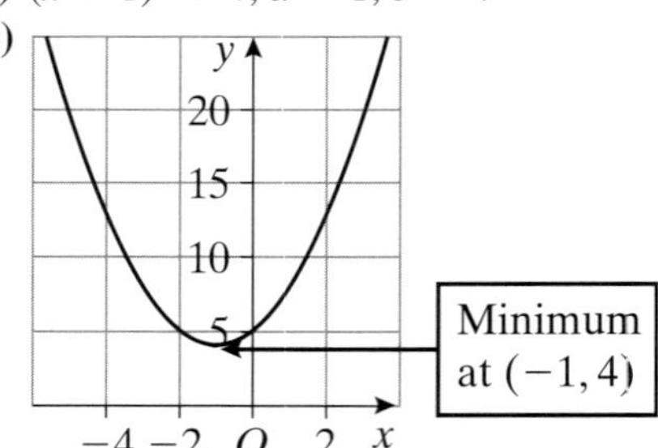

8 a) $(x + 12)^2 + 6$; $a = 12, b = 6$
b)

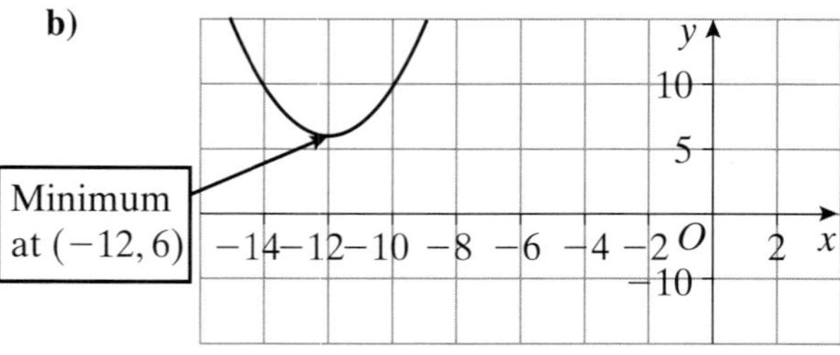

9 a) $y = (x + 6)^2$

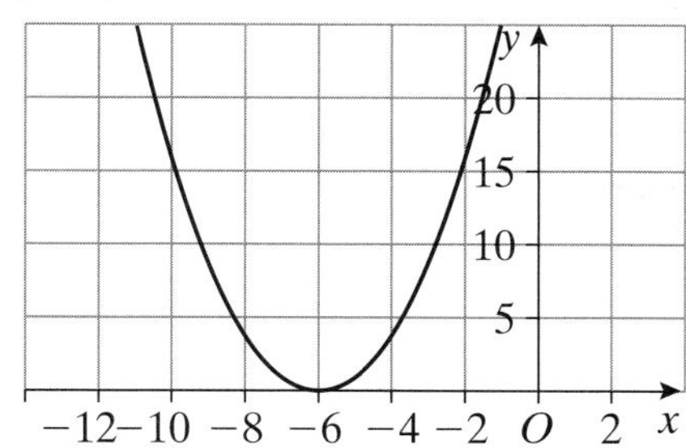

b) Line of symmetry is $x = -6$

Review Exercise 25

1 a) $(x - 2)(x - 4)$
  b) 2, 4
2 $6.5, -3.5$
3 $4.236, -0.236$
4 b) $2.19, -3.19$
5 a) $(x + 4)(x - 3) = 78$
  b) (ii) $9, -10$
    (iii) 13 cm, 6 cm
6 b) $9.93, -3.93$. Radius of circle is 9.93 cm.
7 c) $2\frac{2}{3}, 2$
8 b) 12.361
9 a) "is identically equal to"
  b) $a = 4, b = -21$
10 a) $(x - 5)^2 + 7$ $a = -5, b = 7$
  b)

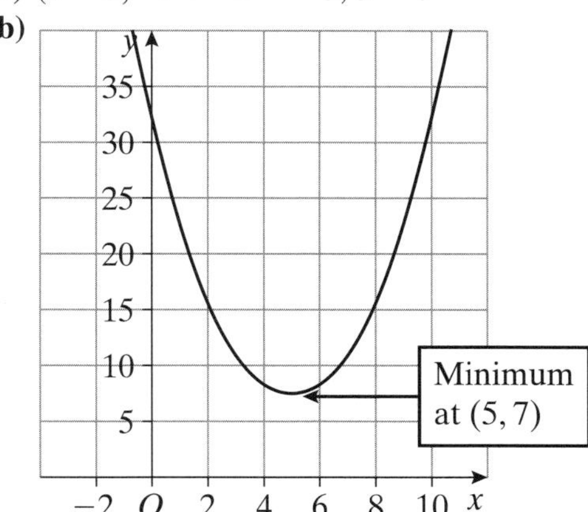

11 a) $k = m^2$
  b) (i) $-m^2$ **(ii)** m
12 a) (i) $(x - 14)(2x - 7)$ **(ii)** $14, 3.5$
  b) (i) $\dfrac{7}{n + 7}$
    (ii) $\dfrac{7}{n + 7} = \dfrac{2}{5}$ gives $n = 10.5$
      But n must be an integer.
  d) $n = 14$ so 7 white balls out of 21, giving $\frac{1}{9}$

Internet Challenge 25

1 Parabola
2 The other three are the circle, ellipse and hyperbola.
3 Yes, a parabola.
4 If the orbit is closed it must be an ellipse (or a circle). Some comets probably have open orbits; these could be parabolas or hyperbolas.
5 Good method but requires some skill!

Chapter 26: Advanced algebra

Starter 26

By counting, the numbers of squares/rectangles are 9, 36, 30. Thus $k = 4$.

To prove the formula, select one corner of a square/rectangle at random. On an m by n grid, there are $m + 1$ possible choices for the x coordinate and $n + 1$ for the y coordinate, giving $(m + 1)(n + 1)$ possibilities altogether.

Now choose a second corner, not in the same row or column as before; this can be done in mn ways. Thus there would seem to be $m(m + 1)n(n + 1)$ choices altogether.

However, each different square/rectangle gets counted four times in this way. Thus the number is $(m + 1)(n + 1) \div 4$ and the result is proved.

Exercise 26.1

1 $3\sqrt{2}$ **2** $4\sqrt{2}$ **3** $5\sqrt{2}$
4 $3\sqrt{5}$ **5** $5\sqrt{6}$ **6** $2\sqrt{6}$
7 $3\sqrt{11}$ **8** $6\sqrt{3}$ **9** $6\sqrt{3}$
10 $7\sqrt{2}$ **11** $3\sqrt{3}$ **12** $7\sqrt{2}$
13 $4\sqrt{11}$ **14** $4\sqrt{2}$ **15** $12 + 4\sqrt{3}$
16 $8 + 7\sqrt{2}$ **17** 22 **18** $22 + 11\sqrt{5}$
19 $\dfrac{2 + 3\sqrt{5}}{5}$
20 a) $4 + 2\sqrt{7}$ **b)** $18 + 2\sqrt{7}$ **c)** $6 + 6\sqrt{7}$
21 $-2 \pm \sqrt{11}$ **22** $\dfrac{-1 \pm \sqrt{5}}{2}$ **23** $\dfrac{-3 \pm \sqrt{13}}{2}$
24 $\dfrac{-4 \pm \sqrt{10}}{2}$ **25** $\dfrac{5 \pm \sqrt{17}}{2}$

Exercise 26.2

1 $\dfrac{8x + 3}{15}$ **2** $\dfrac{7x + 2}{24}$
3 $\dfrac{5x + 2}{10}$ **4** $\dfrac{5x}{6}$
5 $\dfrac{13x + 4}{20}$ **6** $\dfrac{11x + 2}{12}$
7 $\dfrac{2x + 5}{x(x + 1)}$ **8** $\dfrac{5x + 7}{(x + 1)(x + 2)}$
9 $\dfrac{5x + 5}{(x + 3)(2x + 1)}$ **10** $\dfrac{3x + 10}{(x + 3)(x + 4)}$
11 $\dfrac{x + 19}{(x - 2)(x + 5)}$ **12** $\dfrac{4x + 9}{(x + 1)(x + 2)}$
13 3   **14** 2
15 3   **16** -2
17 3   **18** 5
19 $4, \frac{2}{3}$ **20** $2, -\frac{5}{14}$

Exercise 26.3

1 $\dfrac{x + 3}{2x + 1}$ **2** $\dfrac{3x + 5}{2x + 1}$
3 $\dfrac{4x + 6}{3}$ **4** $\dfrac{x^2 + 5}{2x}$
5 $\dfrac{x + 2}{x}$ **6** $5(x + 3)^4$

7 $\dfrac{x + 10}{5}$

8 $\dfrac{3x + 2}{x}$

9 $\dfrac{x - 5}{x}$

10 $\dfrac{4}{(2x + 1)^2}$

11 $\dfrac{x + 8}{2}$

12 $\dfrac{x}{2}$

13 $\dfrac{x}{x + 2}$

14 $\dfrac{x + 2}{x + 7}$

15 $\dfrac{x + 2}{x + 4}$

16 $\dfrac{x + 5}{x + 3}$

17 $\dfrac{x + 3}{x + 4}$

18 $\dfrac{x + 4}{x + 2}$

19 $\dfrac{x + 5}{x - 3}$

20 $\dfrac{1}{x - 4}$

Exercise 26.4

1 $x = 2$ and $y = 2$ or $x = -1$ and $y = -1$
2 $x = 3$ and $y = 10$ or $x = -2$ and $y = 5$
3 $x = 3$ and $y = 19$ or $x = -1$ and $y = 3$
4 $x = 2$ and $y = 20$ or $x = \frac{1}{5}$ and $y = \frac{1}{5}$
5 $x = 4$ and $y = 17$ or $x = 0$ and $y = 1$
6 $x = 2$ and $y = 0$ or $x = -1$ and $y = -3$
7 $x = 3$ and $y = 1$ or $x = -1$ and $y = -3$
8 $x = 2$ and $y = 2$ or $x = -\frac{2}{5}$ and $y = -2\frac{4}{5}$
9 $x = 3$ and $y = -1$ or $x = 1$ and $y = -3$
10 $x = 1$ and $y = -6$ or $x = 6$ and $y = -1$
11 $x = 5$ and $y = 13$ or $x = -3$ and $y = -3$
12 $x = 6$ and $y = 1$ or $x = -5$ and $y = -10$
13 $x = 1$ and $y = 2$ or $x = 2$ and $y = 4$
14 $x = 2$ and $y = -3$ or $x = 3$ and $y = -1$
15 $x = 5$ and $y = 3$ or $x = 0.6$ and $y = -5.8$
16 $x = 1$ and $y = 2$ or $x = -2$ and $y = -1$

Exercise 26.5

1 $x = \dfrac{5}{3 - m}$

2 $x = \dfrac{d - b}{a - c}$

3 $x = \dfrac{2k}{2 - k}$

4 $y = \dfrac{1 - 2d}{d - 1}$

5 $t = \dfrac{bc - a}{1 - c}$

6 $x = \dfrac{n + 2}{3 - k}$

7 $x = \dfrac{ab}{1 - 5b}$

8 $x = \dfrac{3}{a - 2}$

9 $x = \dfrac{ka}{1 - k}$

10 $u = \dfrac{vf}{v - f}$

Exercise 26.6

1 b) $b = 1.5$ c) 1688
2 a) $a = 12\,000, b = 0.7$
   b) £692
3 a) (i) 76 500   (ii) 68 850
   b) $p = 85\,000, q = 0.9$ c) 29 638
4 a) £165 000   b) 6%   c) £56 million
5 a) 100 °C   b) 31.75 °C   c) 22 °C

Review Exercise 26

1 a) 10   b) 3   c) 2
2 $\sqrt{22}$

3 a) (i) 3.5   (ii) 1
   b) 3
4 a) 4   b) 2   c) $83\frac{1}{3}\%$
5 a) $\dfrac{3x}{(x - 2)(x + 4)}$ b) $8, -1$
6 $10\frac{1}{2}$
7 a) $(x + 1)(2x + 5)$ b) $\dfrac{11x + 15}{(x + 1)(2x + 5)}$
8 a) $23 - 6x$ b) $32x^5 y^{15}$ c) $\dfrac{2(n - 1)}{n - 2}$
9 a) 7   b) $\dfrac{2x}{2x + 3}$
10 $y = \dfrac{2k}{4 + 3k}$
11 $x = \dfrac{ay}{y + 1}$
12 $x = 2$ and $y = 5$ or $x = -1.4$ and $y = -5.2$
13 a) If $y = 6$ then $x^2 = -11$ so Bill must be wrong.
   b) $x = 3$ and $y = 4$ or $x = -1.4$ and $y = -4.8$
14 a) $p = 1600, q = 0.5$
   b) £6400

Internet Challenge 26

1 Pythagoras' theorem
2 Circumference of a circle
3 Area of a trapezium
4 Voltage = Current $\times$ Resistance
5 Volume of a cone
6 Quadratic equation formula
7 Energy = mass $\times$ (speed of light)^2
8 Surface area of a sphere
9 Distance s in terms of initial speed u, acceleration a and time t
10 Periodic time for a pendulum of length l
11 Euler's formula for faces, edges and vertices of a polyhedron
12 Conversion from degrees Fahrenheit to degrees Celsius
13 Kinetic energy
14 Potential energy
15 Optics formula, u = object distance, v = image distance, f = focal length
16 Electrical resistance (resistors in parallel)
17 Simple interest
18 Area of a triangle
19 Gravitational force of attraction
20 Work done by a force F moving over a distance d

Chapter 27: Further trigonometry

Starter 27

By calculation, the height is 22.0 metres, to 3 significant figures. Scale drawings will scatter around this value.

Exercise 27.1

1 $a = 6.14$ cm, $b = 5.70$ cm
2 $c = 11.7$ cm, $d = 11.9$ cm
3 $e = 5.71$ cm
4 $f = 7.15$ cm
5 $g = 5.15$ cm

6 $h = 1.70$ cm
7 $i = 6.77$ cm
8 $j = 4.71$ cm
9 $p = 33.2°$
10 $q = 48.7°$
11 $r = 41.9°$
12 $s = 56.2°$

Exercise 27.2
1 75.4° or 104.6°   2 37.4°
3 53.4° or 126.6°   4 44.4° or 135.6°
5 18.8°   6 70.9° or 109.1°
7 76.9° or 103.1°   8 31.1°

Exercise 27.3A
1 6.27 cm   2 5.85 cm
3 6.16 cm   4 21.9 cm
5 9.94 cm   6 4.63 cm
7 60.6°   8 56.6°
9 $r = 129.0°$, $s = 29.4°$ 10 24.8°

Exercise 27.3B
1 9.11 cm   2 10.1 cm
3 16.3 cm   4 13.9 cm
5 20.2 cm   6 5.48 cm
7 112.3°   8 70.7° or 109.3°
9 58.5°   10 50.8°

Exercise 27.4
1 86.8 cm²
2 31.6 cm²
3 a) 48.6°   b) 19.0 cm
4 a) 117.3°   b) 21.3 cm²
5 a) 97.2°   b) 35.2 cm²
6 a) 31.6°   b) 16.7 cm
7 a) 452 cm²   b) 13.0 cm²   c) 374 cm²
8 45.3 cm²

Exercise 27.5
1 a) 43.9 cm   b) 34.7°
2 a) 28.3 cm   b) 28.7 cm   c) 60.5°
3 a) 6.3°   b) 99.0°   c) 5.8°
4 a) 24 cm   b) 44.9°
5 a) 35.4 cm   b) 36.2 cm   c) 12.7°

Exercise 27.6
1 a) 2830 cm³   b) 5 cm   c) 2720 cm³
2 a) 15 cm   b) 1880 cm³   c) 1650 cm³
3 704 cm³

Review Exercise 27
1 152 m²
2 a) 28.9 cm²   b) 9.40 cm
3 a) 11.7 cm   b) 42.5°
4 18.3 cm
5 177 m
6 a) 56.4 cm²   b) 7.84 cm
7 b) 41.6 m   c) 22.04 m
8 a) 9.11 cm   b) 19.2°
9 22.2°
10 a) 1700 cm³   b) $h = \sqrt{\dfrac{S^2}{4\pi^2 d^2} - d^2}$
   c) 1012.5 cm²

Internet Challenge 27
1 Area $= \sqrt{s(s - a)(s - b)(s - c)}$ where the triangle has sides a, b, c and semiperimeter s.
2 26.98 cm²
3 85.45 cm², 87.00 cm², 44.90 cm² so triangle B is largest.
4 The formula was proved by Heron of Alexandria in the 1st century AD, but is thought to be rather older than this.
5 Various proofs exist, including one based on the cosine rule.

Chapter 28: Graphs of curves

Starter 28

x	0	30	45	60	90	120	135	150	180
$\sin x$	0	0.5	0.71	0.87	1	0.87	0.71	0.5	0

x	210	225	240	270	300	315	330	360	390
$\sin x$	-0.5	-0.71	-0.87	-1	-0.87	-0.71	-0.5	0	0.5

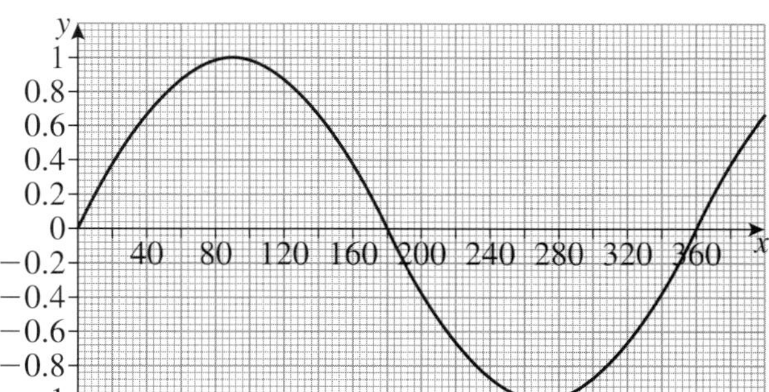

Exercise 28.1
1 $y = x^2 - 5$

x	-3	-2	-1	0	1	2	3
y	4	-1	-4	-5	-4	-1	4

2 $y = x^3 + x$

x	-2	-1	0	1	2	3
y	-10	-2	0	2	10	30

3 $y = 2x^2 + 3$

x	-2	-1	0	1	2	3
y	11	5	3	5	11	21

4 $y = 3x^2 + x$

x	-2	-1	0	1	2	3
y	10	2	0	4	14	30

5 $y = x^2 - 4x$

x	-1	0	1	2	3	4	5
y	5	0	-3	-4	-3	0	5

6 $y = 2^x$

x	-2	-1	0	1	2	3
y	0.25	0.5	1	2	4	8

7 $y = x + \dfrac{1}{x}$

x	-2	-1	-0.5	0	0.5	1	2
y	-25	-2	-2.5	not defined	2.5	2	2.5

8 $y = 10^x$

x	-1	0	1	2	3
y	0.1	1	10	100	1000

Exercise 28.2

1 a), b) $y = x^2 - 3$

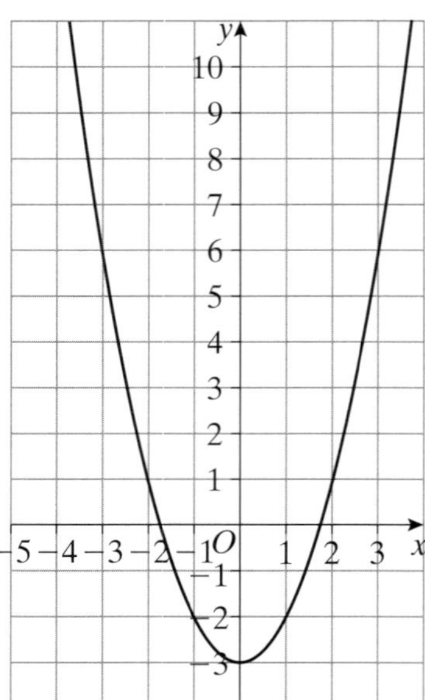

c) -0.75 **d)** $(0, -3)$

2 a)

x	-3	-2	-1	0	1	2	3
y	-3	-4	-3	0	5	12	21

b) $y = x^2 + 4x$

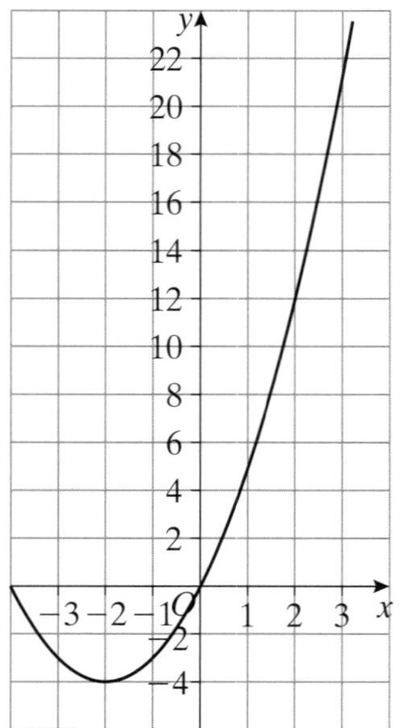

c) $(-2, -4)$

3 a)

x	-3	-2	-1	0	1	2	3
y	15	5	-1	-3	-1	5	15

b) $2x^2 - 3$

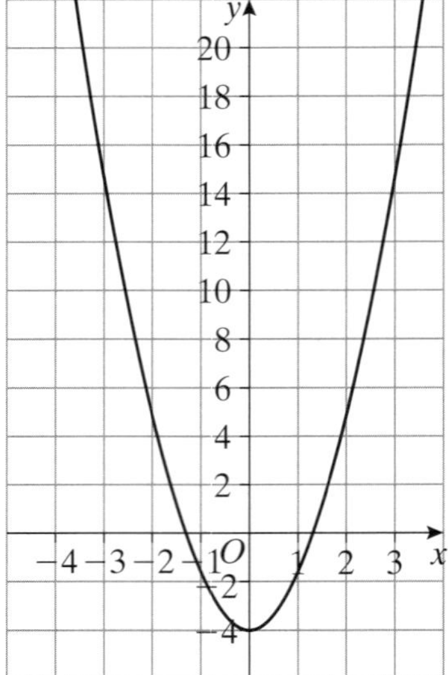

c) $1.2, -1.2$

4 a)

x	-3	-2	-1	0	1	2	3
y	-15	0	3	0	-3	0	15

b) $y = x^2 - 4x$

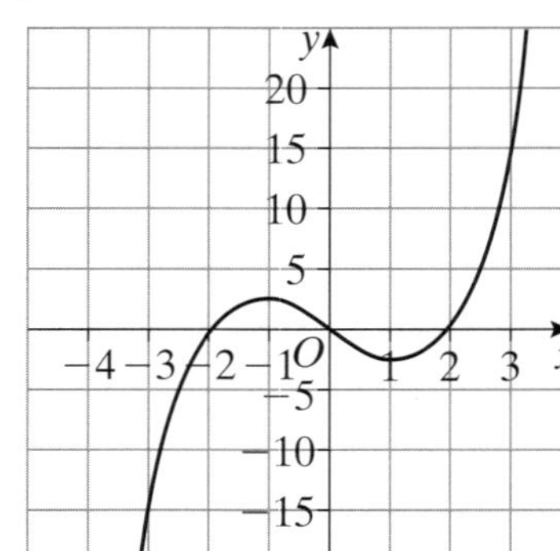

c) $-2, 0, 2$ **d)** 2.46

5 a)

x	-1	0	1	2	3	4
y	0.5	1	2	4	8	16

b) $y = 2^x$

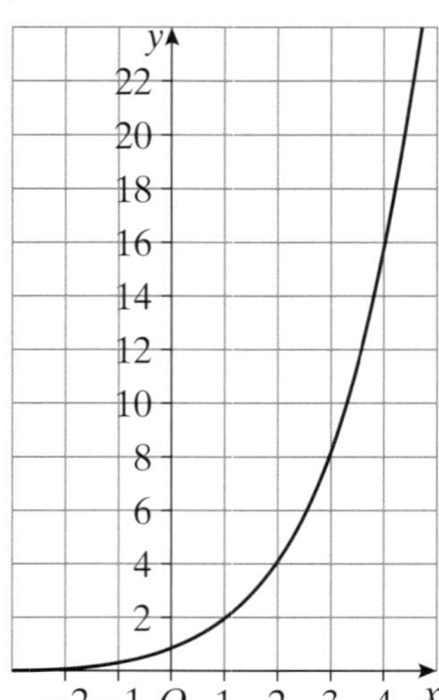

c) 3.58

6 a)

x	-3	-2	-1	0	1	2	3
y	-1	4	7	8	7	4	-1

b) $y = 8 - x^2$

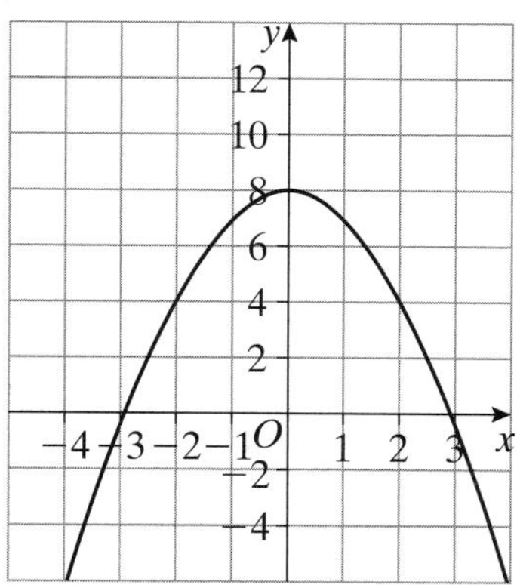

c) $(0, 8)$　　　**d)** $-2.8, 2.8$

7 a)

x	-3	-2	-1	0	1	2	3	4
y	-4	-6	-12	not defined	12	6	4	3

b), c), d) $y = \dfrac{12}{x}$

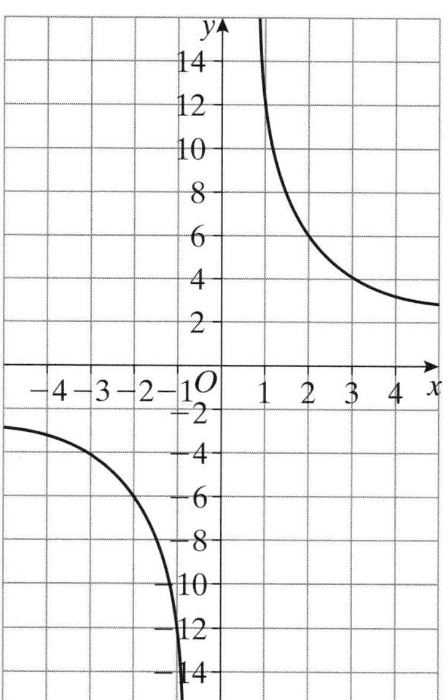

e) 1.33

8 a)

x	-3	-2	-1	0	1	2	3	4
y	6	0	-4	-6	-6	-4	0	6

b) $y = x^2 - x - 6$

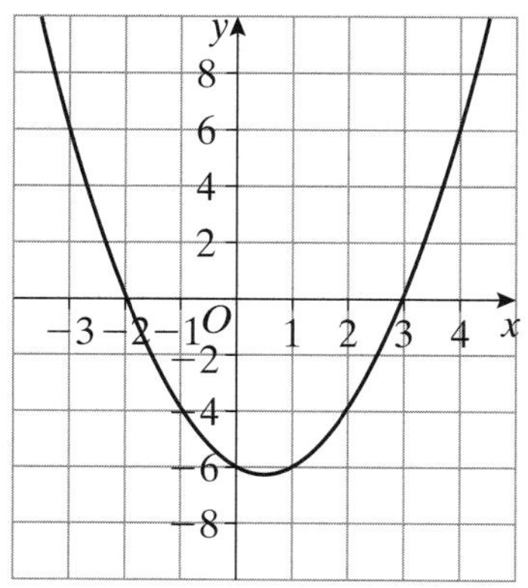

c) $-2, 3$　　　**d)** $(0.5, -6.25)$

9 a) $1.7, 5.3$　　　**b)** $0.6, 6.4$
10 1.25

Exercise 28.3
1 a) $y = \sin x$

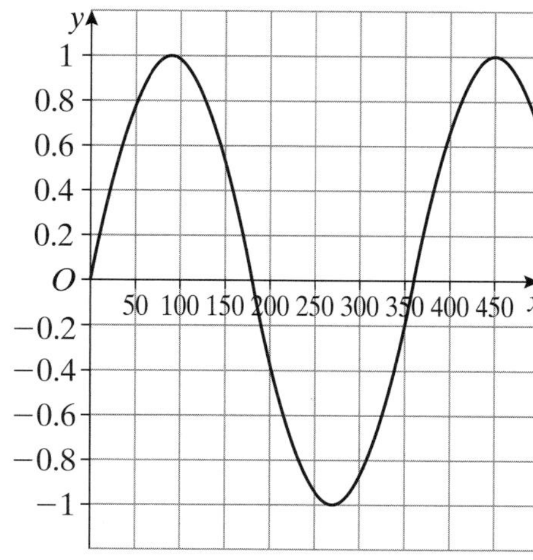

b) $y = \cos x$

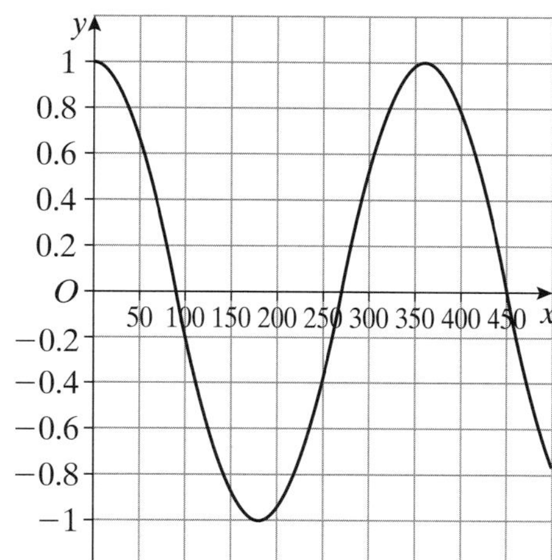

c) $y = \tan x$

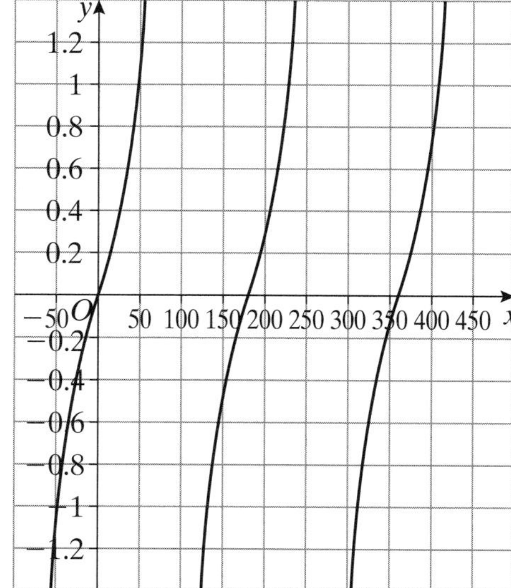

2 a) True　　　**b)** False
　c) False　　　**d)** False
　e) True

3 a) $y = \cos x$　　　**b)** $y = \tan x$
　c) $y = \sin x$　　　**d)** $y = \cos x$
　e) $y = \sin x$

Exercise 28.4

1 a), b), c)

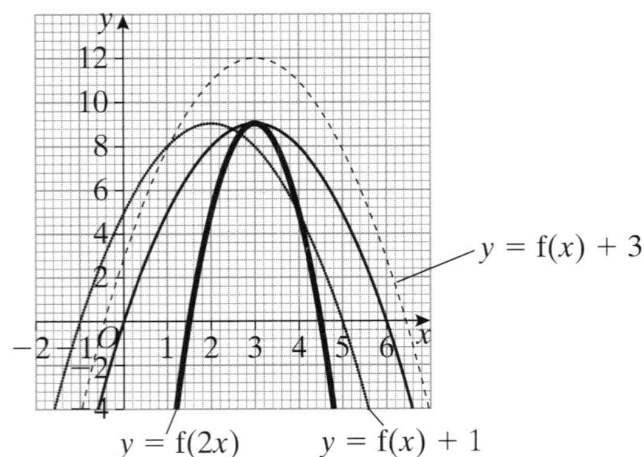

2 a), b), c)

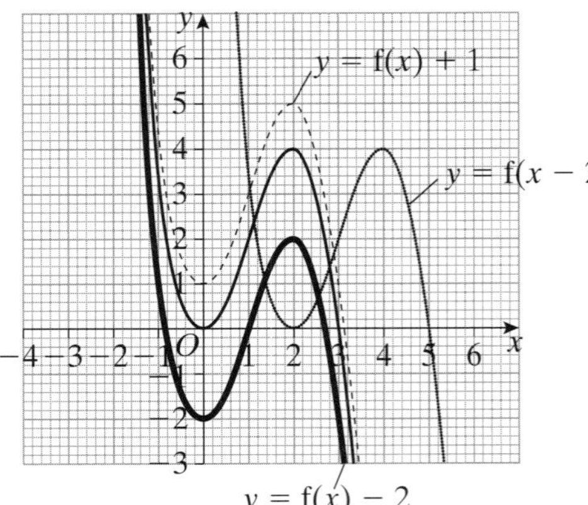

3 a), b) $y = f(x + 1) + 2$ $y = f(x + 1)$

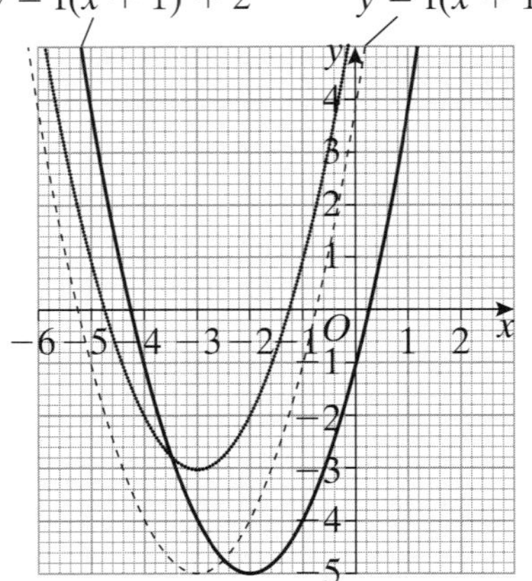

4 a), b) $y = f(-x)$

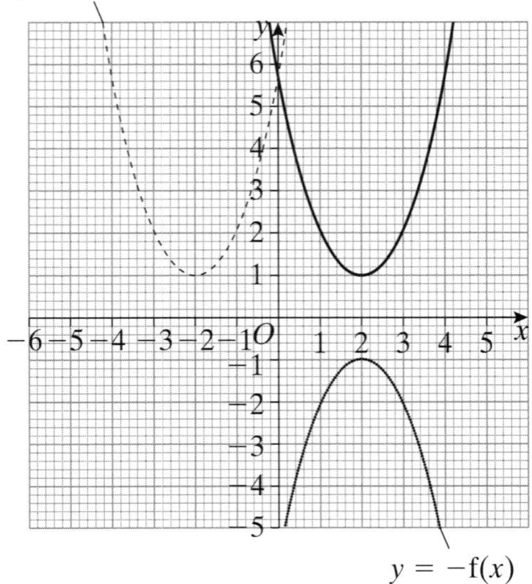

5 $a = 3$ because the wave oscillates between -3 and 3 instead of -1 and 1.

$b = 2$ because the graph completes a full wave in only 180°, not 360°.

6 $a = -2$ because the graph is translated 2 units to the right.

$b = 1$ because the graph is translated 1 unit upwards.

7 $a = 4$ because the wave oscillates between -4 and 4 instead of -1 and 1.

$b = 2$ because the graph completes a full wave in only 180°, not 360°.

8 $a = 2$ because the wave oscillates 2 units above and below its centreline.

$b = 2$ because the graph oscillates above and below 2, not 0.

Review Exercise 28

1 a)

x	-3	-2	-1	0	1	2	3
y	18	8	2	0	2	8	18

b) $y = 2x^2$

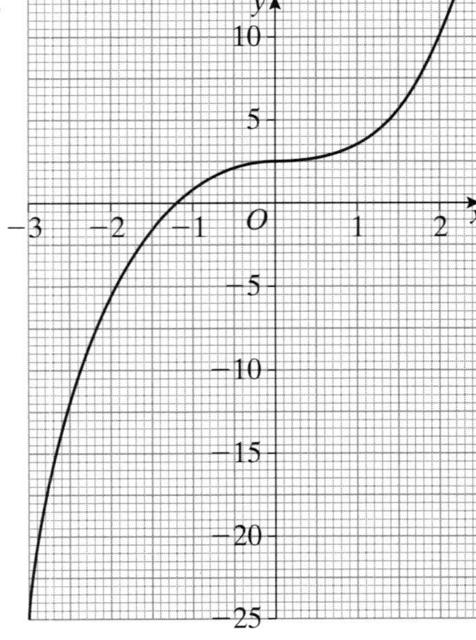

c) (i) 12.5   **(ii)** 2.45 and -2.45

2 a)

x	-3	-2	-1	0	1	2
$y = x^3 + 2$	-25	-6	1	2	3	10

b) $y = x^2 + 2$

c) (i) -1.3 **(ii)** 1.8

3 a) (i) $(5, -4)$
(ii) $(2, -9)$
(iii) $(2, 4)$
(iv) $(1, -4)$
b) $f(x) = (x - 2)^2 - 4$
4 a) (i) $(90, 1)$ **(ii)** $(270, -1)$
b)

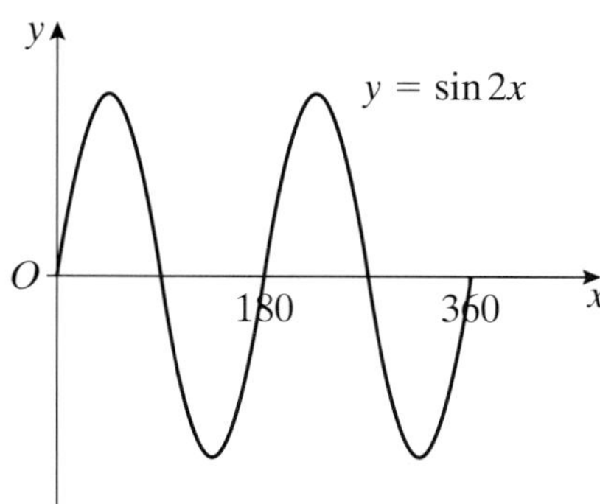

5 a) $(1, 6)$
b) $(3, 10)$
c) $(-3, 6)$
6 a) (i) $y = \sin x° + 1$
(ii) $y = 2 \sin x°$
b) Stretch $\times 3$ parallel to y axis and stretch $\times \frac{1}{2}$ parallel to x axis.
7 a) Reflection in the x axis.
b) $p = 2, q = -4$
c) Translation 2 units to the left and 4 units down.
8 a)

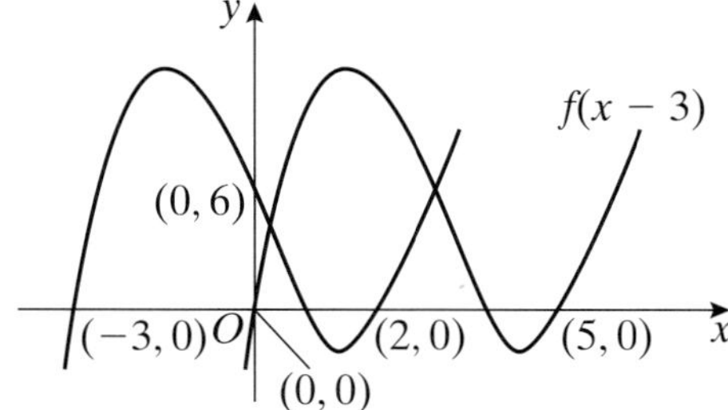

b)

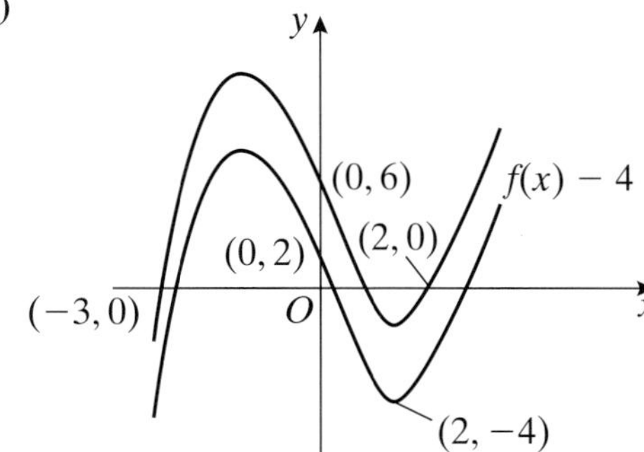

9 a) $a = 50$ **b)** $b = 50$ **c)** $k = 2$
10 a) $a = 2, b = -1$
b) $0°, 234°, 360°, 594°, 720°$
c) 3

Internet Challenge 28

conchoid   equiangular spiral   double folium
trifolium   Archimedean spiral   cardioid
limacon of Pascal   rose curve   lemniscate

Chapter 29: Vectors

Starter 29

There are many ways of completing a knight's tour; here is one way:

Partly complete

2	7	4	11				
13	10	3	8				
6	1	12	15				
17		9	4				
	5	16					

Complete

2	7	4	11	28	39	46	43
13	10	3	8	45	42	29	40
6	1	12	15	38	27	44	47
17	22	9	4	53	48	41	30
64	5	16	21	26	37	54	49
23	18	59	62	57	52	31	34
60	63	20	25	36	33	50	55
19	24	61	58	51	56	35	32

Exercise 29.1

1 a) $\begin{bmatrix} 2 \\ 3 \end{bmatrix}$ **b)** $\begin{bmatrix} 3 \\ -2 \end{bmatrix}$

c) $\begin{bmatrix} -2 \\ -2 \end{bmatrix}$ **d)** $\begin{bmatrix} 0 \\ 3 \end{bmatrix}$

e) $\begin{bmatrix} 1 \\ -4 \end{bmatrix}$ **f)** $\begin{bmatrix} -4 \\ 1 \end{bmatrix}$

g) $\begin{bmatrix} 1 \\ -5 \end{bmatrix}$ **h)** $\begin{bmatrix} 3 \\ 2 \end{bmatrix}$

i) $\begin{bmatrix} 7 \\ -2 \end{bmatrix}$ **j)** $\begin{bmatrix} -2 \\ 4 \end{bmatrix}$

2

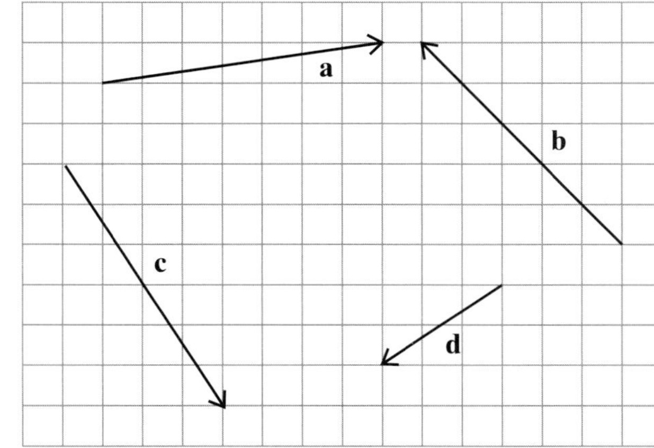

1 $\begin{bmatrix} -1 \\ 13 \end{bmatrix}$

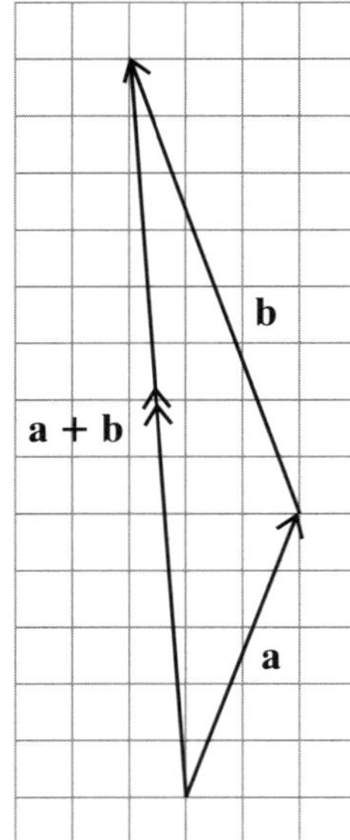

2 $\begin{bmatrix} -5 \\ 10 \end{bmatrix}$

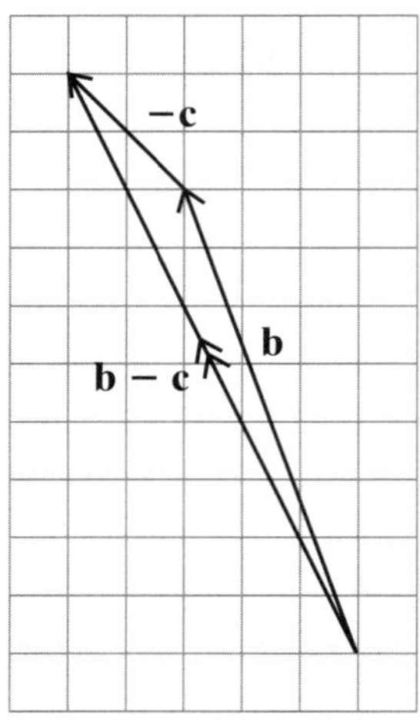

3 $\begin{bmatrix} 4 \\ 3 \end{bmatrix}$

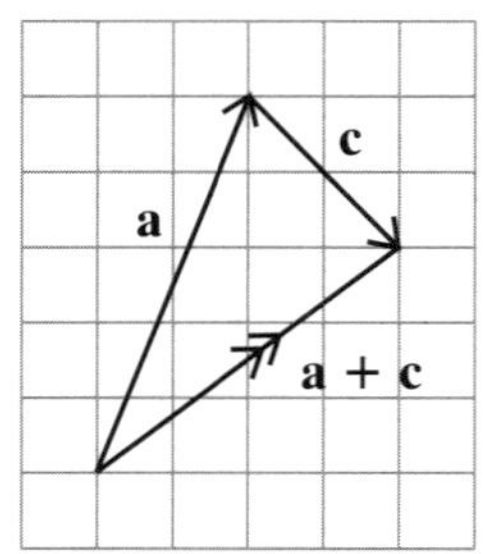

4 $\begin{bmatrix} 5 \\ -10 \end{bmatrix}$

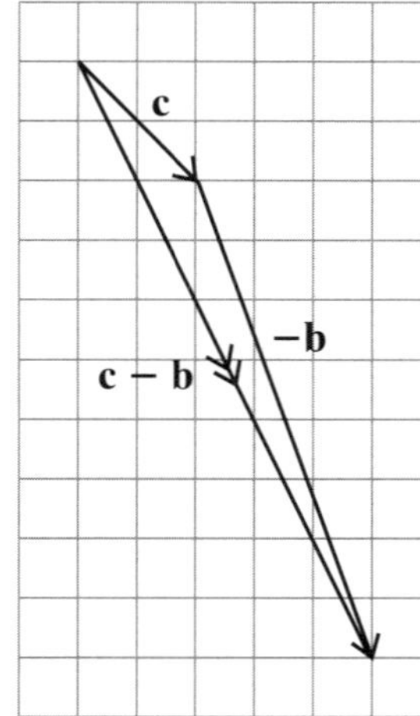

5 $\begin{bmatrix} -3 \\ 15 \end{bmatrix}$

6 $\begin{bmatrix} -3 \\ 1 \end{bmatrix}$

7 $\begin{bmatrix} -3 \\ -5 \end{bmatrix}$

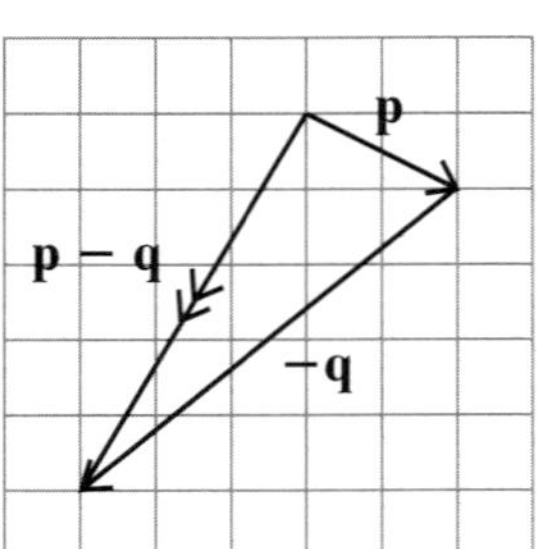

8 $\begin{bmatrix} 13 \\ 9 \end{bmatrix}$

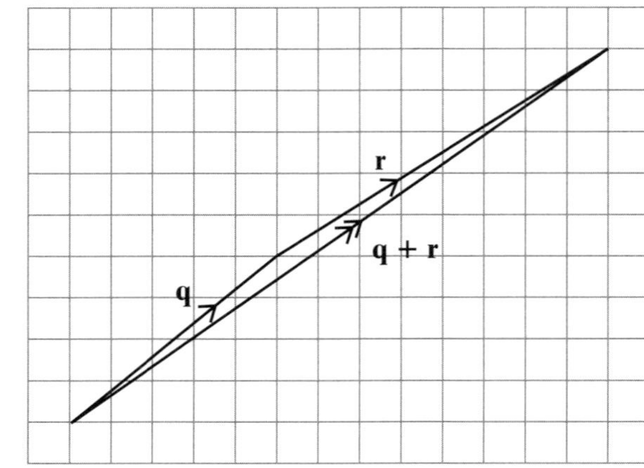

9 $\begin{bmatrix} 6 \\ 6 \end{bmatrix}$

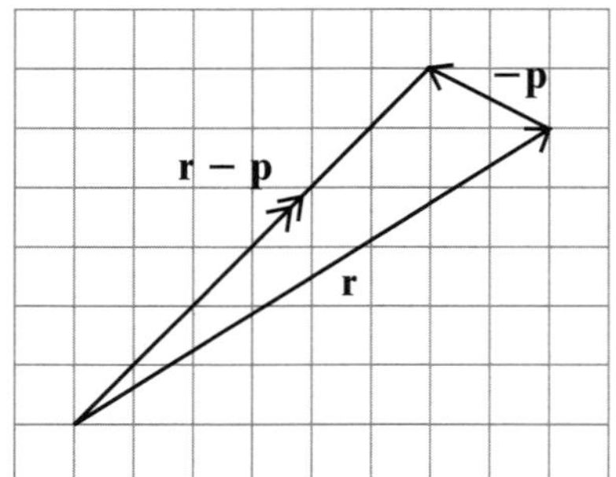

10 $\begin{bmatrix} 3 \\ 1 \end{bmatrix}$

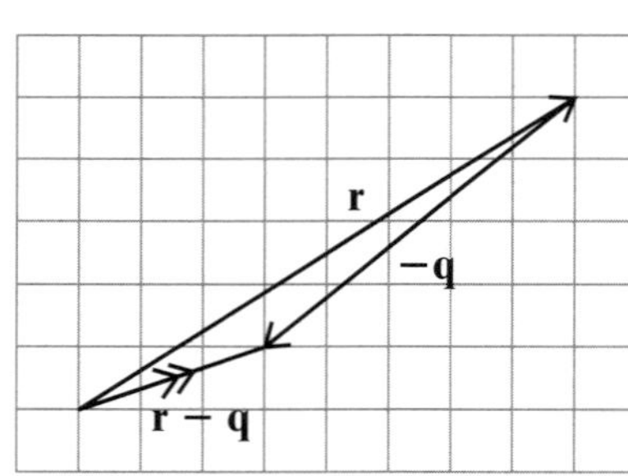

11 $\begin{bmatrix} 15 \\ 8 \end{bmatrix}$

12 $\begin{bmatrix} -1 \\ -2 \end{bmatrix}$

13 $x = 3$

14 $x = 9, y = 7$

15 $x = -3, y = 15$

Exercise 29.3

1 $\begin{bmatrix} 12 \\ -6 \end{bmatrix}$

2 $\begin{bmatrix} 1 \\ 11 \end{bmatrix}$

3 $\begin{bmatrix} 1 \\ 1 \end{bmatrix}$

4 $\begin{bmatrix} -8 \\ -22 \end{bmatrix}$

5 $\begin{bmatrix} 21 \\ 17 \end{bmatrix}$

6 $\begin{bmatrix} 12 \\ -8 \end{bmatrix}$

7 $\begin{bmatrix} -10 \\ 25 \end{bmatrix}$ **8** $\begin{bmatrix} -9 \\ 3 \end{bmatrix}$ **9** $\begin{bmatrix} 12 \\ -17 \end{bmatrix}$

10 $\begin{bmatrix} -5 \\ 27 \end{bmatrix}$ **11** $\begin{bmatrix} 15 \\ -17 \end{bmatrix}$ **12** $\begin{bmatrix} -13 \\ 17 \end{bmatrix}$

13 $x = -3$
14 $x = 1, y = 22$
15 $x = 3, y = -1$

Exercise 29.4

1 a) $\mathbf{q}$ **b)** $\mathbf{p} + \mathbf{q}$ **c)** $2\mathbf{q}$ **d)** $\mathbf{p} + 2\mathbf{q}$
2 a) $-\mathbf{a}$ **b)** $\mathbf{a} + \mathbf{b}$ **c)** $2\mathbf{b}$ **d)** $-\mathbf{a} + 2\mathbf{b}$
3 a) (i) $-\mathbf{p} + \mathbf{q}$ **(ii)** $3\mathbf{p}$
     (iii) $3\mathbf{q}$ **(iv)** $-3\mathbf{p} + 3\mathbf{q}$
  b) $\overrightarrow{PQ} = -\mathbf{p} + \mathbf{q}$, $\overrightarrow{BC} = -3\mathbf{p} + 3\mathbf{q} = 3 \times \overrightarrow{PQ}$ so BC and PQ are parallel.
4 a) (i) $\begin{bmatrix} 4 \\ 7 \end{bmatrix}$ **(ii)** $\begin{bmatrix} 4 \\ 7 \end{bmatrix}$
  b) AB and DC are the same length, and parallel.
  c) (i) $\begin{bmatrix} 6 \\ 3 \end{bmatrix}$ **(ii)** $\begin{bmatrix} 6 \\ 3 \end{bmatrix}$
  d) Parallelogram
5 a) (i) $-\mathbf{a} + \mathbf{b}$ **(ii)** $-\frac{1}{2}\mathbf{a} + \frac{1}{2}\mathbf{b}$
     (iii) $\frac{1}{2}\mathbf{a} + \frac{1}{2}\mathbf{b}$
  b) SR is parallel to PQ, and equal in length, since PQRS is a parallelogram.
  c) $\mathbf{a} + \mathbf{b}$
  d) Since $\overrightarrow{PR} = 2 \times \overrightarrow{PE}$ it follows that E is the midpoint of PR. But E is also the midpoint of QS (given). Thus the diagonals of a parallelogram bisect each other.
6 a) (i) $\begin{bmatrix} 12 \\ 6 \end{bmatrix}$ **(ii)** $\begin{bmatrix} 4 \\ 2 \end{bmatrix}$
  b) Since $\overrightarrow{AB} = 3 \times \overrightarrow{DC}$ it follows that AB and DC are parallel.
  c) Trapezium
7 a) $\overrightarrow{QR} = \overrightarrow{QP} + \overrightarrow{PS} + \overrightarrow{SR} = -2\mathbf{a} + 2\mathbf{b} + 2\mathbf{c}$
  b) $\mathbf{b} + \mathbf{c}$
  c) $\mathbf{b} + \mathbf{c}$
  d) They are parallel, and the same length.
  e) Parallelogram

Review Exercise 29

1 $x = 6, y = 11$
2 a) $\overrightarrow{PQ} = \begin{bmatrix} -6 \\ 8 \end{bmatrix}$ and $\overrightarrow{QP} = \begin{bmatrix} 6 \\ -8 \end{bmatrix}$
  b) 10
3 a) (i) $\begin{bmatrix} -4 \\ -3 \end{bmatrix}$ **(ii)** 5
  b) $\begin{bmatrix} -2 \\ 6 \end{bmatrix}$
  c) $(-6, 2)$
4 a) $\frac{1}{2}\mathbf{p} + \frac{1}{2}\mathbf{q}$
  b) $\overrightarrow{RS} = \frac{1}{2}\mathbf{q} = \frac{1}{2}\overrightarrow{OQ}$ so RS and OQ are parallel.
5 a) $-\mathbf{a} + \mathbf{b}$ **b)** $\frac{1}{3}\mathbf{a} + \frac{2}{3}\mathbf{b}$

6 a) $2\mathbf{a} + 4\mathbf{c}$
  b) $\overrightarrow{OM} = 3\mathbf{a} + 6\mathbf{c} = 1\frac{1}{2} \times \overrightarrow{OP}$ so OP is parallel to OM. Since both OP and OM pass through O and are parallel, OPM is a straight line.
7 a) $\mathbf{a} + \frac{1}{2}\mathbf{b}$ **b)** $-\frac{2}{3}\mathbf{a} + \frac{1}{2}\mathbf{b}$
8 a) (i) $-\mathbf{a} + \mathbf{b}$
     (ii) $2\mathbf{b} + 2\mathbf{c}$
     (iii) $-2\mathbf{a} + 2\mathbf{b} + 2\mathbf{c}$
     (iv) $-\mathbf{a} + \mathbf{b}$
  b) KN and LM are parallel and equal in length.
9 a) (i) $-6\mathbf{a} + 6\mathbf{b}$
     (ii) $6\mathbf{a}$
  b) $-3\mathbf{a} + 12\mathbf{b}$
  c) $\overrightarrow{EY} = -4\mathbf{a} + 16\mathbf{b}$ so $\overrightarrow{EY} = \frac{4}{3}\overrightarrow{EX}$
10 a) (i) $\mathbf{a} + \mathbf{b}$
     (ii) $2\mathbf{a} - \mathbf{b}$
  b) $2\mathbf{a} + \frac{1}{2}\mathbf{b}$
  c) $\mathbf{a} + \mathbf{b}$

Internet Challenge 29

1 12 distinct solutions (ignoring rotations/reflections)
2 1, illustrated below.

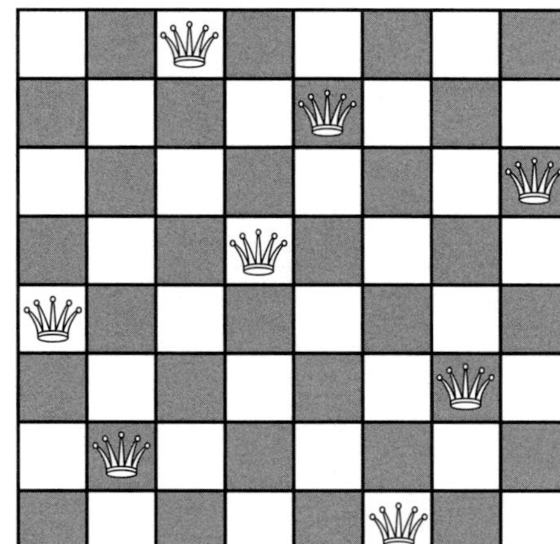

3 A Latin square is an n by n square grid in which each row (or column) contains the same n distinct symbols; any particular symbol occurs exactly once in each row and column (like a Sudoku). The eight Queens puzzle resembles the Latin square in some ways, although it is not a true Latin square.
4 32 knights (an easy solution is to put them all on squares of the same colour).
5 14 bishops
6

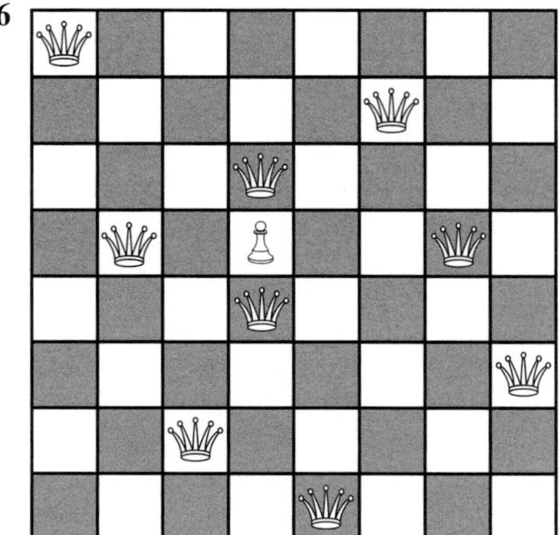

Chapter 30: Mathematical proof

Starter 30
Since $a = 2b$ then $2b - a$ is zero.
Therefore the division at **Step 6** is not permitted.

Exercise 30.1
1 **a)** Yes, SAS   **b)** No   **c)** No
   d) Yes, ASA   **e)** No   **f)** No
   g) Yes, SAS   **h)** Yes, ASA
2 **a)** Prove by showing that triangles OPT and OQT are congruent (RHS).
   b) Angle OPT + angle OQT = $90° + 90° = 180°$
     Thus OPTQ is a cyclic quadrilateral.
3 **a)** QR = RS, angle PQR = angle TSR, angle PRQ = angle TRS
     Thus triangles PQR, TSR are congruent (ASA).
   b) PQ = ST and PR = RT (corresponding sides in congruent triangles)
   c) The diagonals of a parallelogram bisect each other.
4 OA = OB, AM = MB, side OM is common.
   Thus triangles OMA and OMB are congruent (SSS). Angles OMA and OMB are equal, but sum to $180°$, so each must be a right angle.
5 PQ = QR, angle PQS = angle QRT, QS = RT. Thus triangles PQS, QRT are congruent (SAS).

Exercise 30.2
1 Let the numbers be n and $n + 1$.
   Their sum is $2n + 1$ which is odd.
2 Let the numbers be $2n$ and $2m$.
   Their product is $4mn = 2 \times 2mn$, hence even.
3 Let the numbers be $2n + 1$ and $2m + 1$.
   Their product is $(2n + 1)(2m + 1) = 4mn + 2n + 2m + 1 = 2 \times (2mn + n + m) + 1$, hence odd.
4 Let the numbers be n, $n + 1$ and $n + 2$.
   Their sum is $n + n + 1 + n + 2 = 3n + 3$
$$= 3 \times (n + 1),$$
   hence a multiple of 3.
5 Let the numbers be $2n + 1$ and $2m + 1$.
   Then
$$(2n + 1)^2 - (2m + 1)^2 = [4n^2 + 4n + 1]$$
$$- [4m^2 + 4m + 1]$$
$$= 4n^2 + 4n + 1 - 4m^2 - 4m - 1$$
$$= 4n^2 + 4n - 4m^2 - 4m$$
$$= 4(n^2 + n - m^2 - m),$$
$$\text{hence a multiple of 4.}$$
6 **a)** $4 \times \frac{1}{2}ab = 2ab$
   b) (i) $c^2 + 2ab$ **(ii)** $a^2 + b^2 + 2ab$
   d) Pythagoras' theorem
7 **b)** Setting $x = 3$ gives $301 \times 299 = 89\,999$, so not prime.
8 Let the consecutive odd numbers be $2n - 1$ and $2n + 1$.
   Then
$$(2n + 1)^2 - (2n - 1)^2 = [4n^2 + 4n + 1] - [4n^2 - 4n + 1]$$
$$= 4n^2 + 4n + 1 - 4n^2 + 4n - 1$$
$$= 8n, \text{ hence a multiple of 8.}$$

Exercise 30.3
1 (For example) 3 and 5 are prime but $3 + 5 = 8$ is not.
2 It could simply be a rhombus.
3 (Any square number is a counter-example.) The factors of 4 are 1, 2, 4, making 3 factors in all – not even.

---

4 (For example) 4 and 6 have an LCM of 12, but $4 \times 6 = 24$

5
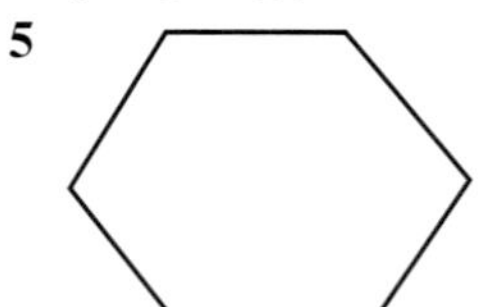

6 (For example) $64 = 8^2$ and $64 = 4^3$
7 (For example) when $x = 11$, then $1 + 10x - x^2$ is -10
8 (For example) (-2) squared is 4, which is not less than 1.

9
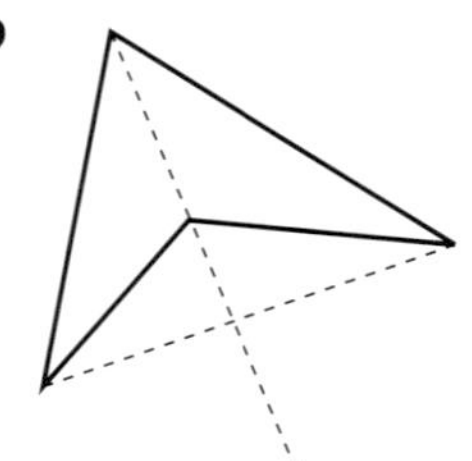

10 (For example) when $x = 41$ then $41^2 + 41 + 41$ is a multiple of 41, hence not prime.

Review Exercise 30
1 **a)** Alternate angles (twice) and common side QS, so ASA.
   b) Angle QRS = angle SPQ and so is also obtuse. Thus the two angles add up to more than $180°$, but for a cyclic quadrilateral would add to exactly $180°$.
2 **a)** $5n$
   b) (i) $5n + 5(n + 1) = 5n + 5n + 5$
$$= 10n + 5$$
$$= 10n + 4 + 1$$
$$= 2(5n + 2) + 1, \text{ hence odd.}$$
     (i) $5n \times 5(n + 1) = 25n(n + 1)$
     If n is even, this product is even. If n is odd, then $n + 1$ is even, so the product is even.
3 Comparing the triangles, CD = AD, DG = DE and angle CDG = angle ADE.
   Thus triangles are congruent, using SAS.
4 $(n + 1)^2 - (n - 1)^2 = [n^2 + 2n + 1] - [n^2 - 2n + 1]$
$$= n^2 + 2n + 1 - n^2 + 2n - 1$$
$$= 4n, \text{ hence a multiple of 4.}$$
5 **a)** Congruent since SSS.
   b) $50°$
6 **a)** Result may be shown by multiplying out both sides of the equation.
   b) Using **a)** the difference between the squares of any two odd numbers may be written in the form $4(a - b)(a + b - 1)$. If $a - b$ is even then the whole expression is a multiple of $4 \times 2 = 8$. If $a - b$ is odd then so is $a + b$, thus $a + b - 1$ is even, and again the whole expression is a multiple of $4 \times 2 = 8$.
7 Using RHS, triangles OYM and OXM are congruent. Thus YM = XM.
8 If $n = 2$ then $n^2 + 3 = 7$ so John is not correct.

9 a) 42.2 cm^2

b) Both triangles have the same perpendicular height, so their areas are proportional to their base lengths.

c) Use $\frac{1}{2}ab \sin C$ for each triangle (same angle in each case) and substitute into result b)

Internet Challenge 30

1 Euclid   **2** Newton

3 Einstein   **4** Goldbach

5 Fermat   **6** Hooke

7 Mendel   **8** Simpson

9 Gödel   **10** Riemann

Chapter 31: Introducing coordinate geometry

Starter 31

Coordinate geometry is the study of shapes drawn on grids.

You will use algebra to describe straight lines and circles.

Exam questions on this topic appear scary, but they are easier than they look at first.

Good luck with the rest of this chapter!

Exercise 31.1

1 a) 5   **b)** 17   **c)** $\sqrt{80}$ (approx 8.94)

2 a) $AB = \sqrt{37}$, $BC = 5$, $CA = \sqrt{34}$

b) Dee is wrong.

3 a)

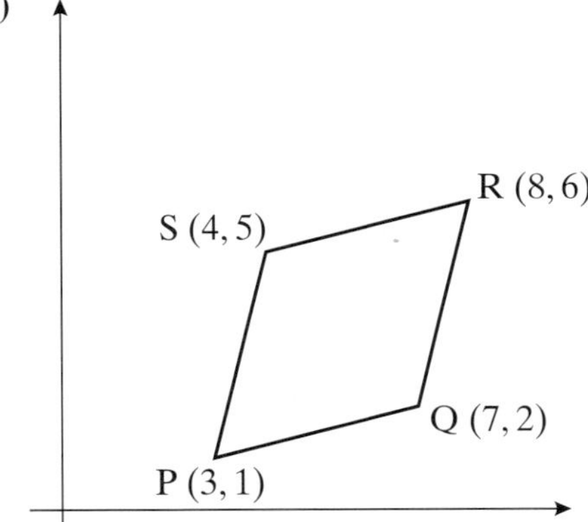

b) $PQ = QR = RS = SP = \sqrt{17}$

c) PQRS is a rhombus.

4 a) 11   **b)** $\sqrt{243}$ **c)** $\sqrt{38}$

The longest side is BC.

Exercise 31.2

1 6

2 $x^2 + y^2 = 25$

3 16

4 $x^2 + y^2 = 49$

5 a) $\sqrt{29}$ **b)** $x^2 + y^2 = 29$

6 a) 5   **b)** $(4, 3)$ and $(-4, 3)$

7 b) $(4, 6)$ and $(-6, -4)$

8 a) Solving as a quadratic gives $x = -3$ twice, so only one point.

b) The line is a tangent.

9 a) Graph B is a circle.

b) Graph A is a parabola, graph C is a straight line.

10 a) The diameter is $2\sqrt{34}$

b) $x^2 + y^2 = 34$

c) $(3, 5)$ and $(5, 3)$

Exercise 31.3

1 a) (i) 3   **(ii)** $\frac{1}{2}$

      (iii) -5 **(iv)** $-\frac{7}{3}$

b) (i) $-\frac{1}{3}$ **(ii)** -2

      (iii) $\frac{1}{5}$ **(iv)** $\frac{3}{7}$

2 $y = 3x - 3$

3 $y = -2x + 10$

4 $y = -\frac{1}{4}x + 3$

5 $y = -2x + 5$

6 a) Lines A and C   **b)** Lines B and D

7 a) $-\frac{1}{2}$ **b)** $y = -\frac{1}{2}x + 4\frac{1}{2}$

c) 2   **d)** $y = 2x - 8$

8 $y = 5x - 3$

9 $y = -x + 9$

10 $-\frac{1}{2}$

Review Exercise 31

1 10

2 $\sqrt{69}$ (approx 8.31)

3 a)

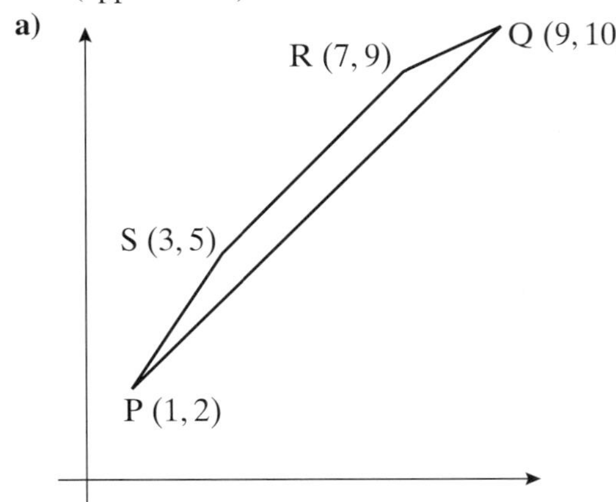

b) $PQ = \sqrt{128} = 8\sqrt{2}$, $RS = \sqrt{32} = 4\sqrt{2}$

c) $2 : 1$

d) $QR = \sqrt{5}$, $PS = \sqrt{13}$

e) The trapezium is not isosceles, since QR and PS are not the same length.

4 a) (i) $(0, 0, 3)$ **(ii)** $(8, 0, 0)$ **(iii)** $(8, 5, 0)$

b) $\sqrt{89}$ **c)** $\sqrt{98}$

5 a) $\sqrt{20} = 2\sqrt{5}$

b) $(2, -4)$ and $(4, -2)$

6 a) O is the midpoint of PQ, so must be the centre of the circle.

b) 17   **c)** $x^2 + y^2 = 289$

7 $y = 2x - 2$

8 $y = -\frac{1}{2}x + 9$

9 $x = 2$, $y = 5$ and $x = -5$, $y = -2$

10 a) $y = 2x + 6$ **b)** $y = -\frac{1}{2}x + 6$

c) Since opposite angles of a rectangle add up to $180°$, it is a cyclic quadrilateral.

11 Substituting for y in $x^2 + y^2 = 9$ gives $x^2 + 5x + 8 = 0$. Using the formula, there is no solution for x, so the line does not cross the circle.

12
$$x^2 + (y - 2)^2 = y^2$$
$$x^2 + y^2 - 4y + 4 = y^2$$
$$x^2 + 4 = 4y$$
$$y = \tfrac{1}{4}x^2 + 1$$

1

Area A	Internal points L	Boundary points B	$L + \frac{1}{2}B - 1$
5	1	10	5
$6\frac{1}{2}$	2	11	$6\frac{1}{2}$

2 Check students' work.

3 The value of $L + \frac{1}{2}B - 1$ is equal to the area, A.

4 This is Pick's theorem, discovered by Georg Alexander Pick.

5 Here is a neat proof that an equilateral triangle cannot be drawn by connecting three points drawn on rectangular spotty paper. It requires an appreciation that quantities such as sin 60 are irrational, that is, they cannot be written in fractional form.

Let the equilateral triangle have sides of length x.

If the triangle can be drawn on spotty paper, then x^2 is a whole number (using Pythagoras' theorem).

But the area is an irrational multiple of x^2 (using $\frac{1}{2}ab$ sin C).

Thus the area is an irrational multiple of a whole number.

Thus the area is irrational.

Now count the number of internal points, L, and boundary points, B, and use $L + \frac{1}{2}B - 1$ to obtain the area, A.

This is clearly rational, contradicting the above requirement.

Thus it is not possible to draw an equilateral triangle on spotty paper.

Chapter 32: Further probability and statistics

Starter 32

1 £20   **2** £30   **3** £4   **4** £6   **5** £12   **6** £10

Ben's average spending per present is less than Simon's on Monday. Ben's average is also less than Simon's on Tuesday. Yet overall, Ben's average is more than Simon's!

Exercise 32.1

1 9 kg   **2** 10 kg   **3** 30 runs   **4** 115.5

5 Billy is right; his method works because there are equal numbers of rods in each batch.

6 43 years   **7** 90 marks   **8** 6

Exercise 32.2

1 a)

1974	1975												1976
Dec	Jan	Feb	Mar	Apr	May	Jun	Jul	Aug	Sep	Oct	Nov	Dec	Jan
17	13	14	6	4	1	0	0	0	3	2	5	8	11
	14.7	11	8	3.7	1.7	0.3	0	1	1.7	3.3	5	8	

b), c)

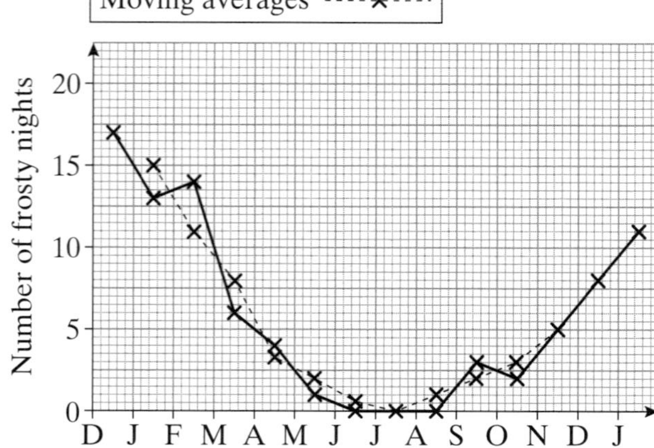

The graphs show that the number of frosty nights is low in the summer months and high in the winter; the data shows clear seasonal variation.

2 a)

	Week 1					Week 2					Week 3				
	Mon	Tue	Wed	Thu	Fri	Mon	Tue	Wed	Thu	Fri	Mon	Tue	Wed	Thu	Fri
	6	2	3	2	12	4	5	3	2	10	4	3	2	4	11
			5	4.6	5.2	5.2	5.2	4.8	4.8	4.4	4.2	4.6	4.8		

b), c)

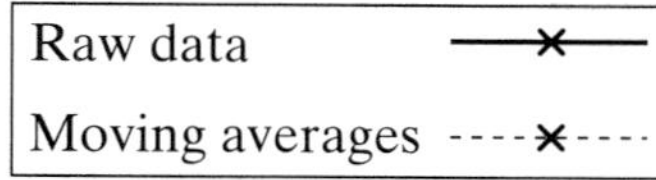

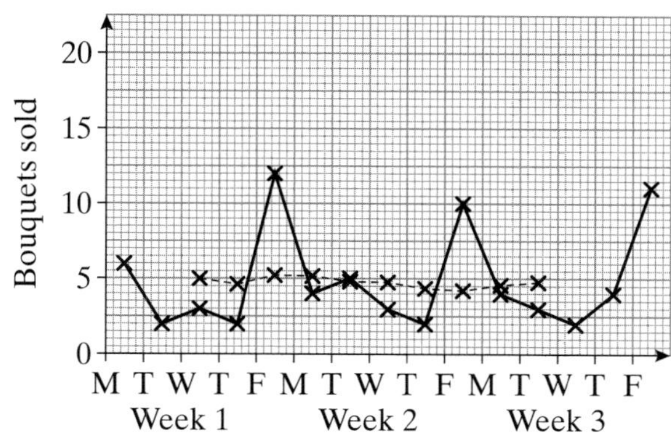

The sales on different days are widely different, especially on Fridays when they peak. The moving averages graph shows that the average sales are approximately constant.

3 a)

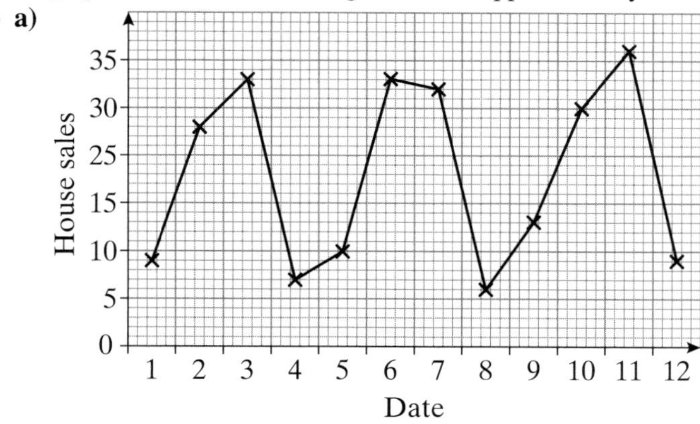

The raw data shows strong seasonal variation, with most sales in the 2nd and 3rd quarters.

b)

Year	1996				1997				1998			
Quarter	1st	2nd	3rd	4th	1st	2nd	3rd	4th	1st	2nd	3rd	4th
	9	23	28	7	10	28	29	6	13	25	31	9
		16.75	17	18.25	18.5	18.25	19	18.25	18.75	19.5		

c) d)

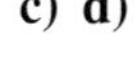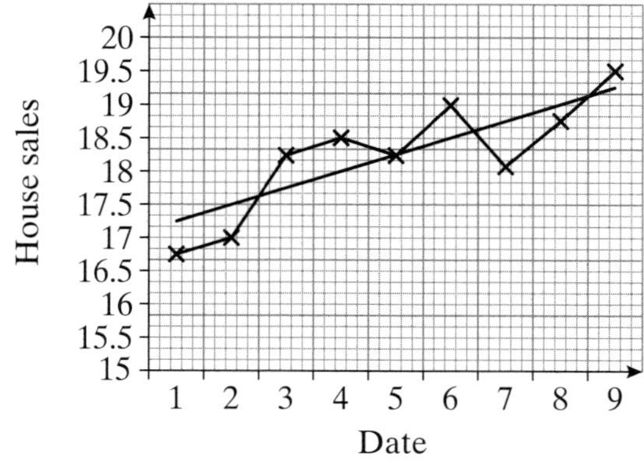

Exercise 32.3

1 a)

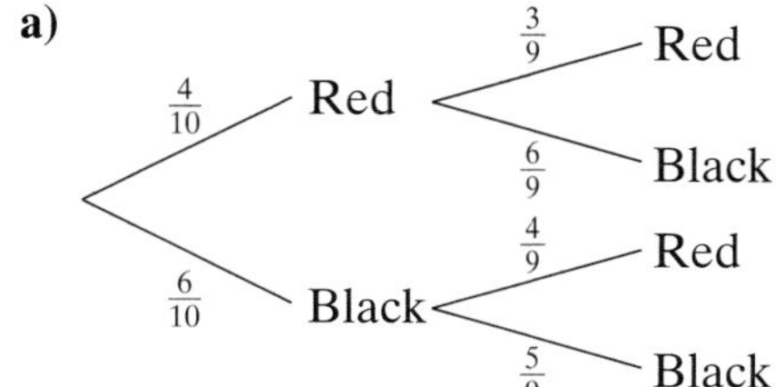

b) $\frac{30}{90} = \frac{1}{3}$

c) $\frac{42}{90} = \frac{7}{15}$

2 a)

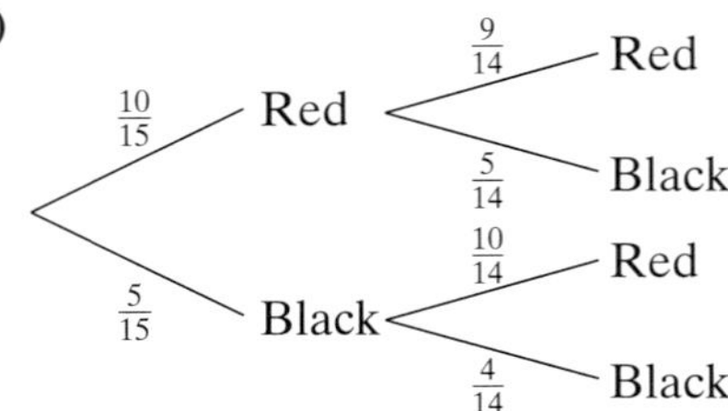

b) $\frac{72}{132} = \frac{6}{11}$ **c)** $\frac{78}{132} = \frac{13}{22}$

3 a)

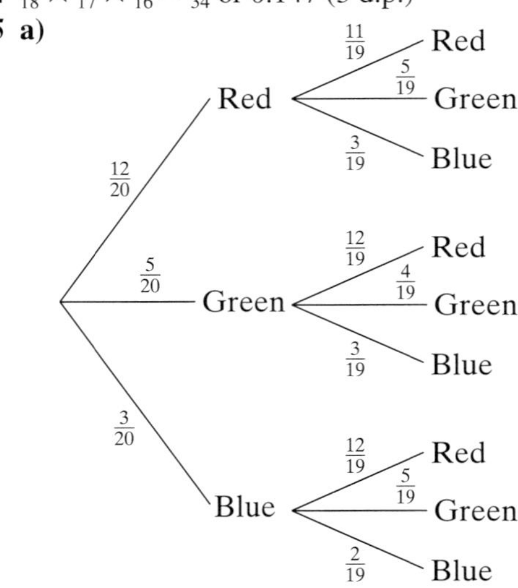

b) $\frac{100}{210} = \frac{10}{21}$

4 $\frac{10}{18} \times \frac{9}{17} \times \frac{8}{16} = \frac{5}{34}$ or 0.147 (3 d.p.)

5 a)

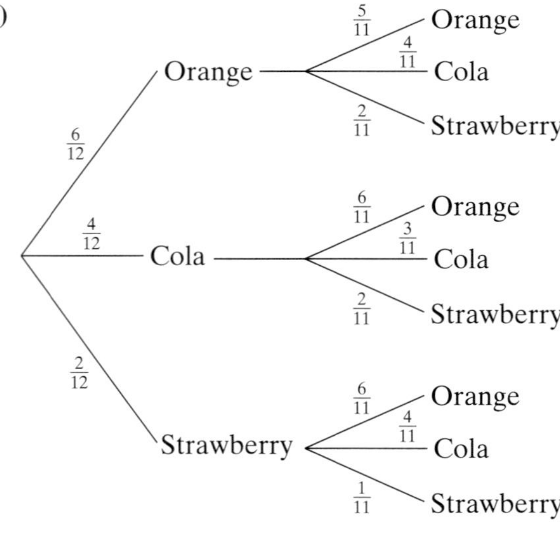

b) $\frac{132 + 20 + 6}{380} = \frac{158}{380} = \frac{79}{190}$

6 a)

b) $\frac{12}{132} = \frac{1}{11}$

c) $\dfrac{24 + 12 + 24 + 8 + 12 + 8}{132} = \dfrac{88}{132} = \dfrac{2}{3}$

7 a) $\frac{6}{16} \times \frac{5}{15} = \frac{1}{8}$ or 0.125

b) $\frac{10}{16} \times \frac{9}{15} \times \frac{8}{14} = \frac{3}{14}$ or 0.214 (3 d.p.)

Review Exercise 32

1 79.3

2 £380

3 a) 1.8

b) p and q might be equal.
Both lists might contain the same number of items.

4 a) 29 minutes

b) A: 12, B: 20, C: 18

5 20, 19

6 117.75, 117.5, 113

7 $\frac{41}{95}$ (approx 0.432)

8 a) $\frac{42}{90} = \frac{7}{15}$

b) $\frac{1}{3}$

9 $0.05 \times 0.06 = 0.003$, which is not equal to 0.011
Thus Fred is not correct; the events are not independent.

10 $\frac{33}{50} = 0.66$

Internet Challenge 32

2 Graph of raw data:

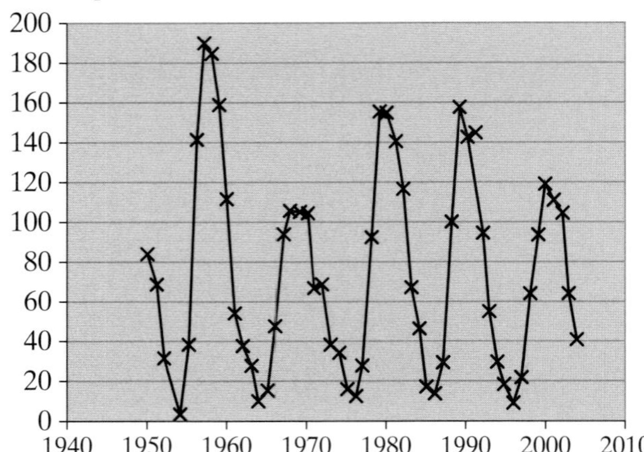

4 Graph of 3-yearly moving averages:

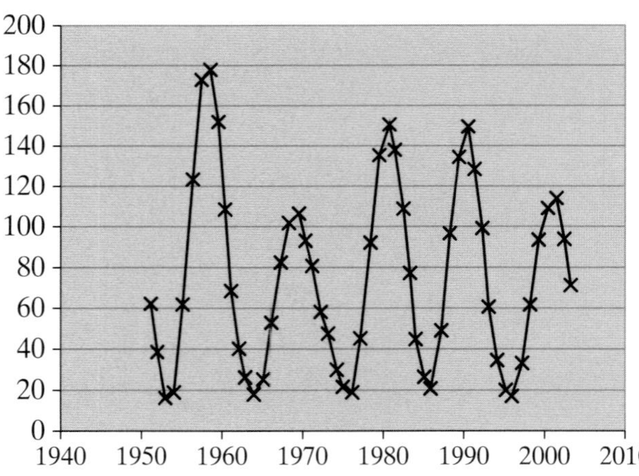

5 The spot numbers appear to increase and decrease with an underlying period of about 11 years.

6 The sunspot cycle length is roughly 11 years.

7 The Maunder minimum was a period from about 1645 to 1715 when the sun was largely spot-free. It was associated with a rather cold period of weather (the Thames routinely froze in the winter, and people held frost fairs on it.)

8 2011